T3-AKE-691

# Strategic Environmental Management for Engineers

O'Brien & Gere
Robert Bellandi, Editor

JOHN WILEY & SONS, INC.

This book is printed on acid-free paper. ♾



Published by John Wiley & Sons, Inc., Hoboken, New Jersey
Published simultaneously in Canada



For general information on our other products and services or for technical support, please contact our Customer Care Department within the United States at (800) 762-2974, outside the United States at (317) 572-3993 or fax (317) 672-4002.

Wiley also publishes its books in a variety of electronic formats. Some content that appears in print may not be available in electronic books. For more information about Wiley products, visit our web site at www.wiley.com

***Library of Congress Cataloging-in-Publication Data:***

Strategic environmental management for engineers / by O'Brien and Gere Engineers ; Robert Bellandi, editor.
p. cm.
Includes bibliographical references and index.
ISBN 0-471-09221-5 (cloth)
1. Environmental management. I. Bellandi, Robert. II. O'Brien & Gere.
GE300.S78 2004
628—dc22 2003023271

Printed in the United States of America

10 9 8 7 6 5 4 3 2 1

*To O'Brien & Gere staff—past, present, and future*

# CONTENTS

# CONTRIBUTORS

**Robert Bellandi,** ELS, has edited two earlier books with O'Brien & Gere Engineers on the subject of hazardous waste remediation and has recently contributed to a review of water resource and water quality management in New York State from 1789 to 1970.

**Timothy J. Barry,** PE, manages the construction of complex wastewater treatment and other environmental and remediation projects for O'Brien & Gere Engineers and its sister firms. He has specialized in advanced wastewater treatment projects.

**Marc J. Dent,** PE, is a managing engineer at O'Brien & Gere Engineers. He has managed environmental site assessments and compliance audits, developed environmental management systems, and managed Phase II investigations, remedial investigations, and remedial design projects.

**Mark R. Greene,** PhD, of O'Brien & Gere Engineers, heads wastewater and groundwater treatment projects. His areas of expertise include biological processes, solid/liquid separation, and process modeling for chemical manufacturers, pharmaceutical plants, food and beverage production, petroleum refiners, manufactured gas plants, military bases, Superfund sites, and landfills.

**Hilary G. Grimes** is a doctoral candidate in Environmental Manufacturing Management at Clarkson University, Potsdam, New York. Her areas of interest include product stewardship, the consumer–producer interface, and environmental decision making.

**Susan E. Powers,** PhD, PE, is a professor in the Department of Civil and Environmental Engineering and Director of the Center for the Environment at Clarkson University. Her research work includes the fate of organic fluids in subsurface environments and assessment of the environmental impacts of energy sources and fuels. She teaches classes, including industrial ecology, which incorporate many aspects of the material covered in this text.

**Thomas P. Seager,** PhD, is a research engineer at the Recycled Materials Resource Center at the University of New Hampshire. He is an experienced engineering educator and has written extensively on topics in industrial ecology.

**Peter S. Segretto,** PhD, is corporate director of Environmental Affairs at Alcan Corporation, Cleveland. Before that, he was assistant professor of geography at the University of South Florida. He has been involved in developing and documenting environmental management systems since 1995 and has been a RAB Certified Lead Auditor for ISO 14001 since 2000.

**Thomas L. Theis,** PhD, is director of the Institute for Environmental Science and Policy and professor of civil engineering, at the University of Illinois at Chicago. He is an expert in systems analysis for problems in pollution prevention, life-cycle assessment, and environmental chemistry.

**Michael J. Tracy,** PE, is a senior project engineer with O'Brien & Gere Engineers. He has been involved in all aspects of the design of wastewater conveyance, pumping stations, and treatment plants. He holds degrees in environmental and civil engineering and physics.

**Ross Whaley,** PhD, has been a University Professor at the State University of New York College of Environmental Sciences and Forestry for the past 3 years. The focus of his scholarly pursuits has been the political economy of sustainable development. Before that, Whaley was president of SUNY ESF for 16 years. He recently was appointed chairman of the Adirondack Park Agency by New York Governor George Pataki.

**Anahita A. Williamson** is a doctoral candidate in civil and environmental engineering, as part of the Environmental Manufacturing Management program, at Clarkson University in Potsdam, New York. Her research interests include using life-cycle assessment as a tool for environmental analysis and adapting process technologies to incorporate industrial ecology concepts

**Bill Vitek,** PhD, is an associate professor of philosophy at Clarkson University. He is co-editor, with Wes Jackson, of *Rooted in the Land: Essays on Community and Place.*

# ACKNOWLEDGMENTS

The authors and editor have benefited from the support and involvement of many individuals in the production of this book. Foremost among these supporters were their colleagues at O'Brien & Gere Engineers and Clarkson University. Al D'Agostino of O'Brien & Gere created the figures for this book. Jim Harper of John Wiley & Sons offered critical guidance and support throughout the development of the manuscript. Shannon Egan expertly guided the manuscript through the production process. Thank you, all!

Terry L. Brown, PE, president of O'Brien & Gere Limited deserves special recognition for his patronage and for his determination to complete this book. It is his position that the next generation of engineers must have the tools to foster "sustainable engineering" of our world. This book is a contribution to that future.

Several accomplished individuals donated their time and expertise to review the authors' work, and we are particularly grateful for their generous contributions. The chapters here have benefited from their expert knowledge, critical comments, and encouragement. Their points of view will give the readers additional valuable insights; the authors and editor, however, are solely responsible for any limitations in this book. We thank the following individuals:

Gordon P. Becker, Alcan Aluminum Corporation
Patricia Calkins, Xerox Corporation
Adrienne T. Cooper, University of South Carolina
Mary Ann Curran, USEPA
John R. Ehrenfeld, Massachusetts Institute of Technology
Daniel Gagnier, Alcan, Inc., and former chair of ISO TC-207
Scott A. Grieco, O'Brien & Gere Engineers
William H. Jairrels, Alcan Aluminum Corporation
David P. Lawrence, Lawrence Environmental
Reid Lifset, Yale University
Archie McCulloch, Marbury Technical Consulting
James Mihelcic, Michigan Technological University
Justin R. Moses, O'Brien & Gere Engineers
Parry M. Norling, formerly of the RAND Institute and the DuPont Company
Karen Palmer, Resources for the Future
Brenda D. Pulley, Alcan Aluminum Corporation
Donald James Wuebbles, University of Illinois
Julie Beth Zimmerman, USEPA Office of Research and Development, National Center for Environmental Research

# INTRODUCTION

# You *Can* Get There from Here: Green Engineering—Precursors, Principles, and Prospects

BILL VITEK AND ROSS WHALEY

*Engineering is the art of modeling materials we do not wholly understand, into shapes we cannot precisely analyze so as to withstand forces we cannot properly assess, in such a way that the public has no reason to suspect the extent of our ignorance.*

—A. R. Dykes, British Institution of Structural Engineers, 1976

## Precursors

This book provides the student of engineering with the principles, tools, and motivation to assess future professional actions in the broad and long-term view of environmental consequences. This is *sustainability*, and strategic environmental management is a bulwark of sustainability.

The ideas in this book are innovative; hence, they require explanation and context. We present a brief history of American environmentalism and the engineer's role in this history. The ideas here also carry forward engineering's long tradition of providing innovative technical solutions to the challenges of living well in a world that can be harsh, crowded, and limited by material and energy flows, ethical goodwill, and political stability.

### Engineering

To relate engineering practices to sustainability is to risk the charge of redundancy. At the very least it sounds like a linguistic safety factor. What else is the practice of engineering—in its most basic intentions and inventions—than an attempt to sustain and improve the lives of people and societies? The charge is true in one respect. Engineering—from its earliest expression in building aqueducts (600 BC) or paved roads (312 BC) to military defenses, to the creation of drinking and wastewater and electric power systems, to nearly every "modern" convenience—can be easily seen as intending to improve our capacity to live in the world. Nicholas Murray Butler, the 1931 recipient of the Nobel Peace Prize, friend of President Theodore Roosevelt, and president of Columbia University from 1901 to 1945, described the field thus:

> [E]ngineering . . . is the direction of the sources of the power of nature for the use and convenience of man. It is the link, the bridge between man and nature; a bridge over which man passes into nature to control it, guide it, understand it, and the bridge over which nature and its forces pass to get into man's field of interest and service. (Davenport and Rosenthal, 1967, p. 133)

The fact that engineered solutions have not always solved the targeted problems or that they sometimes did so in ways that created unintended and unforeseen problems of equal or worse consequence is simply a reflection of the difficulty of the tasks engineers have undertaken at the behest of society. Perhaps the tools and assumptions available to the working engineer were inadequate. As we shall see below, green engineering practices (now in their infancy) offer new tools, additional assumptions, and larger boundaries.

History makes clear that working engineers have always created new tools, assumptions, and boundaries in pursuit of solutions to social, political, and economic conditions—in short, in pursuit of sustaining and improving life. A few examples may suffice. Most of us know West Point as America's first military academy, but few know that it was created to provide the U.S. government and its military with a corps of engineers. George Washington brought this problem to light during the Revolutionary War, and in 1802 Congress appropriated money to create the West Point Academy, modeled after the École Polytechnique in Paris (Perrucci and Gerstl, 1969, p. 101).

The engineering profession likewise responded to the immense challenges of creating America's transportation and electrical infrastructures in the 19th century. Engineers created lighter stronger bridges with materials and techniques that were new at the time, as best evidenced in John Roebling's work with suspension bridges, notably the Brooklyn Bridge (Petroski, 1994, chapter 8). We may best know Thomas Edison as the inventor of the lightbulb; but Edison, Samuel Insull, and S. Z. Mitchell were largely responsible for the innovation, management, and financing necessary to electrify America. As cities and then suburbs became crowded, the demand for freshwater and waste disposal became paramount to preserve public health. Engineers responded. Ellis Sylvester Chesborough's work in Chicago created the largest comprehensive sewage system at the time by literally raising the elevation of the entire city to create a gravity flow sewage network. On the freshwater side, "he designed and constructed a 2 mile-long tunnel to draw water from far out in Lake Michigan" (Reynolds, 1991, p. 14).

The 20th century is equally impressive and representative of the working engineer's ability to respond to a challenge with solutions that required new materials, processes, and design parameters. From plastics to the transistor and integrated circuit to high-efficiency energy systems and complex communications networks, the engineering profession has had a profound effect on the world in which we live. Even when engineered solutions have caused secondary technical, social, and environmental problems, a fundamental goal of engineering design has remained a commitment to creating a world to sustain and enhance human life.

*Sustainability* marks an important break from our earlier history of control and working against nature (technology conquers nature) to one of working with and learning from nature (technology models nature in its design, productivity, and material/energy flows). See the following box with the 12 principles of green engineering.

**The 12 Principles of Green Engineering**

**Principle 1:** Designers need to strive to ensure that all material and energy inputs and outputs are as inherently nonhazardous as possible.

**Principle 2:** It is better to prevent waste than to treat or clean up waste after it is formed.

**Principle 3:** Separation and purification operations should be designed to minimize energy consumption and materials use.

**Principle 4:** Products, processes, and systems should be designed to maximize mass, energy, space, and time efficiency.

**Principle 5:** Products, processes, and systems should be "output pulled" rather than "input pushed" through the use of energy and materials.

**Principle 6:** Embedded entropy and complexity must be viewed as an investment when making design choices on recycle, reuse, or beneficial disposition.

**Principle 7:** Targeted durability, not immortality, should be a design goal.

**Principle 8:** Design for unnecessary capacity or capability (for example, "one size fits all") solutions should be considered a design flaw.

**Principle 9:** Material diversity in multicomponent products should be minimized to promote disassembly and value retention.

**Principle 10:** Design of products, processes, and systems must include integration and interconnectivrty with available energy and materials flows.

**Principle 11:** Products, processes, and systems should be designed for performance in a commercial "afterlife."

**Principle 12:** Material and energy inputs should be renewable rather than depleting.



Early proponents of sustainability summed up the formerly prevailing ethic thus: "If brute force doesn't work, you're not using enough of it" (McDonough and Braungart, 2002, p. 30). American environmentalism's transition to pollution treatment and prevention in the last quarter of the 20th century demonstrated a grudging recognition that controlling nature did not always work. Sometimes we have to control the way that humans use nature.

The need for sustainable development worldwide is now asking engineers to rethink entire processes, to create and experiment with new materials, and to answer novel questions—all in response to shifting cultural and political goals and values. In this respect, green engineering practices begin to describe what can be considered a revolutionary change of assumptions about the natural world: the human ability to understand and control nature, and the engineer's job at the interface of nature and culture. Sustainable development and, now, green engineering practices are an outgrowth of global environmentalism. Strictly speaking, sustainability did not originate in the United States but can be seen as a logical outgrowth of American environmentalism—a movement bound up with the engineering profession.

## Environmentalism

For the purposes of this introduction, the history of American environmentalism can be divided roughly into three periods:

- Sources
- Sinks
- Systems

The earliest concerns for the natural environment focused on sustainable *sources* of natural resources. Later attention turned to *sinks* for pollutants. Currently, we are turning our attention to *systems* as a more comprehensive way to solve problems and to minimize unintended consequences of our actions. Engineers have made, and are making, important contributions in each of these areas. What has changed most—and is the subject of this book—are the assumptions that simultaneously delimit and expand the boundary conditions of engineering design problems in the direction of complex living systems.

### *Sources*

In the late 1800s, writers, philosophers, and politicians began to recognize that while this still-young country was blessed with a rich abundance of natural resources, the supply was not inexhaustible. A century ago, the presidency of Theodore Roosevelt (1858–1919) was marked by a concern for seemingly dwindling water supplies and timber for future generations. The solution to this concern was subsumed under the term *conservation*. With advice from Gifford Pinchot (1865–1946), chief forester of the U.S. Forest Service, Roosevelt set aside lands as national forests and introduced the idea of science-based management on these lands. In supporting the often-controversial policies of Roosevelt, Pinchot said:

> I stand for the Roosevelt policies because they set the common good of all of us above the private gain of some of us; . . . because they oppose all useless waste at present at the cost of robbing the future; because they demand the complete, sane, and orderly

> development of all natural resources; because they insist upon the equality of opportunity and denounce special privilege. . . . (Pinchot, 1910, p. 888)

The early practice of conservation offered two new approaches to how we managed the country's resources. The first was the intervention of government in ownership, regulation, and management. This was a dramatic reversal to the *laissez faire* philosophy of minimizing the role of government in our economic life. The second was attention to scientific management of natural resources. The rationale here was that improved technology would expand the amount of renewable natural resources supplied over time and improve the efficiency of their manufacture and use so they would be available in perpetuity. The idea of resources being available in perpetuity certainly is open to question, but the significant advances in the management, processing, and use of renewable natural resources through the conservation movement cannot be denied.

Roosevelt and Pinchot fostered changes in attitude, the most important being that the natural world could not be depended upon to shell out endlessly the goods and materials for a growing human population. Engineers responded to this conservation ethic, but a central assumption remained *the control of nature for the benefit of human beings.* In 1938 Aldo Leopold—forester, philosopher, and author of *A Sand County Almanac* (1949)—gave a lecture to the University of Wisconsin College of Engineering titled "Engineering and Conservation." Leopold wasted no time in challenging the prevailing ethic that the "engineer believes, and has taught the public to believe, that a constructed mechanism is inherently preferable (and superior) to a natural one," and that oftentimes an engineered solution pushes "trouble downstream and seeks benefit for the locality at the expense of the community." He then contrasted the engineering view of nature with the ecological, and stated, "the real difference lies in the ecologist's conviction that to govern the animate world it must be led rather than coerced" (Flader and Callicott, 1991, pp. 252–253).

Nonetheless, until the early 1960s, the primary engineering challenge remained to improve the efficiency of the use of natural resources and labor and to tame a wild country so that its economic growth could proceed in an orderly manner. Examples included the Tennessee Valley Authority, the electrification of rural America (a process that coincided with the transformation to farms of ever larger size, productivity, and efficiency), and hydroelectric-irrigation projects in the West.

### *Sinks*

The gains made in extending the availability of natural resources came at a cost. The protection of plants from insects led to the use of DDT (dichlorodiphenyltrichloroethane) and other pesticides that ultimately proved harmful to wildlife and human health. Attention to surface water supplies often neglected water quality, the integrity of aquatic ecosystems, or subsurface supplies. Manufacturing that improved efficiency in the use of labor and raw materials was at times insensitive to increases in air or water pollution. Rachel Carson brought this dramatically to the fore by creating a fable in which a community experienced a "silent spring" because the birds had been poisoned by the accumulation of human-spread chemicals in the soil, water, and air (Carson 1962). Carson's book, *Silent Spring,* then presented evidence that the fable might well be a precursor of things to come. Indeed, when Carson wrote in the 1960s, the evidence was mounting that both the quantity and toxicity of human-created substances could overwhelm the natural world's capacity to absorb, store, or cleanse them.

The American public reacted with a new wave of environmentalism in the late 1960s and 1970s. This movement resulted in a host of new laws—for the most part, still our blueprint today—aimed at evaluating the environmental impacts of public projects, curbing declines in air and water quality, regulating the use of chemicals, and protecting threatened and endangered species of plants and animals. The Clean Air Act expanded regulations to include automobile emissions as well as stationary sources of pollution, set deadlines for compliance, and authorized research aimed at reducing pollution. Each revision was a new challenge for engineers. The Clean Water Act and its amendments presented an analogous set of challenges to find new treatment technologies. These laws created standards, data management and reporting procedures, and harsh penalties including criminal sanctions for noncompliance. The federal government's approach to managing human effects on the environment was described as "command and control."

Considerable environmental benefits resulted from these legislative and enforcement efforts. Some would argue, however, that they were achieved at the cost of economic growth, and that there were alternative approaches to working with the industrial sector that would have achieved the same benefits, but with greater efficiency and effectiveness. Whether or not alternative approaches would have been as effective is beyond the scope of this book. What can be stated with certainty is that the arousal of environmentalism in the 1960s and 1970s marked the shift away from the control of nature expressed by Pinchot, Butler, and countless others. The new ethic ushered in by Leopold and Carson, among others, was a fledgling sense that human control can make matters worse for humans and natural systems. Nature, while resilient, is not immune to the depredations of human-created compounds. Complex natural systems can be harmed in unpredictable ways. And, nature itself may offer lessons that can be put to good use in an industrialized and highly technological culture.

Engineers, of course, played a central role in designing the technological products and systems of the "sinks" era. It was during this time that environmental engineering began to move away from its 19th century sanitary engineering roots ("dilution is the solution to pollution") and toward pollution control and treatment technologies. Products such as secondary wastewater treatment processes, catalytic converters and smokestack scrubbers, and lined landfills were first created and put into commercial use. It was a time of mostly fixing problems caused by an earlier generation of engineering design errors, oversights, and unforeseen and unintended consequences. Threshold limits of synthetic toxic compounds were set. Controls were placed on the synthetic chemicals that could be used in manufacturing and services. This new ethic forced the fall of the assumption that nature can accommodate without limit the waste products of industrial systems.

Carson's book and the pollution control revolution it promoted were one step away from the systems approach advocated by green engineering practices, but the concluding chapters of *Silent Spring* foreshadowed the inevitable evolution toward systems thinking. Although Carson's groundbreaking book is best known for its assault on DDT and pesticides generally, her views go beyond criticism and critique. She does not rule out chemical pesticides as a last resort in controlling pests that attack food crops but recommends that alternatives be considered, including insect sterilization, ultrasound techniques, bacterial control, and integrated pest management. These approaches "are *biological* solutions, based on understanding of the living organisms they seek to control and of the whole fabric of life to which these organisms belong" (Carson, 1962, p. 278). Carson states the same idea more ominously in the last paragraph of her book:

> The "control of nature" is a phrase conceived in arrogance, born of the Neanderthal age of biology and philosophy, when it was supposed that nature exists for the convenience of man. The concepts and practices of applied entomology for the most part date from the Stone Age of science. It is our alarming misfortune that so primitive a science has armed itself with the most modern and terrible weapons, and that in turning them against the insects it has also turned them against the earth. (p. 297)

Both her criticisms and call for a new approach to pest control presaged the change that represents the third and current stage of American environmentalism and the sustainability practices that are the subject of this book.

### *Systems*

We have suggested that the first era of American environmentalism focused on an efficient and continuous source of natural resources. The second era directed us to pitfalls of chemical use beyond the capacity of ecosystems to be a sink for chemical residues. As we look to the future, it is appropriate to ask whether engineering solutions can be approached in a way that will reduce the demand on available virgin natural resources, reduce pollution, protect ecosystems, promote economic development, and enhance the quality of life for all citizens. The quest to achieve a balance of all of these ends require what we call *green engineering practices*. It is a strategic systems-centered boundary-expanding approach to engineering solutions in contrast to the successful but tactical, linear, and focused approaches of the past.

Green engineering is a logical outgrowth of the international attention being devoted to *sustainable development*, a term coined by the International Union of Conservation Naturalists in its landmark document, *Conservation Strategy.* The ideas in this document later became the foundation of *Our Common Future*, a report ("The Brundtland Report," so-called because Gro Harlem Brundtland chaired the commission when the report was issued in 1987) prepared by the United Nation's World Commission on Environment and Development (1988). The commission defined sustainable development as "development that meets the needs of the present without compromising the ability of future generations to meet their own needs." The commission attempted to refute the idea that economic development and environmental protection were always at odds and that we must choose one over the other. In the report's interpretation economic development, environmental protection, and social equity were inextricably linked.

Some would claim that the definition proposed in the Brundtland Report is general and offers little guidance. It did, however, offer a rallying point for international attention to the potential scarcity of natural resources over time, the need to reduce waste that pollutes the natural environment and affects human health, and the need to deal with the problem of income distribution and poverty . In 1992, delegates to the United Nations Conference on Environment and Development held in Rio de Janeiro refined the conceptual basis of sustainable development and created an action plan. One product of this conference was the Rio Declaration, which added protection of the environment to previous goals of development—peace and security, economic development, human rights, and supportive national governance (Dernbach, 2002). The same conference proposed Agenda 21, a comprehensive plan of action.

Essential to the idea of sustainable development is the recognition of limits. In response to the notion of limits in the natural world, Goodland and Daly (1996, p. 1002) proposed their definition of *sustainable development* as "development without growth in throughput of matter and energy beyond regenerative and absorptive capacities." In practice, their definition requires an awareness of limits; the primacy of relationships in light of these limits, and creative and appropriate action within relationships and in response to these limits. Green engineering practices are simply the application of these boundary conditions to specific problems. They do not compose a subdiscipline but an essential approach to all engineering. They differ from other engineering approaches in that they work within—rather than trying to transcend—the ecological limits of sources, sinks, and natural systems, as well as the ethical-political limits imposed by a view of justice that seeks to better the world for all people. Within these admittedly daunting constraints, the practicing engineer is free to express the creativity, daring, and commitment to excellence that have marked the best engineering projects and designs throughout the ages.

The implication of limits to green engineering is captured well by Garret Harden in what he admits is rather "folksy" language. Following are 5 of his 12 "paramount positions" that seem particularly appropriate to the limits that must be addressed by green engineering;

- "The world available to the human population is limited to earth." That is, there are real limits to resource availability on the planet Earth. There is little evidence to date that this will be solved through interplanetary sources of material and energy.
- "There is no such thing as a free lunch." A perpetual motion machine independent of additional resources or energy has proved beyond our ability and will continue to be so.
- "The First Law of Human Ecology: We can never merely do one thing." Some economists have referred to this as the law of *unintended consequences*.
- "The Second Law of Human Ecology: There is no away to throw to." The by-products of our production and consumption activities must go somewhere.
- "Cultural carrying capacity and the standard of living are inversely related." Limits apply to as diverse activities as enjoying solitude or attending the theater. As long as there are limits, there will be questions of distributive justice (Hardin, 1991, pp. 52, 55).

Others have built on this framework as we examine the implications for green engineering practices. Hawken et al. (2000) use the term *natural capitalism* to describe an approach to producing goods and services that embraces four principles:

1. A radical increase in the productivity of resource use
2. A shift to biologically inspired production (biomimicry)
3. A shift in the business model from selling products to providing the service that the products supplied (e.g., carpets, copiers, and automobiles)
4. Reinvestment in natural and human capital (Lovins 2001)

In 2003, dozens of engineers and scientists met for the first conference on "Green Engineering: Defining the Principles," sponsored by Engineering Conferences International and held at the Sandestin Resort, Destin, Florida. The summary from that conference provides a practicable list for the full implementation of green engineering techniques:

1. Engineer processes and products holistically, use systems analysis, and integrate environmental impact assessment tools.
2. Conserve and improve natural ecosystems while protecting human health and well-being.
3. Use life-cycle thinking in all engineering activities.
4. Ensure that all material and energy inputs and outputs are as inherently safe and benign as possible.
5. Minimize depletion of natural resources.
6. Strive to prevent waste.
7. Develop and apply engineering solutions, while being cognizant of local geography, aspirations, and cultures.

8. Create engineering solutions beyond current or dominant technologies; improve, innovate, and invent (technologies) to achieve sustainability.
9. Actively engage communities and stakeholders in development of engineering solutions.

The scientists and engineers also stated that practitioners have *a duty to inform society of the practice of green engineering*. In other words, it is not sufficient if you, as an engineer, adopt techniques that will promote a sustainable world, it is also your obligation to inform your clients, employer, and those affected by your work of your green approach and the hoped-for result.

## Principles

This book gives shape and definition to these larger and sometimes lofty principles. In the chapters that follow, you will be introduced to the tools and trade of green engineering, including an introduction to and demonstration of concepts and applications already current in the engineering professions (environmental management systems, green construction, life-cycle management, extended product responsibility, and total cost accounting), descriptions of emerging disciplines that take sustainability as their central goal (green chemistry and industrial ecology), as well as the introduction of new tools to manage our environmental impacts (pollution potential, zero liquid discharge). There are plenty of how-to explanations and examples, honest assessments of the many remaining challenges and unknowns, and a guarded optimism about the future of green engineering. The twin goals of these chapters are to:

- Introduce innovative ideas and solutions and demonstrate their current application, practicality, and potential
- State unequivocally that green engineering has emerged as an approach that cuts across all engineering disciplines

An overview of the chapters to follow will support these claims. The discussion begins with Tom Seager's chapter on industrial ecology and the multiple dimensions of sustainability. Industrial ecology is an emerging branch of science concerned with the interrelationships of human industrial systems and their environments. It is sometimes referred to as *the science of sustainability*. To the extent that engineering practice is the implementation of scientific discoveries into useful products and processes, industrial ecology may rightly be called the science of green engineering. Seager summarizes this new discipline including historical and intellectual antecedents; he contrasts two methods of engaging in industrial ecology research—life-cycle analysis and systems analysis; and he identifies and defines industrial ecology's multiple metrics (economic, thermodynamic, environmental, sociopolitical, and aggregated). The chapter includes a welcome glossary of key terms.

The book turns next to three decision-making tools available to assist engineers in understanding the environmental effects of manufacturing processes and production. In Chapter 2 Susan E. Powers and Anahita A. Williamson provide a history of life-cycle assessments (LCA), an approach that first began in the 1960s. They give an overview of LCA's philosophical assumptions—a systems approach focused on mitigating environmental impacts from cradle to grave, numerous examples and case studies from industries that have used LCA, and a discussion of LCA's limitations. Despite the challenges of conducting an LCA, many of the world's largest and most productive industries continue to employ and improve the technique, relying on their engineering employees to fashion the quantitative techniques of this complex but valuable industrial-environmental tool.

Hilary Grimes' chapter on extended product responsibility (EPR) is the second tool discussed. EPR is one part quantitative systems analysis, one part public policy (mandated or voluntary), and one part market conditions. Taken together the goal of EPR is to manage the environmental consequences of products from raw materials to producer to consumer to recycler to producer. The technique is as complicated and difficult as it sounds, especially as different countries (and their potential product markets) promulgate different requirements. Grimes focuses on the American experience. She characterizes the current conditions of U.S. EPR; describes corporate activities, tools, and experiences with EPR; discusses the social and political pressures that drive consumer demands, potential cost savings, and market differentiation; and presents corporate case studies.

While life-cycle assessment and extended product responsibility are already relatively well known in industry and academia and have at least a short history of application, Chapter 4 by Tom Seager and Tom Theis break new ground. They introduce a new environmental metric that they call *pollution potential* and define as "a measure of the change in chemical composition of the environment." The advantage of this metric, they claim, is that it avoids risk-based environmental metrics that rely on biological measures of health or increased incidence of disease. Their metric is, instead, a thermodynamic measure expressed as the energy per mole required to remove a pollutant from the environment in an ideal thermodynamic process. (Dig out your thermodynamics notes!) Using chlorofluorocarbons (CFCs) as a case study, Seager and Theis demonstrate the value of the pollution potential metric as a tool that evaluates all environmental effects on a single scientific scale, allowing comparison of different environmental priorities. It is too early to tell if or how this new metric will influence public policy or decisions on the production floor or in the corporate boardroom, but it is an excellent example of green engineering at the boundaries of consideration, exploring new ways to measure, quantify, and evaluate environmental impacts.

The book then turns to exemplars and applications. Imagine that you are a plant manager who is given the task of monitoring the diverse material and energy inputs of your plant, as well as the outputs, including air, water, and chemical discharges. The task is daunting, especially when state and federal regulatory bodies change their regulations, and local community activists can insist that you do more to improve environmental quality. Peter Segretto and Marc Dent provide help in Chapter 5. They outline in detail the ISO 14001 method of "specifying requirements for an environmental management system" (EMS). Such a system enables an organization to formulate a policy and objectives taking into account legislative requirements and information about significant environmental impacts. It applies to environmental aspects over which a corporation can be expected to have influence. It does not itself state specific environmental performance criteria. Segretto and Dent illustrate the EMS process of "plan, do, check, act" by showing how facilities of Alcan, Inc., implemented specific elements of ISO 14001. Alcan has operations in 38 countries in primary aluminum, fabricated aluminum, as well as flexible and specialty packaging. In 2002, it had revenues of $12.5 billion, and it had 53,000 employees in 2003. This was obviously not a small undertaking, and the results are still being tabulated, but initial results demonstrate an improved record across the environmental profile.

Managing a whole system for environmental quality is followed by an equally difficult task. Chapter 6 by Timothy Barry and Michael Tracy provides an overview of industrial water use in North America, including the amount and type of wastewater contaminants and the national and international regulations that affect industrial water use. The automotive industry is this chapter's case study, and it highlights DaimlerChrysler's manufacturing facility in Toluca, Mexico, home of a state-of-the-art "total water recycling" facility.

The final chapter highlights the application of green engineering principles to an entire scientific discipline (chemistry). Not long ago "green chemistry" would have sounded like a cruel oxymoron. A great deal of the environmental criticism for over 50 years has focused on the chemical industry's creation of chemical compounds that have proved detrimental to human and environmental health. As Mark Greene demonstrates in Chapter 7, green chemistry is an emerging field, complete with a methodology, principles, and metrics. Green chemistry "utilizes alternative feedstocks, develops, selects, and uses less environmental harmful solvents, finds new synthesis pathways, improves sensitivities in reactions, generates less waste, and avoids the use of highly toxic compounds."

Greene claims that green chemistry is not yet a discipline in itself but rather a way of applying knowledge in kinetics, catalysis, reaction engineering, materials and interfaces, process design and control, separations, and thermodynamics to lessen the impact that chemical products and processes have on the environment. (Dust off your old chemistry notes.) This chapter highlights the 12 principles of green chemistry and demonstrates the Synthetic Methodology Assessment for

Reduction Techniques (SMART) module. And it begins to blunt traditional environmental criticisms of chemistry's role in industry. Green chemistry is already making inroads in academia. Chemistry programs across the United States are beginning to recognize the need for a scientific base on which to build green engineering practices.

## Prospects

This textbook offers the latest data practices across the physical green engineering perspective and as practiced in some of the world's most innovative and productive industries. Its authors are leaders in their respective fields, and the many case studies provide clear and convincing evidence that the practices of green engineering are here to stay. But, as the saying goes, you do not have to take our word for it. The American Society for Engineering Education (ASEE) affirms in its Statement on Sustainable Development Education:

> Engineering students should learn about sustainable development and sustainability in the general education component of the curriculum as they are preparing for the major design experience. For example, studies of economics and ethics are necessary to understand the need to use sustainable engineering techniques, including improved clean technologies. In teaching sustainable design, faculty should ask their students to consider the impacts of design upon U.S. society, and upon other nations and cultures. Engineering faculty should use systems approaches, including interdisciplinary teams, to teach pollution prevention techniques, lifecycle analysis, industrial ecology, and other sustainable engineering concepts. . . .
>
> ASEE believes that engineering graduates must be prepared by their education to use sustainable engineering techniques in the practice of their profession and to take leadership roles in facilitating sustainable development in their communities. —ASEE website

Sustainability is also included in the Accreditation Board for Engineering and Technology (ABET) Engineering Criteria 4, stating that the curriculum must include a major design experience that incorporates realistic constraints, including "most" of the following considerations: "economic, environmental, sustainability, manufacturability, ethical, health, and safety, social, and political" (*Engineers Forum on Sustainability,* March 2003, p. 1). If your are using this book in a class or are using it at work, it is probably because your university or company or engineering professional society has been encouraged by one or a number of signals (economic, environmental, social justice, legal, etc.) to consider making the shift to green engineering practices. We are confident that this book will make the case for green engineering unassailable, exciting, and challenging.

This does not mean that there won't be dead ends, failures, and unintended and unforeseen consequences along the way. The practicing "sustainable" or "green" engineer will not be able to avoid such experiences, but he or she will be practicing an approach to engineering that cooperates as much as possible with natural systems and forces and will seek to understand—in Wes Jackson's phrase—the "genius of the place" in designing products and processes. The practicing engineer will incorporate earlier environmental discoveries that resources are not limitless nor are sinks infinitely capable of handling our wastes and will think in terms of systems interacting, overlapping, and—unfortunately—sometimes conflicting.

Green engineering practices, in providing a link and a bridge between humans and nature, recognize an obvious link and interdependence between human beings and the natural world. In seeking to enlist nature for the use and convenience of humans, the sustainable engineer must not ignore the limits of living systems or their collective creativity and genius. Nor must the sustainable engineer create solutions for human interest and service without likewise limiting the damage to or even enhancing the power of living systems to provide the feedstocks of energy and materials—as well as nature's regenerative capacities—necessary for modern life. To put it another way, the universe, at least in our small corner of it, has created conditions conducive to life. Green engineering works with these conditions to advance and improve the human condition.

## References

An excellent source for a broad call to rethink the entire college curriculum in the direction of sustainability can be found in David Orr's *Earth in Mind: On Education, Environment, and the Human Prospect*. Island Press: Washington, DC, 1994. Or visit http://www.oberlin.edu/envs/ajlc/ to see his ideas in action.

See http://www.undp.org/capacity21/ for information on the progress of implementing Agenda 21 from the 1992 United Nations Conference on Environment and Development in Rio de Janeiro.

Anastas, P. T. and J. B. Zimmerman. Design through the 12 Principles of Green Engineering. *Environmental Science & Technology*. March 1, 2003, pp. 95A–101A.

Carson, R. 1987 (1962). *Silent Spring*. Houghton Mifflin: Boston.

Davenport, W. H. and D. Rosenthal, eds. 1967. *Engineering: Its Role and Function in Human Society*. Pergamon: New York.

Dernbach, J. C. 2002. *Stumbling Toward Sustainability*. Environmental Law Institute: Washington DC.

Flader, S. and J. Baird Callicott, eds. 1991. *The River of the Mother of God and Other Essays by Aldo Leopold*. University of Wisconsin Press: Madison.

Goodland, R. and H. Daly. 1996. Environmental sustainability: Universal and non-negotiable. *Ecological Applications*, **6**(4):1002–1017.

Hardin, G. 1991. Paramount positions in ecological economics. In *Ecological Economics: The Science and Management of Sustainability*, R. Costanza, ed. Columbia University Press: New York.

Hawken, P. A. Lovins, and L. Hunter Lovins. 2000. *Natural Captialism: Creating the Next Industrial Revolution*. Back Bay Books: Newport Beach, Calif.

Leopold, A. 1970 (1949). In *A Sand County Almanac*. Ballantine Books: New York.

Lovens, A. 2001. Natural Capitalism: A presentation to the Australian Broadcasting System. www.abc.net.au/science/slab/natcap.

McDonough, B. and M. Braungart. 2002. *Cradle to Cradle: Remaking the Way We Make Things*. North Point Press: New York.

Perrucci, R. and J. E. Gerst, eds. 1969. *The Engineer and the Social System*. Wiley: New York.

Petroski, H. 1994. *Design Paradigms: Case Histories of Error and Judgment in Engineering*. Cambridge University Press: Cambridge, England.

Pinchot, G. 1910. *The Fight for Conservation*. Doubleday: New York.

Reynolds, T. S., ed. 1991. *The Engineer in America: A Historical Anthology from Technology and Culture*. University of Chicago Press: Chicago.

World Commission on Environment and Development. 1988. *Our Common Future*. Oxford University Press: New York.

## CHAPTER 1

# Understanding Industrial Ecology and the Multiple Dimensions of Sustainability

THOMAS P. SEAGER

Industrial ecology is a branch of science concerned with the interrelationships of human industrial systems and their environments. It is sometimes referred to as *the science of sustainability* because it is motivated by the hypothesis that industrial systems can be harmonized with the environment by imitating some of the salient features of natural systems, such as material efficiency and organism symbiosis. Its aim is to understand the natural and economic processes that relate to the well-being of future generations.

Industrial ecology has not yet established reliable modes of inquiry or a uniform framework for dialog, partly because sustainability is difficult to characterize and partly because industry–environment interactions are difficult to model quantitatively. Experts disagree on how to measure sustainability or what characteristics of industrial or natural systems are most worthy of maintaining for future generations. Many sustainability metrics have been proposed; they encompass different perspectives and analytical techniques. Each captures different qualities depending on the priorities of their designers. Considered together, however, they constitute a fragmented approach. A uniform framework has yet to be established or proposed.

This chapter traces some of the historical and intellectual antecedents of industrial ecology. It povides a concise lexicon of the biological analog, contrasts the two most promising analytical methods by which industrial ecology research may be carried out—life-cycle assessment (LCA) and systems analysis—and proposes a taxonomy by which the quantitative criteria of sustainability can be classified as economic, thermodynamic, environmental, ecological, sociopolitical, or aggregated.

This chapter introduces students to the concepts and methods of industrial ecology. Its goal is to teach students to assess critically the interrelationships and trade-offs between industry and the environment and the linkages among different approaches to sustainability.

**Glossary**

**Aggregation** is the process of summing or averaging measures with identical units that are associated with activities or things at different locations in time and space or are attributable to different chemicals or processes. Typically, life-cycle inventories are aggregated by linear summation of mass and energy resource consumption. Weighted aggregation schemes adjust for the quality of different resources, as well as quantity. Most LCA researchers assume a linear relationship between aggregated inventory values and environmental or ecological impacts, although this may not be the case.

**Dematerialization** is the use of plastics, composites, and high-strength alloys to reduce the mass of products and cited as an important trend in the auto industry wherein cars are becoming lighter, more fuel efficient, but also increasingly difficult to recycle as the materials from which they are manufactured become more diverse and complicated.

**Disaggregation** Whenever data are aggregated, some information is lost. In economic input–output analysis, the aggregation of economic activities into 500 sectors means that it may be difficult to identify which aspect of the life cycle makes the greatest contribution to the inventory, or impossible to make distinctions between different activities that have been lumped together.

**Design for Environment (DfE)** is a process of incorporating environmental specifications, especially those that are associated with the use or end-of-life aspects of the typical product life cycle, into product development or design criteria.

**Dose–response** refers to the incidence of disease or disability in a population as a function of the concentration of a chemical pollutant that has accumulated in or passed through the body. It is generally assumed that larger doses result in greater incidence (response), but this is not always the case. Also, some metals that are required in small amounts for normal metabolic activities may result in adverse effects when their dosages are too small, as well as too large. Consequently, dose–response curves may be U-shaped or upside-down U-shaped (with dose as the independent variable on the *x* axis and response the dependent variable on the *y* axis).

**Energy intensity** is a measure of the energy required to sustain an economic activity, typically normalized in units of joules per dollar or joules per kilogram.

**Emergy** is a life-cycle energy accounting of the energetic inputs required to produce a marginal increase in economic demand or population of a species in an ecosystem.

**Exergy** is a thermodynamic property of state that measures the maximum amount of work extractable from a thermodynamic system under ideal conditions, relative to ambient environmental conditions.

**Exposure** refers to the concentration of chemical pollutant that is mobile and bioavailable in the environment. Exposure does not always result in dosage. For example, not all of a pollutant that is present in the air in the lungs is transferred to the bloodstream. Some of the pollutant is likely to be exhaled.

**Global warming potential (GWP)** is a measure of the heat trapped over a specific time period (20-, 100-, or 500-year periods are typically reported) by a certain gas released to the atmosphere at a steady rate during that period, relative to an equivalent mass release of carbon dioxide. Units are expressed in equivalent kilogram of carbon dioxide per mass pollutant.

**Human toxicity potential** is a toxicity- and exposure-weighted aggregated measure of pollutant releases.

**Industrial ecology** is a branch of science concerned with the interrelationships of human industrial systems and their environments.

**Industrial ecosystem** is a model of a community or interacting system of firms that is based on a natural analog.

**Industrial metabolism** is the process by which mass and energy (exergy) flows are handled or transformed by the economy.

**(Economic) input–output analysis** is a method of estimating the overall demands on the economy related to a marginal increase in demand for a specific product or service.

**Life cycle** is a metaphor for tracking the resource requirements and the use and disposal of industrial products or services.

**Life-cycle assessment** refers to studying the origin, use, and eventual fate of industrial materials together to identify sustainability impacts or opportunities for resource savings.

- *Scoping* identifies the subject of study and boundaries of the analysis.
- *Inventory* is an aggregated accounting of the material and energetic resource demands, as well as residuals or by-products, required to produce an industrial product or service.
- *Impact assessment* is the process of identifying hazardous or deleterious environmental and ecological end points that are associated with a life-cycle inventory.
- *Improvement assessment* is the process of identifying opportunities for reducing life-cycle impacts.

**Materials intensity** is a measure of the mass required to sustain an economic activity, typically measured in terms of kilograms per dollar or kilograms per person.

**Ozone depletion potential** is a measure of the globally and annually averaged reduction in stratospheric ozone attributable to a steady-state mass flux release of a pollutant at the earth's surface, relative to an equivalent mass-flux release of chlorofluorocarbon-11 (CFC-11).

**Total equivalent warming impact (TEWI)** is a GWP-weighted aggregated life-cycle measure of the total equivalent kilograms of carbon dioxide associated with an industrial product or economic activity.

**Toxic release inventory** is a mass-weighted aggregation of toxic releases to the environment compiled yearly from a manufacturing industry survey.

**Toxicological risk assessment (TRA)** is a method of determining the probability or expected incidence of an adverse ecological or human health reaction to chemical pollution. Results may be expressed in terms of increased incidence of disease or disability, such as cancer or infertility, or expressed in a quotient relative to a threshold, reference, or background level of risk. A hazard quotient (HQ) of greater than 1 is usually considered unacceptable.

**Tropospheric ozone formation potential (TOFP)** or photochemical ozone creation potential is a measure of the air pollutants contribution to smog-forming chemical reactions. TOFP is reported as a quotient in comparison to a reference pollutant (ethylene), which is assigned a TOFP of 100.

**Triple bottom line** is a metaphor that extends profit-maximizing criteria to include environmental and social factors in business management.

**Supraoptimal** is a neologism coined to communicate the sense that expanding the boundaries of a system may lead to optimal solutions or designs that were inaccessible without cooperation between what were previously separately optimized subsystems.

## Introduction

Scientific approaches to the environment have continuously been changing in surprising ways. While most branches of science tend to evolve into increasingly specialized subbranches, the range of knowledge encompassing environmental topics has continuously expanded, as has environmental awareness. The state of the art has progressed:

- From remedial action on highly visible problems such as smoke, sewage, and hazardous waste
- To end-of-pipe pollution prevention treatments such as scrubbers, digesters, and clarifiers
- To waste minimization and green chemistry

Today, environmental engineers take courses in biological and ecological sciences such as organic chemistry, microbiology, or limnology in addition to the standard physical science and mathematics regimen that is required of all engineers. As the locus of environmental problem solving has moved up the pipe from the outfall to the treatment plant to the chemical reactor, the range of issues confronted by all engineers has become increasing complex.

Now, environmental issues are moving even further. The cutting edge of environmental strategy is no longer found on the factory floor. It is found in design meetings, marketing offices, and in corporate boardrooms. This new paradigm demands changes in engineering practice. Engineers of all types are increasingly called on to incorporate environmental specifications in new-product or process designs. They must consider not only how products are made and how they are used, but also what happens to them after they are used up. Should they be recycled? Remanufactured? Reused? Moreover, what was previously focused just on products and pollution must now also consider resources. At some point, engineers will no longer be working on what comes from the pipe; they will be more concerned about *what is going in*. The difference is indicative of a move from concern for the environment to concern for *sustainability*, which is a broader concept that captures an overall sense of environmental and societal well-being.

Environmental engineering, in the meantime, may be on the cusp of another major expansion: incorporating tools that come from schools of liberal arts and business. This book is one example. If environmental engineers must increasingly be called on to join with business executives, managers, or marketers, should they not also be able to apply principles of management, economics, or communications just as they must apply the principles of physics, chemistry, or biology? It is a formidable multidisciplinary challenge. For this reason, environmental problems today are more likely to involve teams of people with a variety of expertise. No one textbook, discipline, or even degree program can possibly prepare a student for the myriad issues related to sustainability. This chapter is an attempt to introduce the issues related to sustainability and to begin to classify them.

This chapter focuses on ideas that apply to quantitative problem solving, that is, the basis of engineering design. Many facets of sustainability, however, are difficult to quantify. To be effective, a metric must not only be reliably related to performance objectives, it must be easy to communicate to other members of a sustainability team. It must be something that inexpert individuals can appreciate and feel confident in, and it must have transparent methods of calculation. Furthermore, independent researchers must be able to reproduce the results. Sustainability brings new audiences, new problems, new engineering challenges, new methods, and new measures to engineering.

## What Is Industrial Ecology?

*Industrial ecology* is a term employed by researchers and practitioners from many scientific disciplines. It has not yet established a consistent definition or a uniform analytical framework. Engineers in particular are far from universally embracing the concept, often believing that industrial ecology provides little quantitative basis for design. Many engineers are sympathetic to the ethic that motivates industrial ecology —the production of goods and services in our economy should be accomplished with fewer thermodynamic resources, fewer adverse environmental effects, and greater economic efficiency—but engineers are (by and large) at a loss to put the concept of a natural analog into practice.

In industry, engineers may be employed in two ways (besides research) that relate to the environment: product development and environmental health and safety (EHS). It is far more common to see the principles of industrial ecology employed in product development teams (such as in *design for environment* specifications or gateways) than in EHS, although the latter seems more directly related to environmental objectives. In part, this is because EHS teams are organized around end-of-pipe solutions, legacy issues (e.g., old hazardous waste sites), emergency responses such as spill containment, or chemical fate and transport. In the industrial ecology ideal, there is no *end* to the pipe; industrial materials are constantly recirculated by the economy just as natural materials are constantly recirculated in nature.

Unfortunately, the industrial ecology ideal remains a distant vision. To date, industrial ecology has not yielded to concise explanation or reliable investigative methods. It has been variously described as a metaphor, paradigm shift, broad umbrella of concepts, socioeconomic trend, agenda, and science that is still being defined by its proponents. A myriad of definitions, descriptions, and new terms have appeared, disappeared, or reappeared in the recent literature only to confuse experts and neophytes alike. Establishing a common vocabulary is an important first step to founding a new science, but it takes time for scientists to become comfortable with new interpretations or meanings for otherwise familiar words.

The neologism *industrial ecology* was originally popularized by a *Scientific American* article that proposed industrial systems would function more efficiently and with fewer environmental effects if they were modeled after natural ecosystems; with such an approach, the consumption of energy and materials was optimized, waste generation minimized, and the effluents of one process served as the raw materials for another process (Frosch and Gallopoulos, 1989). This supposition is an extension of what was previously called *industrial metabolism* and characterized as the energy-and-value-yielding process essential to economic development. It is analogous to the metabolic processes that are essential to life (Ayres, 1989).

Ironically, the seminal authors were trained not in ecology but physics. The earliest articles emphasized the physical aspects of both natural and industrial systems such as material and energetic efficiency, rather than biodiversity, species succession, or evolution. Nevertheless, the first textbooks in the field almost completely favored the term *ecology* over *metabolism*. That change by itself may be considered a broadening of the metabolic analogy, suggesting a larger scale of analysis. For a brief time, some authors may have used *ecology* and *metabolism* interchangeably. However, just as in natural ecology, each term has come to imply a different perspective for study. *Metabolism* focuses on the material and energy transfers necessary to sustain an organism, or part of an organism (such as a cell), or organization (such as a bee hive or a household); *ecology* emphasizes the interactions between different components that imbue a larger structure called a *system*.

The word *industrial* has also been subject to debate and requires clarification. By and large, it is meant to apply to all kinds of human activities that relate to material and energy consumption, including agricultural, manufacturing, transportation, mining, fishing, and service industries, not just factories. Arguing for a broad definition of *industrial* might lead to the logical conclusion that the natural and industrial are not separate systems but inexorably linked and, therefore, must be considered together. Perhaps a merging of natural and industrial ecology may seem the inevitable consequence of a holistic application of the natural analog, but a theoretical basis for modeling the linkages has not yet been established, and the complexities involved represent an enormous barrier to collective study. An assumption in industrial ecology, rarely questioned or discussed, is the hypothesis that a boundary can be drawn between human activities and natural ones and that materials, energy, or information flows that cross the boundary can be accounted for as if each were a separate system. At global scales, such an approach is dubious, although it may be applicable at the small scale of a specific factory or product life cycle.

Nevertheless, it is clear that an analogy has led to a new field of study and a new perspective from which to model industrial processes. Therefore, to facilitate clarity, understanding, and economy of thought, this chapter defines these terms thus:

*Industrial ecology* is a branch of science concerned with the interrelationships of human industrial systems and their environments.

*Industrial metabolism* is the process by which mass and energy (exergy) flows are handled or transformed by the economy.

*Industrial ecosystem* is a model of a community or system of economic entities that is based on a natural analog.

These definitions are broad enough for a wide range of applications and consistent with the biological and etymological roots that inspired coining of the terms.

To apply the concepts, however, engineers require quantitative criteria and objectives. The focus of engineering is design, and the word *design* implies a **purpose**. Every design alternative must be judged against measurable criteria; otherwise, it would be impossible to determine which design best suits the purpose at hand. In industrial ecology, metrics and measures abound, but they are difficult to relate to objective scales. What does it mean to say that one technology results in release of gases to the atmosphere that may warm the planet in comparison to another that may contaminate the groundwater beneath a landfill? While understanding the consequences of design is informative, many specific measures are incommensurate—meaning that design depends entirely on a client's subjective preference, say, for the risk of global warming compared to the risk of groundwater contamination. Industrial ecology, therefore, requires new broad-based metrics and multiple perspectives that help to inform these judgments.

## Hypotheses of Industrial Ecology

Industrial ecology research is motivated by the intuitive sense that modeling industrial systems after natural ecosystems could lead to more efficient and profitable use of energy and raw materials by minimizing waste and adverse environmental effects. This hypothesis is manifested at two scales: the macro and the micro. The most commonly cited is the macroscale of the system. Here, the analogy emphasizes the interactions between different sectors of the economy in the same way the natural ecology emphasizes the interactions between different species in an ecosystem and suggests that the material or energetic residuals of one industry can be employed to benefit another. On the macroscale, the analogy calls attention to the reduction or elimination of waste by discovering symbiosis and is supported by the mathematically defensible position that efficiency gains may be realized by removing barriers separating independently optimized subsystems and adopting a more holistic approach that again optimizes the connected systems in concert (see also Paton, 2001).

On the other hand, the microscale application of the analog emphasizes biomaterials, biocatalysis, or bioprocesses with the potential to increase the thermodynamic efficiency, resource renewability, or sustainability of the economy by mimicking technologies or employing renewable resources found in nature. The theory and methods of macroscale applications are far more advanced than microscale applications, but there is not yet any reliable way to test whether application of the principles of industrial ecology really leads to more sustainable choices. This is because sustainability is difficult to define in a manner that can be measured, even by multiple metrics and because some potential important effects may have no measure. For example, whether the investigative methods of industrial ecology can evaluate the environmental or ecological implications of biotechnologies such as genetic engineering remains an unanswered question.

### Contrasting Analytical Perspectives

Two perspectives provide an analytical basis for industrial ecology: life-cycle assessment (LCA) and systems analysis. Each embodies the notion that environmental problems should be examined with an increasingly holistic rather than reductionist approach. Although the two approaches have significant differences, they are both extremely sensitive to how the **boundaries of study** are defined. In fact, one of the principle motivating hypotheses of industrial ecology is the idea that **as the boundaries expand, *supraoptimal* solutions may emerge**. That is, the most favorable outcome may be found by coordinating the activities of all system components rather than by combining the single best option of each subsystem.

Still, LCA and systems analysis approach this synthesis of separate subsystems in different ways. See Table 1.1.

## Life-Cycle Assessment

Life-cycle assessment is the principal tool by which industrial ecology research is carried out. Two approaches predominate:

- Bottom-up, or processed-based method—developed most intensively by the U.S. Environmental Protection Agency (USEPA, 1993) and the Society of Environmental Toxicology and Chemistry (SETAC, 1993)
- Top-down, or economic input-output based method.

Each typically follows a four-step methodology consisting of (1) scoping, (2) inventory analysis, (3) impact assessment, and (4) improvement assessment. LCA is introduced in this chapter and expanded in Chapter 3.

## Scoping and Inventorying

*Scoping* is a process of identifying the goals that motivate the assessment and determining the proper boundaries of study. The *inventory* analysis is an accounting of the resource requirements of a particular product, process, or industry from extraction of virgin materials to final disposition. The *impact assessment* is conducted to relate the inventory data to specific environmental concerns. Finally, the *improvement assessment* (or interpretation phase) identifies the aspects of the material's life cycle that might be most amenable to mitigation or evaluates the potential for application of new strategies that offer the greatest leverage for environmental benefits.

The purpose of an LCA is to allow designers to gauge the comprehensive environmental implications of a technological activity so that

**Table 1.1** Comparison of Approaches: Systems Analysis and LCA

| Factor | Systems Analysis | Life-Cycle Assessment |
|---|---|---|
| Approach | Flexible: amenable to any scale, from single product or process to an entire industry or geographical region | Metaphor of a product of process lifetime—although this is difficult to define unambiguously |
| Objective | Explicitly emphasizes optimization and decision making | Predominantly comparative |
| Models employed | Prescriptive | Descriptive |
| Data requirements | Less extensive than LCA | More extensive than systems analysis, at least where systems analysis may neglect information that is beyond the boundaries of a focused study or irrelevant to the decision at hand |
| Applicability | Industrial ecology where the emphasis is on examining interrelationships | Industrial metabolism where the emphasis is on examining specific materials flows and processes |

decisions made at one stage of a product life cycle do not create adverse environmental consequences at other stages, although these stages may seem disconnected from a narrowly focused objective. For example, an automotive engineer may be charged with reducing the weight of a vehicle, thereby increasing the gasoline mileage. Increased gasoline mileage results in consumption of fewer thermodynamic resources and pollution for an equivalent economic utility; so the designer may clearly see the benefits. One design solution may be to employ complex specialty materials (with high strength-to-weight ratios) as a substitute for steel or simpler plastics. Specialty materials may be more problematic to produce and recycle, however. The overall benefit of increased mileage, therefore, must be weighed against the potential increased adverse environmental effects of manufacturing, maintenance, and final disposition of the new material. An LCA may be able to determine the relative importance of the resources consumed and pollution created at each stage, thereby justifying one decision as preferable to another.

Both process-based and economic input–output methods are controversial. Process-based LCA usually begins with a process diagram, which facilitates the scoping stage. A process diagram may be greatly simplified, such as that pictured in Figure 1.1 depicting the ethanol cycle: corn production, fermentation, and combustion in the transporta-

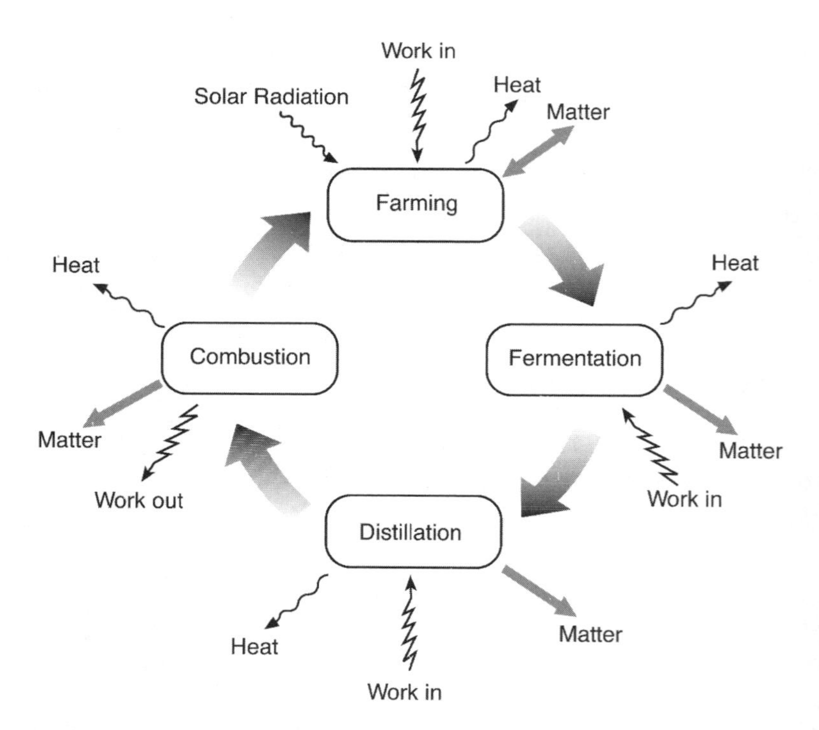


**Figure 1.1 Simplified ethanol process diagram.** The life cycle of ethanol (for corn production) may be depicted in a greatly simplified diagram that shows the basic production and consumption steps. (Adapted from Berthiaume et al., 2001. Reprinted with permission from Elsevier.)

tion industry as fuel. Alternately, it could be more detailed, such as that pictured in Figure 1.2 showing the production, use, and reclamation of an automobile catalytic converter. Researchers typically start with a final demand (fuel) and work backward through the chemical and physical transformation steps to the extraction of raw materials.

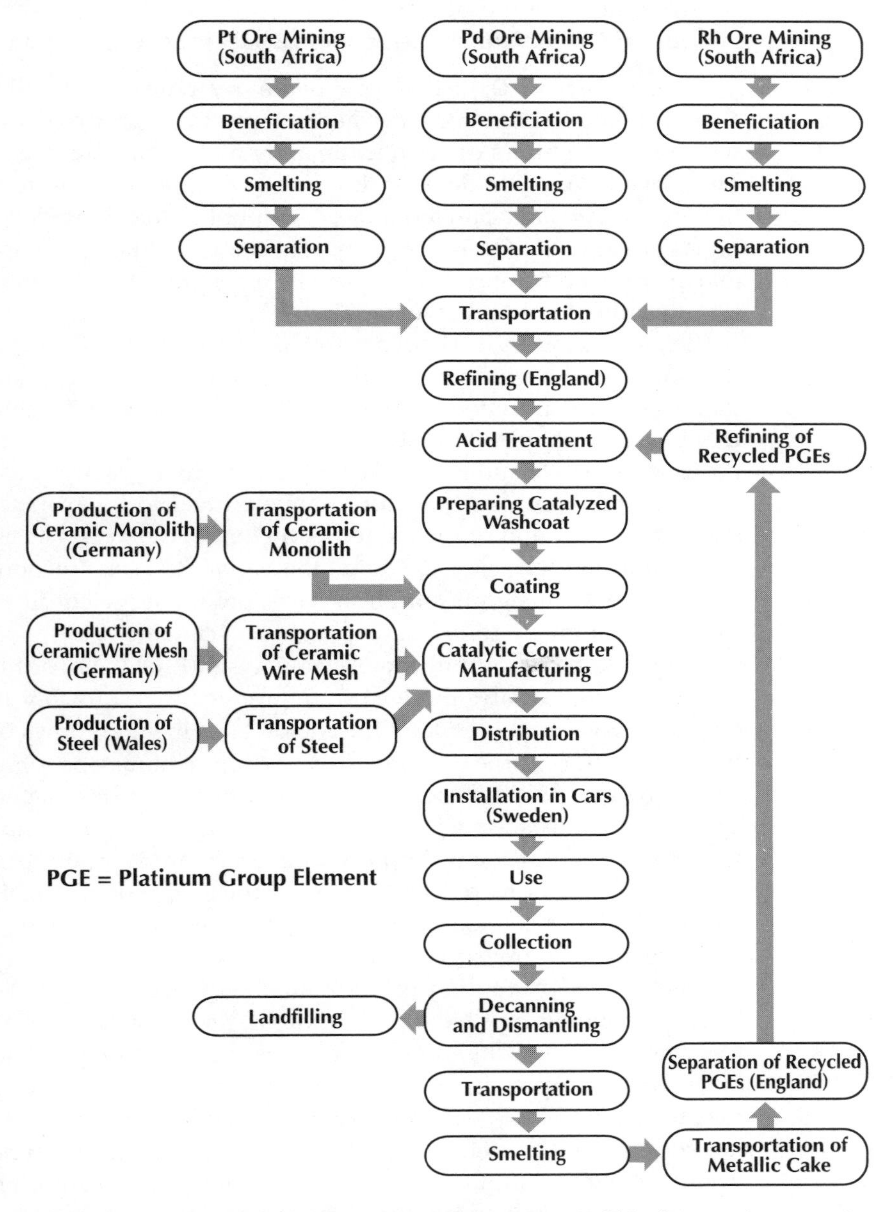


**Figure 1.2 Life-cycle diagram of a catalytic converter.** This diagram shows greater detail than the previous one, including information regarding the geographical location of some of the processes consider. (*Source:* Amatayakul and Ramnas, 2001. Reprinted with permission from Elsevier.)

Current LCA techniques have been criticized as unreliable, unscientific tools subject to quantitative and qualitative errors (Ayres, 1995). The limitations identified with current practices include the following:

- Lack of adequate inventory data
- Difficulties in identifying the boundaries of the system
- Disparate underlying assumptions that make LCA results difficult to compare or reproduce
- Environmental impact assessments that are not directly comparable

In many instances, a life-cycle inventory requires compilation of chemical information for intermediate products or by-products for which manufacturers or suppliers maintain little quantitative account. Even when such data do exist, it is often regarded as proprietary and therefore unavailable, unverifiable, or unpublishable. Boundary definitions vary among LCA researchers, resulting in studies or comparisons that are difficult to reproduce, especially when the alternatives compare significantly different raw materials.

In some instances, it is apparent that a specific materials cycle may have no logical beginning or end and that the boundaries of an inventory analysis may rely on arbitrary (though informed) judgments. In the ethanol example, corn stalks and leaves are assumed to be composted completely and recycled as soil inputs; in the catalytic converter example, platinum and palladium are assumed to be extracted from spent converters and resmelted for use in new converters. The life-cycle inventories in each case depend on the *assumed* recycle fractions. Such problems present significant obstacles to producing scientific research that different investigators can reproduce.

Economic input–output life-cycle analysis (EIOLCA) uses broader boundaries and creates more easily reproducible results, but it is less detailed (Lave et al., 1995). EIOLCA correlates material flows between up to 500 sectors of the economy (the environment may be represented as one or more sectors) for material and energetic interdependence. The entire U.S. economy is represented to ensure a wide definition of life-cycle boundaries, but imported materials are typically treated as domestic production (for lack of foreign data). One model is freely available on the Internet (eiolca.net) and greatly speeds the time required to estimate life-cycle inventories. Results, however, depend on nationwide averages, based on a survey of U.S. manufacturers that is updated every 5 years. Economic input–output analysis is, therefore, not a reliable tool for evaluation of plant-specific processes (because the plant's performance may deviate significantly from nationwide average results) or for assessment of new technologies (because the most advanced technologies will not be well represented by aggregated data from previous years). Moreover, input–output tables do not provide environmental data on use or disposal phases. (Use phases data must be constructed from a use series of "new" purchases of manufactured goods, such as gasoline, oil, water, or tires for automobiles, based upon the resource demands, for example, gas mileage, of a particular product.)

A hybrid approach combines the advantages of process-based and EIOLCA methods. In the hybrid approach, proprietary data must be gathered from factory or chemical process data, working backward from the final demand (e.g., a consumer product such as a laundry detergent) to commodity products (raw materials) that are available on the open market from many different suppliers (e.g., electricity). The life-cycle inventories attributable to commodity materials may then be estimated from input–output analysis. Figure 1.3 depicts a hybrid process diagram for ethanol production from sugar crops (sugar cane, sugar beets, corn). Economic input–output analysis can compile crop production life-cycle inventories, such as those depicted in Table 1.2 for sugar crop production. Process-based LCA, based on the technologies or specifics of the processes in question, can complete life-cycle inventories for conversion of sugar crops to ethanol. Note that EIOLCA is not specific enough to distinguish between sugar beet or sugar cane production but represents average quantities for all sugar crops. On the other hand, it greatly simplifies treatment of iron, steel, and other infrastructure or capital requirements that would be difficult to amortize in a process-based LCA, and it greatly speeds analysis.

The hybrid approach encourages manufacturing engineers to:

- Use data that are readily available from production experience, laboratory experiments, or pilot-scale studies for proprietary or new technologies.
- Complete the LCA inventory for processes outside their industry with census data accessible through standardized economic input–output tables.

This approach may resolve the difficulty of producing widely disparate results that depend largely on what information is excluded from the study or which underlying assumptions are applied.

## Assessment Stages

No matter which method is employed to build a life-cycle inventory, the impact assessment and improvement assessment stages remain problematic. In economic input–output analysis, inventory data are aggregated such that the researcher often finds it difficult to identify which stage of the life-cycle activity (raw materials extraction, processing, manufacturing, use, collection, reclamation, or disposal) is most important or how decisions at one stage may affect the environmental resources consumed at other stages. In this case, *disaggregation* of input–output data is required. This process may be just as time-consuming and ambiguous as process-based LCA. On the other hand, even when life-cycle inventories are constructed separately for each stage, environmental effects are likely to be difficult to compare or assess based on simplified criteria. Resource consumption and pollution are typically dislocated in time and space. Static comparisons do not account for these factors.

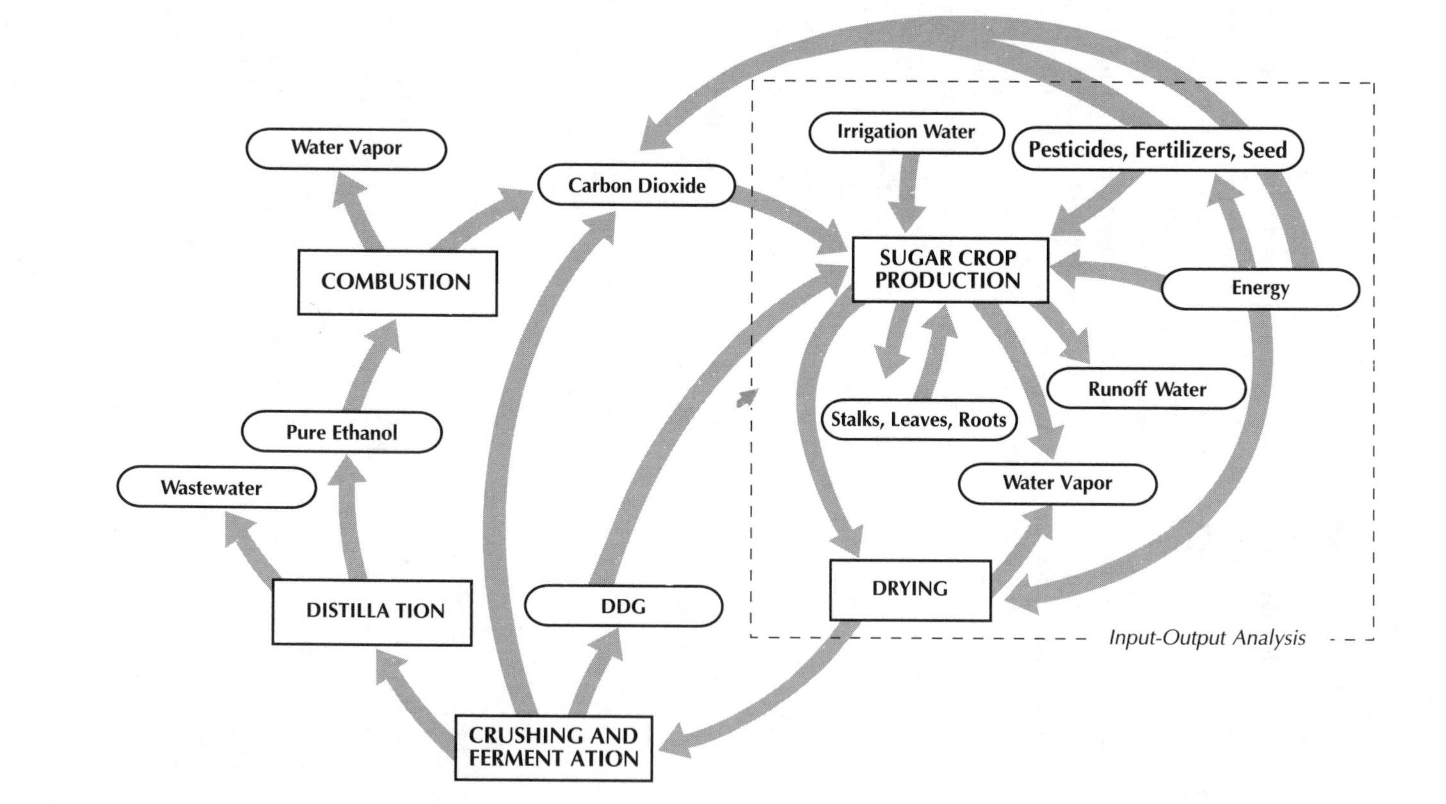


**Figure 1.3 Hybrid LCA diagram for ethanol production for sugar crops.** In the hybrid approach, commodity production (such as sugar crops) is modeled with economic input–output (EIO) data, but proprietary or novel technologies are modeled with process data. The combination of the two LCA methods results in a life-cycle inventory that incorporates the broad boundaries of the entire domestic economy (albeit in a general way) and specific process data. (DDG is dried distiller grain.)

**Table 1.2** Economic Input–Output Inventory of Sugar Crop Production

| Effects | Total for All Sectors | Measure |
|---|---|---|
| Economic purchases | 1.77 | $ million |
| Electricity used | 0.26 | MkWh |
| Energy used | 8.4 | Terajoules |
| Conventional pollutants released | 47.9 | Metric tons |
| OSHA safety[a] | 0.0021 | Fatalities |
| Greenhouse gases released | 583 | Metric tons $CO_2$ equiv. |
| Fertilizers used | 0.009 | $ million |
| Fuels used | 8.1 | Terajoules |
| Ores used—at least | 32.1 | Metric tons |
| Hazardous waste generated RCRA[b] | 58 | Metric tons |
| External costs incurred median | 0.11 | $ million |
| Toxic releases and transfers | 0.53 | Metric tons |
| Weighted toxic releases and transfers | 5.2 | Metric tons |
| Water used | 0.0018 | Billion gallons |

*Source:* Carnegie Mellon University Green Design Initiative. 2003. Economic Input–Output Lifecycle Assessment (EIO-LCA) model (Internet). Available from http://www.eiolca.net/ (accessed 28 July 2003).
[a]OSHA = Occupational Safety and Health Administration.
[b]RCRA = Resource Conservation and Recovery Act.

A streamlined LCA may partially overcome the difficulties of completing an extensive impact assessment. The methodology is greatly simplified. A subjective score is assigned to each component of a matrix describing the type of effect (e.g., energy consumption and solid waste generation) versus portion of the life cycle. The advantage of this method is the reduced time frame for completion—a few days rather than several months—and the ease in communicating the results of the analysis to administrators and the public. This method does little, however, to simplify the scoping stage in which the boundaries of study are defined and the subjective aspects make it nearly impossible to corroborate research results. [A subjective process is not necessarily invalid. The criticism here is that streamlined LCA is not scientifically reproducible by independent researchers because subjective studies depend on value judgments that vary among different organizations and individuals, yet there is nothing invalid about basing decisions on a subjective analysis consistent with one's own values (Graedel, 1998).] Table 1.3 depicts a streamlined impact assessment of alternative automobile propulsion technologies. One of the advantages of streamlined impact assessment is the ease with which social issues may be incorporated, such as energy independence. The subjective nature of the streamlined assessment may result in confusion or controversy regarding the basis of the scoring, however. For example, could global warming and fossil fuel depletion be considered "social" issues?

Many challenges must still be met to improve the utility of LCA as a decision, design, or policy-making tool. Future methods are likely to incorporate more information on economic costs, uncertainty, and

**Table 1.3** Streamlined Assessment of Alternative Automotive Technologies[a]

| | Reformulated Gasoline | Reformulated Diesel | Compressed Natural Gas | Ethanol from Biomass | Battery Electric Vehicle | Hybrid Electric Vehicle | Fuel Cell |
|---|---|---|---|---|---|---|---|
| Vehicle emissions | | | | | | | |
| Ozone, $NO_x$, VOC | +2 | −1 | +3 | +2 | +4 | +3 | +3 |
| Particulate matter | 0 | −1 | 0 | 0 | +2 | +1 | +1 |
| Air toxics | +1 | −3 | +2 | +1 | +3 | +3 | +3 |
| Fuel related | | | | | | | |
| Fuel cycle emissions | −1 | −1 | +1 | −1 | −1 | 0 | 0 |
| Fuel cost | −1 | +1 | +1 | −4 | +1 | 0 | 0 |
| Vehicle performance | | | | | | | |
| Range | 0 | +1 | −2 | −1 | −4 | +1 | +1 |
| Vehicle costs | 0 | 0 | −1 | 0 | −2 | −2 | −2 |
| Social issues | | | | | | | |
| Infrastructure cost | −1 | −1 | −2 | −3 | −1 | 0 | 0 |
| Energy independence | −1 | +1 | +2 | +4 | +1 | +1 | +1 |
| Global warming | −1 | +2 | +1 | +5 | 0 | +2 | +2 |
| Fossil fuel depletion | −1 | +1 | 0 | +5 | +1 | +1 | +1 |

*Source:* Reprinted with permission from Lave et al., © 2000 American Chemical Society.

[a] +5 is considered much better than a conventional gasoline automobile; −5 is considered much worse.

chemodynamic or toxicological properties of pollutants in the environment; they probably will also more closely involve public values or social concerns. Incorporating cost data will enhance the utility of LCA as a decision tool for business (Norris, 2001; Emblemsvag and Bras, 2001). Incorporation of uncertainty in the inventory and assessment stages will improve the credibility and applicability of LCA to environmental decision making. Chemodynamic and toxicological data will better relate LCA to other environmental decision methods such as risk assessment (Bennet et al., 2002; Hofstetter and Hammit, 2002; Nishioka et al., 2002). Additionally, public involvement will help to bring inexpert information to the scoping and boundary definition stages and to assess the relative importance of disparate or incommensurate impacts that call for comparison of different alternatives (Anex, 2002). For example, should untreated wastewater released to the deep ocean be assessed in the same way as treated wastewater released to a freshwater river or lake? Should power plant emissions in heavily populated areas be given the same weight as those in remote locations? How should the long-lived but locally contained and slowly decaying risks of radioactive waste associated with nuclear power be compared with the risks of global warming associated with carbon dioxide, which are widespread, perhaps decades distant from now, but worsening and perhaps irrevocable? The answers to these questions may lie outside the purview of LCA itself, although LCA is a tool for gathering information that will inform the answers. The principal advantage of LCA is the broad framework that guides the investigative process beyond local boundaries, thus forcing consideration of factors that may previously have been ignored. LCA is informative as a vehicle for exploring an intuitive appreciation of a more holistic approach.

The photographic industry shows how even nonquantitative approaches can be insightful. Figure 1.4 represents a management perspective typical of current photographic industry practices. The emphasis is on reducing waste and pollution from the source of production. Raw material requirements and pollution emissions are benchmarked to production of "finished" product—for example, photographic film, paper, or chemicals.

Figure 1.5 illustrates a more comprehensive view consistent with the product life-cycle analogy. Typically, photographers capture extra exposures of critical scenes (the "just-in-case" shots) or take exposures that later prove to be unsatisfactory (e.g., the "I look fat in that" photos) or take additional, unnecessary exposures "just to use up the roll." Current industry practice is to develop and print all of the exposures, even to create double prints, before returning the images to the photographer. Although not all photographs are equally valued by the photographer, they all consume equivalent thermodynamic, economic, and environmental resources. The fact that most photographic prints go into long-term storage (the shoe box) without ever being revisited shows that a great portion of the film and paper that is perceived by the manufacturer to be final product *may in fact be perceived by the photographer as waste.*

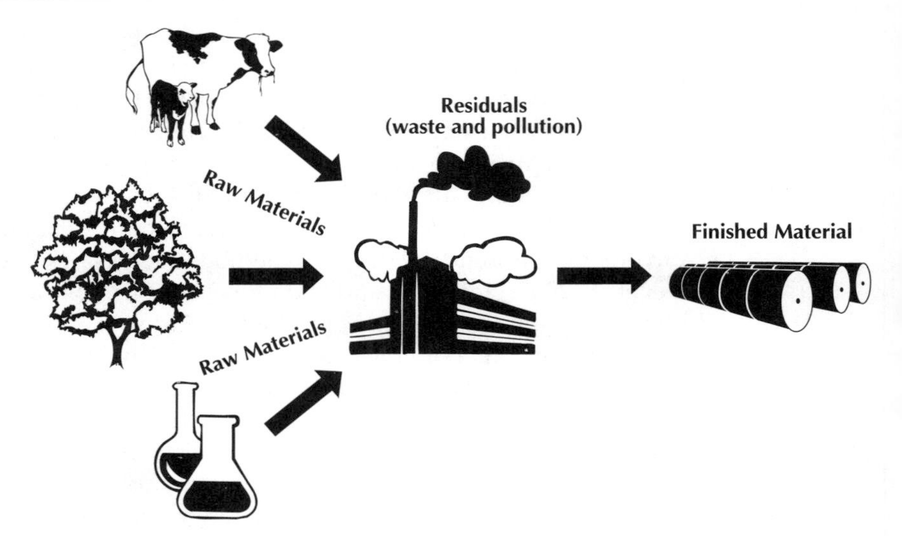


**Figure 1.4 Compliance perspective.** Waste and residuals are typically measured at the boundaries of factory, normalized to final product: In this case photographic film, paper, chemicals, or other materials. (Adapted from Seager and Theis, 2002.)

This example illustrates the conflict between suppliers, such as film manufacturers, and consumers that is typical in our economy. Photographers are not interested in buying film *per se*. The photographer is only interested in what *film does:* help to produce photographs. From this perspective, photographers have little interest in redundancy at intermediate stages of the photographic process (extra film or double prints of unwanted images) but might be better off paying only for the final photographs they really value; this is often the basis on which professional photographers are hired. On the other hand, manufacturers profit only by selling film and chemicals. To the extent that photography is inefficient, the additional materials consumed create extra profit opportunities for materials manufacturers.

The interests of the materials suppliers and the product consumers are at odds. One way to realign these interests is to decouple the business value from the materials so that the final consumer contracts to purchase *services*, rather than *things*. For example, digital technologies may make photographic processes much more environmentally and economically efficient. Digital cameras typically allow photographers to delete unsatisfactory exposures by displaying them on a small screen at the back of the camera. This obviates the need for developing and printing unwanted photographic prints. Images stored in the camera can then be uploaded on the Internet to a printing service that sends regular photographic prints in the mail. In a business model borrowed from the cellular phone industry, it could be possible to contract with a digital image printer to pay a fixed fee for the right to receive a specified number of photographic prints each month. The cost of the camera could be included in the monthly fee in such a way that it is advertised to the consumer as "free." The vendor has the advantage of locking in

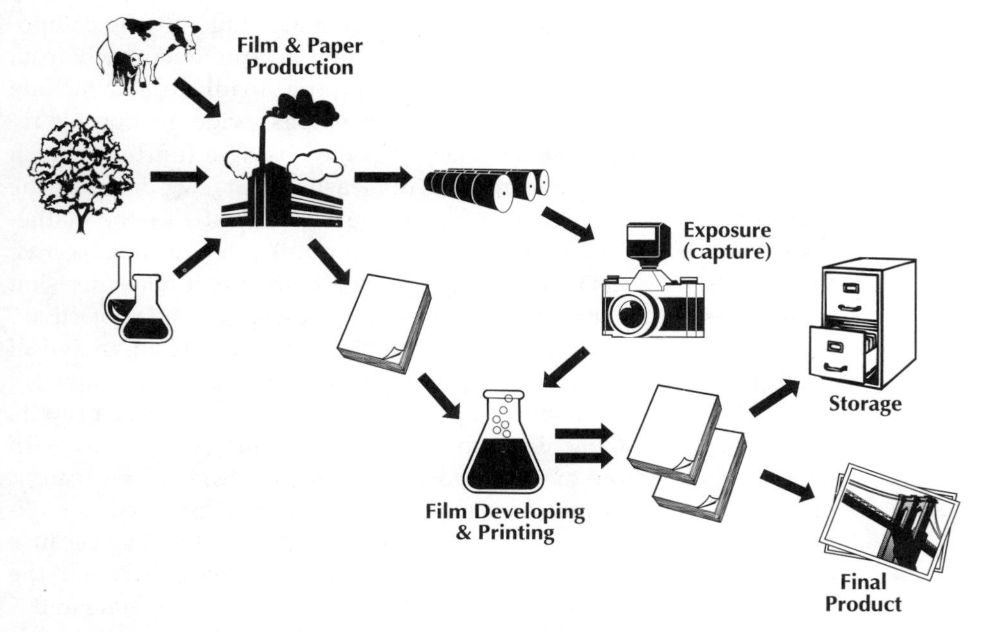


**Figure 1.5 Life-cycle perspective.** Most photographs go into storage or disposal, not on display. In this sense, much of the film, paper, and other materials required to produce an unsatisfactory photograph may be perceived by the photographer as waste, although it is counted by the manufacturer as product. The life-cycle perspective suggests that waste and residuals be measured along the complete product chain, not just at the factory, and normalized in terms of final services to the consumer, not material production. (Adapted from Seager and Theis, 2002.)

a fixed stream of revenues, capturing a customer with whom the vendor will have constant contact for 2 years or more, and presumably will be able to offer photographs (and other services such as T-shirts, coffee mugs, holiday cards, frames) valued more highly by the customer with less material, energetic, and environmental expense. The photographer is spared the large capital expense of the camera, the waste of spoiled rolls of film, unwanted exposures, double prints, and the trouble of trips to the store to drop off film and pick up prints. Service arrangements realign vendor incentives to profit from efficiency rather than waste. Whether this paradigm is *really* preferable from an environmental standpoint depends on the life-cycle issues involved in the enabling technology. Are the batteries used by the digital camera disposable or rechargeable? What are the environmental effects of manufacturing the computer and memory chips in the camera? of using the Internet?

## Systems Analysis

Systems analysis typically requires a mathematical model that characterizes the relationships and constraints governing various systems components. The model is usually the result of a careful examination of the cosmos in question during which quantitative links among components are established. Boundaries may either be drawn narrowly (e.g., around a single manufacturing facility) or more broadly (e.g., to include suppliers, partners, customers, or to encompass wide geographic regions). The focus of systems analysis is the objective function, which must be expressed in uniform units of measurement (e.g., dollars). The goal of a systems analyst is to find a solution that satisfies the mathematical model for the maximum (or minimum) value of the objective function. Systems analysis is a design tool; therefore, it helps decision makers to focus all the elements of a system toward a single objective.

Selection of a *measurable* unifying objective metric is an analytical prerequisite. Current management approaches focus primarily on pecuniary measures: maximizing profits, market share, revenue growth, or minimizing manufacturing costs while maintaining compliance with emissions regulations or self-imposed environmental, health and safety, or other constraints such as public image. It is often assumed in engineering and economics that business decisions are made solely because of a profit-maximizing principle, but in many instances, this is not the case. For example, monopolistic public utilities typically operate under constrained profit margins, instead minimizing costs while maintaining a high standard of service. The explosion of Internet companies in the 1990s was a more dramatic example. Few (if any) were profitable but instead operated under a maximum growth strategy, offering deep discounts or free services to such an extreme extent that they subsidized customer purchases with venture capital in an effort to capture a first-mover advantage and become the *de facto* technology or market standard. Most went bankrupt before a leading position in market share could be translated into a sustainable business enterprise (or profits, for that matter).

Returning to the photographic industry, it is possible to contrast the life-cycle perspective of the previous figures with the perspective depicted in Figure 1.6, which illustrates a systems model of the silver halide and digital photographic imaging chains. Three steps are common to both: capture, processing, and output (or printing). The feasible or hypothetical links between each step on each chain create a myriad of technological possibilities that may efficiently serve a variety of photographer preferences. Both LCA and systems approaches may be adapted to provide contrasting interpretations of the photographic industry model depicted:

- An LCA approach entails separate assessments of each imaging step or link to determine the comparative environmental characteristics of each pathway or identify any opportunities for improvement along any particular pathway.
- The systems approach models the functional relationships at each box and link to determine which pathway best meets the objective criteria.

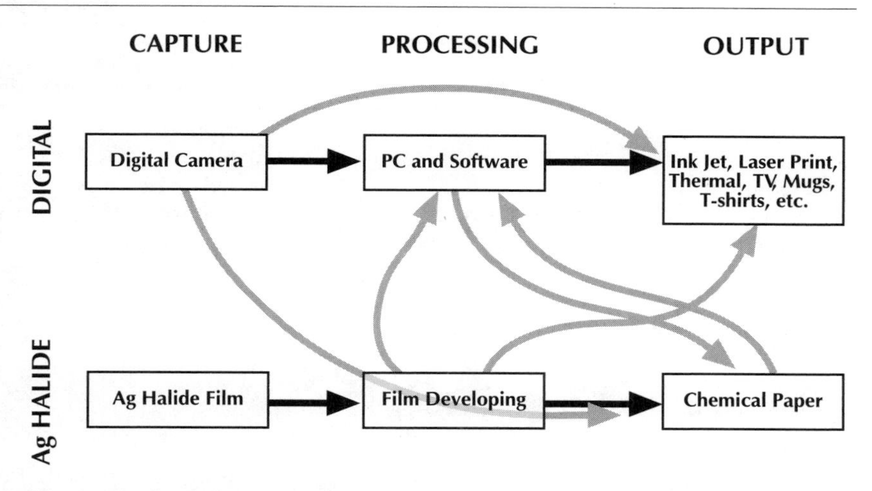


**Figure 1.6 Systems perspective.** New technologies have made a wide variety of process diagrams possible for the capture, storage, and production of images. Is it possible to determine which imaging chain is optimal from a sustainability perspective? (*Source:* Seager and Theis, 2002.)

Because the systems approach is more flexible, it may have advantages as a decision or design tool. The data requirements can be less onerous, and the system boundaries can be defined unambiguously. For example, suppose the focus of a study is to determine the optimum printing technology, comparing ink jet and silver halide photographic printing. As far as each is printed on an identical paper base, the environmental effect of the base is irrelevant to comparison. Only the different coatings, consumable chemicals, and energy requirements need be studied. Although an LCA approach requires consideration of every aspect including the paper on which the image is printed, systems analysis can exclude those factors that remain uninfluenced by the decision without altering the conclusions. In this way, systems analysis allows a more specific definition of system boundaries and thus partially overcomes one of the primary obstacles to LCA: the extensive data requirements called for by the life-cycle inventory. This principle is applicable to other aspects of the imaging chains as well. For example, both digital and film cameras must contain similar elements: camera body and a lens. In some instances, they are interchangeable. To compare the relative environmental effects of each technology, a more focused approach is expedient. Consequently, it is not necessary to complete comprehensive LCAs to draw effective comparisons; an abridged or comparative LCA may be adequate. This, however, dilutes the principal benefit of the life-cycle metaphor—the notion of expanding the boundaries of study; systems analysis is a more appropriate term for such an approach. Table 1.4 summarizes the principal differences between the two perspectives.

**Table 1.4** Summary of Life-Cycle Analysis and Systems Analysis

| | Life-Cycle Assessment | Systems Analysis |
|---|---|---|
| Purpose | Descriptive | Prescriptive |
| Boundaries | Cradle to cradle/grave | Scalable |
| Data requirements | Broad | Focused on decision |
| Emphasis | Materials cycling | Any uniform metric (such as dollars) |
| Applicability | Industrial metabolism | Industrial ecology |

*Source:* Seager and Theis, 2002. Reprinted with permission of Elsevier.

## Metrics of Industrial Ecology

Hypothetically, the optimal photographic image chain could be selected based on some environmental metric. But what metric should be employed? Toxicity measures are too narrowly focused on human health end points and neglect environmental issues such as climate change, ozone depletion, or eutrophication. Moreover, strictly environmental measures may be counterproductive to the extent that social or economic factors may be ignored. In the absence of any ideal measure, to what purpose should industrial systems be designed? How can engineers measure the efficacy of various alternatives? *Perhaps the greatest obstacle to pursuit of a quantitative environmental systems approach for industrial ecology remains the fact that no single metric exists that embodies sustainability.* This partly explains why the original industrial ecology hypotheses have yet to be tested in any comprehensive scientific way.

Most sustainability metrics may be characterized in a taxonomy that includes six broad categories:

1. **Economic** metrics estimate environmental effects or ecosystem services in terms of money so that they may be compared with monetary transactions or industrial accounts. Economic metrics are easy to communicate to diverse audiences. Estimates of the external (or social) costs of pollution and the value of ecosystem services are widely quoted and applied (Costanza, 1998). Both the calculation methods and the validity of the concept of pricing the environment are recognized as controversial, however. The methods and figures reported are extremely uncertain, if meaningful at all. Most importantly, monetization may lead to the erroneous assumption that environmental exploitation can be revocable, reversible, or reparable in the same manner as financial transactions, although in many cases ecological systems can be damaged beyond recovery. (See side bar on economics and thermodynamics in Chapter 4.)
2. **Thermodynamic** metrics are usually intensive measures that are normalized to representative units such as kilogram/person or kilowatt-hour/product. While they indicate resource requirements of industrial activities, thermodynamic metrics usually do

not indicate the specific environmental consequences of resource consumption. Sometimes thermodynamic output metrics are reported relative to thermodynamic inputs in a manner analogous to financial metrics, such as energetic return on investment or *emergy* yield ratio. The purpose is to attempt to indicate whether the resources consumed were invested wisely, resulting in a greater quantity of resources available for future consumption.

3. **Environmental** metrics estimate the extent of chemical changes or hazardous conditions in the environment. They may be simple measures of what is *released* to the environment without considering the fate or mobility of the pollutants after release. They may include chemodynamic considerations such as pollutant degradation, catalysis, or recombination to form new pollutants. They may include potency factors, such as toxicity, reactivity, or rarity. Environmental metrics are measures of environmental changes, not population health, and are not easily or reliably converted into quantitative ecological end points such as death, disability, or disease. (These are ecological metrics; see below.) Environmental metrics are often called *potentials*. The word *potential* in this sense conveys the sense that environmental pollution represents a hazard that may (or may not) be realized, depending on factors that are not expressed in a single number or are assumed to conform to general assumptions in the calculation model. It is possible for environmental metrics to be expressed in chemical or thermodynamic units such as biochemical oxygen demand, concentration, or pollution potential, but environmental metrics are distinguished from thermodynamic metrics by the fact that they are intended to measure environmental loadings or changes rather than resource demands. They are generally measures of the residuals created by industrial processes rather than the raw materials.
4. **Ecological** metrics attempt to estimate the effects of human intervention on natural systems in ways that are related to living things and the interdependent functions of the environment. Ecological metrics are often coupled directly with environmental metrics when the focus of study is the specific health or populations effects (such as cancer) of a particular pollutant. Figure 1.7 shows how small concentrations of cadmium in freshwater (an environmental metric) affects *Daphnia* reproductive rates (an ecological metric). Greater cadmium levels result in fewer young per female in comparable periods of time. Other studies may measure increased mortality or decreased body mass. The selection of an ecological (sometimes biological) measure is known as a hazardous *end point*. To relate environmental metrics to ecological end points, it is important to understand what fraction of the pollutant present in the environment is biologically available, how the pollutant enters the organism, and the relationship between pollutant dosage and organism effect. Not all species are affected in the same way

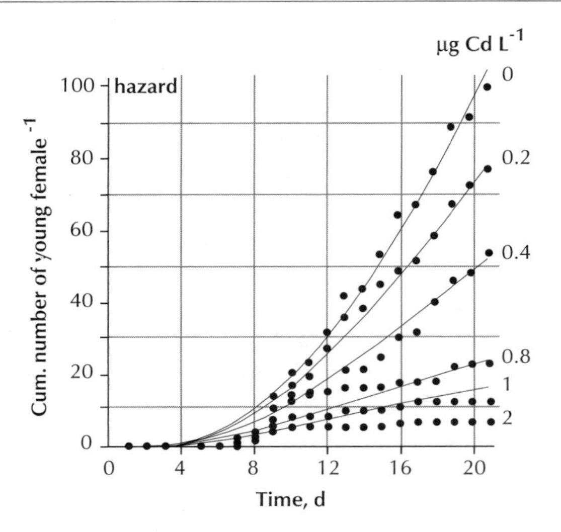


**Figure 1.7 Effect of environmental perturbation on ecological end points.** The accumulated average number of young per female *Daphnia* decreases as cadmium concentration in the water environment increases. In this case, the ecological metric (reproductive rate) may be directly correlated with the environmental (cadmium concentration). (*Source:* Cairns, 1998. This material is used by permission of John Wiley & Sons, Inc.)

by the same pollutant; some are more sensitive or tolerant of different pollutants in the same way that some fish require salt water and others require fresh.

The study of how organisms respond to poisons is called *toxicology*. When the source of the poison is environmental pollution and the dosage or response crosses several different levels of organism interaction (such as the accumulation of hydrophobic pollutants in the fatty tissues of higher predators), the field is called *ecotoxicology*. At the ecosystem level the hazardous end points may be more difficult to identify than they are in an individual organisms. Multiple effects may be caused by a single source. When a number of different ecological metrics are measured independently of each other but may be related to the same overall assessment of an ecosystem, they are called *indicators*. None is a definitive assessment, but taken collectively they may indicate that some ecosystem functions or relationships are seriously degraded. Table 1.5 contrasts example indicators that may be observed at the level of a single species with examples that may be indicative of the health of the ecosystem as a whole.

5. **Sociopolitical metrics** are much less familiar to engineers. With the exception of public hearings or opinion polls, elections, and ballot referendums (which are difficult to imagine being a function of design), engineers are rarely confronted with the social implications of their works or design decisions. Nevertheless, civil engineers (and perhaps environmental engineers in particular) are ostensibly charged with serving the public or social good, even

**Table 1.5** Example Indicators of Single-Species and Ecosystem Response to Environmental Degradation

| Single Species | Ecosystem |
|---|---|
| Accumulation<br>Concentration of pollutant in living tissues | Populations<br>Catch size, total population<br>Presence of preferred species, consumption restrictions, deformity<br>Prices, employment, stocking costs |
| Mortality<br>Life expectancy, death rate | Drinking water<br>Stocks, withdrawals, replenishment<br>Treatment costs, standards violations, consumption restrictions, acute illness incidence, user satisfaction<br>Willingness to pay<br>Depreciation of capital investment or debt service, treatment, and distribution costs |
| Neurotoxicity<br>Nervous system or sensory disorders<br>Behavioral disorders: muscle coordination<br>Brain disorders: memory, intelligence, speech | Recreation<br>Visits, boat registrations, tour attendance<br>Fishing and swimming restrictions, catch per unit effort<br>Employment, fees, gate receipts, stocking and repair costs |
| Reproductive toxicity<br>Birth defects, infant mortality, birth weight<br>Fertility: sperm counts | Productive use<br>Resource stocks, withdrawals, replenishment<br>Resource productivity, livestock losses, fertilizer, irrigation or maintenance costs<br>Compensation for loss of use, increased product cost<br>End-of-life treatment and disposal costs |
| Carcinogenicity<br>Tumor<br>Mutation | Aesthetics<br>Satisfaction, accessible shoreline<br>Odors, turbidity, eutrophy<br>Real estate values, willingness to pay |
| Cardiovascular disease | |
| Immunocompetency<br>Allergies<br>Autoimmune disorders: diabetes<br>Blood cell counts | |

*Source:* Adapted from Cairns 1998; reused with permission of John Wiley & Sons, Inc.

above the interests of their private employers. Who can dispute the far-reaching social and economic effects of transportation projects, dams, or even public sewerage? But how should the social good be *measured?* Are social objectives satisfied by more highway travel lanes? rising property or stock market values? attendance at sports events? There can be no doubt that sustainability will require social cooperation and a certain environmental equity or justice. Sustainability, if nothing else, is motivated by a shared *ethic* that conditions should be better in the future than they are today, and that we (as a society) should not be bettering ourselves at the expense of our children, grandchildren, or great grandchildren. The confusion lies in the term *better* and how to measure this. Could *better* be synonymous with *richer*? Richer in what way: money, fossil fuels, whales?

6. **Aggregated** metrics may combine features or metrics belonging to a variety of other categories, or they may group a number of metrics that belong to a single category. Consequently, aggregated metrics might also be financial, thermodynamic, environmental, ecological, or sociopolitical, or they might be a dimensionless combination of any of these.

## Difficulty of Aggregation

Life-cycle data are voluminous and difficult to interpret. Optimization, as is common in systems analysis, requires that all terms be expressed in a single objective function. In short, the tools of industrial ecology require that data be *aggregated* to communicate the results more easily to decision makers and so that useful conclusions may be drawn. Aggregated data are easier to work with, but they inevitably contain less information than the original. For example, a life-cycle inventory does not necessarily contain data that indicate which stage of the life cycle is most resource intensive: primary extraction, processing, use, or final disposition? Improvement assessment often requires more information than the simple conclusion. The process of working backward—from aggregated to primary data—is called *disaggregation* and can be problematic when working from large databases such as those compiled from input–output analysis. For this reason, it is essential that the methods of aggregation employed in a life-cycle or systems study be completely transparent to other researchers. For example, many industrial processes produce more than one product (called *co-products*), or produce residuals with positive (if incidental) economic value (*by-products*). How should primary resource consumption or pollution generation be allocated among multiple products? One approach is to weight the inventory data by the economic value of the products. This approach has the disadvantage of depending on market prices, which fluctuate and may not fully reflect environmental costs. It does, however, align the life-cycle inventory data with the criteria that are most likely to drive capital investment and production decisions: **return on investment.**

In practice, one of the most problematic issues is the difficulty of properly weighting resource quantities that are dislocated in space and time or are of different quality. Thermodynamic (mass and energy) data are aggregated differently than financial data. There is little agreement on standard approaches.

The most common method of weighting thermodynamic and environmental units is linear (in a simple sum). In fact, resource requirements and environmental loadings come at different stages of the life cycle, at different times, and are engendered in different locations with different environmental priorities; linear aggregation may be a questionable practice. Even seemingly simple tasks such as computing total energy intensity can be complicated by the fact that a variety of logical aggregation schemes may be adopted that lead to different conclusions. One study that reviewed several methods recommended weighting energy values *by price* (rather than heat or exergy content) to correct for the fact that different fuels exhibit different levels of economic utility, for example; ease of storage for periods of high demand, cost of transmission, or environmental control (Cleveland et al., 2000). However, current energy market prices do not fully reflect the social or environmental costs of different fuel types. An environmental LCA aggregated based on market prices may give misleading signals.

Methods of aggregating money flows are also controversial. The most common method of aggregating money flows over time is to use continuous compound discounting at a single discount rate (which is loglinear), despite the fact that no single discount rate can be applied to society, a government project, or even a corporation. All of these organizations apply different discount rates to events with different time horizons or perceived risks (which is appropriate). Consequently, corporate and government debts carry any number of different interest rates that make marginal increases in borrowing difficult to assess on the basis of a single rate. Moreover, interest rates fluctuate over time, belying the assumption of a single interest rate of the life of a particular project.

Even among individuals, a variety of discount (or interest) rates may be simultaneously observed or applicable. For example, many individuals hold multiple lines of credit. The highest interest rates are usually associated with credit card debt, which is usually unsecured. Automobile loans typically carry lower rates than credit cards. Home mortgages carry even lower rates than automobiles. As individuals age, they (hopefully) accumulate wealth; their borrowing instruments typically migrate from higher interest rate contracts (credit cards) to lower (home mortgages). Eventually, individuals reach a creditor point in life at which their implied discount rate is comparable to money market or cash rates of interest—lower even than home mortgage debt. The result is a general tendency for time preferences to reflect greater patience (lower discount rates) as individuals age or accumulate wealth.

Human and animal studies suggest that a hyperbolic function in the form $V = B/(B + t)$, where $t$ is time, $B$ is a fitting parameter, and $V$ is present value, may be a better predictor of behavior than the exponential function $V = F\exp(-it)$ suggested by economic equilibrium theory (Green and Myerson, 1996). In that equation, $F$ is the amount of a monetary value at a time $t$ in the future, and $i$ is the applicable interest (or discount) rate. Compared to an exponential discount function, the hyperbolic places slightly less emphasis on short-term events, about the same on intermediate-term, and more on long-term outcomes. Moreover, the hyperbolic discount function better represents the case where multiple discount rates are applicable [as in large organizations (Ayers and Axtell, 1996)] or exponential discount rates are expected to decline over time (as in wealth-accumulating individuals). The important point to take note of is that entirely different—albeit logically defensible—conclusions may result from selection of different aggregation methods, even in what are assumed the simplest cases.

## Beyond Impacts Aggregation: Multicriteria Decision Analysis

It is difficult to approach environmental decisions; different stakeholder groups have different feelings, assumptions, or values that may be affected. Thus, it is improbable that all groups will prefer a single best alternative among many; different stakeholders are likely to prefer different designs, technologies, or policy alternatives. They may disagree on how to define the boundaries of the problem and on which measures are most important in judging alternatives. The challenge for industrial ecology is to present as nearly a complete picture as possible in a format that allows multiple interpretations. This requires that data be available for aggregation or disaggregation as a decision maker pleases, so that the sensitivity of the results to different priorities or perspectives can be explored. However, because resources are scarce and because some objectives may be mutually exclusive, not all the goals held by all stakeholders can be satisfied in every instance. Some method of compromise or ranking, therefore, must be involved.

*Multicriteria decision analysis* (MCDA) is one approach that can be used to balance the different objectives and interests of different constituencies (Seppälä et al., 2002). There are three broad categories of MCDA methods, each applicable in the context of LCA or systems analysis:

**Optimization models** employ numerical scores to communicate the merit of one option in comparison to others on a single scale. Scores are developed from the performance of alternatives with respect to individual criterion and then aggregated in an overall score. They may be simply summed or averaged, or a weighting mechanism can be used to favor some criteria over others. Optimization models are best applied when objectives are narrow, clearly defined, and easily measured and aggregated. For example, students who seek to maximize their grade point average (GPA) are using an optimization model. A numerical score is typically applied to traditional *A, B, C* grades, and a weighted GPA is computed on the basis of how many credit hours each course is worth. The challenge is for students to allocate their study time among their different subjects to create the highest GPA.

**Goal, aspiration, or reference level models** are used to establish desirable or satisfactory levels of achievement for each criterion. This process seeks to discover options that are closest to achieving, but not always surpassing, these goals. When it is impossible to achieve all stated goals, a goal model can be cast in the form of an optimization problem in which the decision maker attempts to minimize the shortfalls, ignoring excesses. Alternatively, the decision maker may seek to satisfy as many of the goals as possible (even if only just barely) and ignore the fact that some performance metrics may be *very* far from target levels. Goal models are most useful when not all the relevant goals of a project can be met at once. For example, some students may be in danger of failing one or more courses at a time. Even though it is their goal to pass all of their classes, they may not be able to satisfy this goal. Instead of studying a little bit on all subjects and risking multiple failures, they may withdraw from the classes in which they are doing most poorly, and focus on passing the rest. These students are using a goal model.

**Outranking models** compare the performance of two (or more) alternatives at a time, initially in terms of each criterion, to identify the extent to which a preference for one over the other can be asserted. In aggregating such preference information across all relevant criteria, the outranking model seeks to establish the strength of evidence favoring selection of one alternative over another. Outranking models are appropriate when criteria metrics are not easily aggregated, measurements scales vary over wide ranges, and units are incommensurate or incomparable. For example, students selecting courses for the next semester may use many criteria, such as the popularity of the instructor, the course meeting times, whether the course is required, and the distance between buildings or classes. How could such disparate criteria, some quantitative (distance), some semiquantitative (popularity), possibly be aggregated? Instead of an optimization or goal model, students may employ an outranking model that assesses courses in a comparative scorecard. If one course has a better instructor, and more convenient location, and satisfies a graduation requirement, it may be preferable even if it meets at an extremely inconvenient time.

In cases where weighted optimization models are so sensitive to subjective weighting factors that it is possible to recommend any of the available alternatives by adjusting weighting functions, an optimization approach may seem particularly fallacious. The advantage of multiple approaches is that MCDA can consider multiple objectives and compare alternatives in many different ways, rather than trying to reach one ultimate solution. For example, return to Table 1.3 presenting the streamlined impact assessment of alternative automobile technologies. To determine which technology is preferable, any of the three MCDA approaches could be employed.

First, consider an optimization approach in which each of the impact factors is added up in a linear sum. (Note that in Table 1.3, hybrid and fuel cell cars have identical impact factors in every respect and consequently may be grouped together for this analysis). Hybrid/fuel cell cars are rated the best, and ethanol is second. However, suppose that the only considerations were fuel, vehicle, and infrastructure costs (with other factors given zero weighting). Then, reformulated diesel is preferred, while ethanol moves down to the worst.

A goal aspiration approach could operate differently. Certainly, no stakeholder would *prefer* higher emissions, higher costs, lesser vehicle performance, *and* negative social impacts. Suppose the objective were to improve all four impact categories: vehicle emissions, fuel related, vehicle performance, and social issues. Each is evaluated by linear summation of its components so that the total vehicle emission impact for reformulated natural gas cars would be estimated as +3, reformulated diesel as −5, and so forth. No single alternative satisfies the goal of making improvements in every criterion. All of the technologies have some blemish, such as reduced vehicle performance, but compressed natural gas may be *closest* to satisfying this goal because it sacrifices only vehicle performance while improving all other aspects. Other technologies are either negative or neutral in two or more respects. In this approach, impacts are summed within individual criteria and then tested for positivity. The recommended technology is the one with the *most positives*, regardless of *how* positive. Separate criteria are noncompensatory, which is to say that vehicle emissions improvements do not cancel out a loss of vehicle performance. This approach is more sophisticated than simple summation and optimization. It, however, recognizes that criteria are measured on different scales and that positives in one area do not necessarily cancel losses in another.

A simplified outranking approach might ask only which alternative is the best in each subcriterion (Table 1.6). For example, batteries are superior in ozone, particulate matter, and tied with hybrid/fuel cells in air toxics. Each subcriterion in which an alternative dominates all might justify award of one "win" point. (In case of a tie, both winners could get points.) Similarly, "loss" points may be accumulated when a technology is inferior to all others in a subcriterion, as batteries are for vehicle range and costs. By the point system, batteries are awarded *wins* for the criteria already mentioned, plus fuel costs, but *losses* for fuel cycle emissions, range, and vehicle costs for a total record of four wins, three losses, or

**Table 1.6** Sensitivity of Recommended Alternative to MCDA Approach

| | Optimization (All Inclusive) | Optimization (Costs Only) | Goal Aspiration | Outranking |
|---|---|---|---|---|
| Best | Hybrid/fuel cell | Reformulated diesel | Compressed natural gas | Hybrid/fuel cell or ethanol |
| Worst | Reformulated gasoline | Ethanol | Reformulated gasoline | Reformulated gasoline |

a net of plus one. Doing the same for the other alternatives shows that ethanol and hybrid cars each have two more wins than losses and that reformulated gasoline has three more losses than wins. More sophisticated outranking methods allow for awarding partial wins (such as half a win for a tie), or study a series of pairwise comparisons in which alternatives are matched against each other for each criterion like softball teams in a round-robin tournament. The winner might be the alternative with the best winning percentage, or the greatest positive differential.

In MCDA, the multiple metrics of industrial ecology are complementary. Each provides a different perspective. The purpose of MCDA is usually to clarify the complex decision process, not optimally to solve the problem; that is, to compare alternatives based on incommensurate criteria and allow conflicting components of the decision problem to be examined simultaneously. Although MCDA may be focused on synthesizing multiple criteria into a single recommendation, it may also be employed to foster mutual understanding among stakeholders with different viewpoints by showing under what conditions different alternatives may be preferred.

**A stepwise approach to MCDA**

The process of MCDA consists of a series of steps that feed back and forth on each other. These include:

Identification of the problem/issue

Problem structuring (involving values, goals, constraints, external environment, key issues, uncertainties, alternatives, stakeholders)

Building quantitative or semiquantitative models (specifying alternatives, defining criteria and metrics, eliciting values)

Using the model to inform and challenge thinking (synthesis of information, challenging intuition, creating new alternatives, robustness analysis, and sensitivity analysis)

Developing an action plan

(Belton and Steward, 2002)

## MCDA for Sustainability

The MCDA approach may be applied to any type of complex decision process, such as purchasing a car, selecting a college, or deciding for which political candidate to vote. Environmental issues present some peculiar challenges, however:

- Environmental resources are typically shared by groups with different values, and no single group "owns" the legal right to decide the disposition of these resources.
- The criteria by which environmental issues are judged may be nonquantitative, or semiquantitative, and involve incommensurate or incomparable attributes such as pristine views, social costs, health effects, and/or biodiversity.

To handle the special nature of environmental problems, MCDA must be combined with other tools such as public participation and stakeholder value elicitation. Expert knowledge is critical to understanding how alternatives perform in relation to criteria. For example, expert opinion could be the basis for assessing cost or health risks, but the criteria themselves, and the relative importance of each criterion, should be established with stakeholder input (Keeney, 1992; Gregory, 2002).

One critical step in any MCDA process (see box) is identification of objectives and selection of the metrics by which the merit or performance of alternatives in relation to these criteria will be judged. Several examples of sustainability metrics are worthy of at least brief mention. The metrics that are expressed in commensurate units (dollars, joules, or kilograms) are more readily aggregated than metrics that are expressed in many different ways (as is often the case with environmental, ecological, or sociopolitical measures). Chapter 4 introduces a new aggregated environmental metric called *pollution potential* that is not (like toxicological measures) specifically tied to hazardous end points but represents a broad indication of environmental sustainability.

#### *Total Cost Assessment*

Total cost assessment (TCA) is a method of accounting for several different types of liabilities: direct, indirect, contingent, intangible, and external into a single financial metric that may better represent the systemic costs of design or process alternatives (see Kennedy, 1998; CWRT, 1999). Total cost assessment has been advocated as an alternative to traditional accounting. It is consistent with the basic underlying principal that a more inclusive, holistic treatment will lead to better decisions. The concept is analogous to total cost of ownership (TCO), which is typically applied to evaluate capital investment decisions, but the terms are not synonymous. A TCO methodology catalogs direct and indirect costs; TCA encompasses costs from five broad categories:

**Direct costs** are readily attributable to the process or decision at hand. The cost of raw materials, most taxes and shipping fees, many contractors, and capital or durable purchases are easy to include.

**Indirect costs** are incurred during production (and consumption) but are lumped together with other items. It is often difficult to measure how much of these costs should be allocated to specific activities. Electricity consumption is a common example because so many activities are powered by a single meter. Other examples are depreciation of plant and equipment, heating or cooling rooms that house specialized equipment, and real estate taxes.

**Contingent liabilities** include items that cannot be anticipated with certainty but can be expected to occur randomly in a manner consistent with past experience. Examples include legal expenses or product liability expenses, accidents or unplanned outages, product recalls, and returns or failures.

**Intangible costs** are those that are difficult to translate into monetary terms. For example, the time required of salaried employees is not always measured with confidence. And translating salaried time into a financial expression requires application of some uncertain assumptions regarding the value of additional hours, changes in morale, training expenses, or frustration, and how these should be apportioned among many different projects. Customer support requests or product returns may be an indicator of direct costs (in support personnel salaries or warranty servicing costs) but could also indicate a loss of customer goodwill and brand image that are best categorized as intangible.

**External** (or social) costs are those that are not borne by the decision makers (e.g., polluters) but by society. These are generally the most difficult to quantify. Examples include the cost (or benefit) of environmental pollution, police protection, and public views.

*Thermodynamic Metrics*

Thermodynamic metrics come in many different forms and are among the most widely cited in industrial ecology literature. Several are discussed below.

**Materials intensity** and **energy intensity** aggregate the mass and energy requirements per unit of product, service, or value added. The advantage of these measures is that they are generally easy to account for and aggregate because they rely only on the first law of thermodynamics (rather than the second, or entropy law). Data are easily checked for internal consistency by applying mass and energy balances. The disadvantage is that materials and energy vary in quality. Electricity is a higher quality energy source (per joule) than coal or wood. Sand and gravel are low-quality materials compared to refined metals. Simple mass or energy-weighted aggregations do not account for variations in quality.

**Exergy** is a thermodynamic measure of material and energetic resource inputs that combines the first and second laws of thermodynamics. Exergy is analogous to Gibbs free energy but with reference to an unperturbed environmental chemical state. It is a measure of the total thermodynamic work available (by ideal processes) in any thermodynamic system. It may be aggregated along the life-cycle chain to identify the aspects that are most resource intensive, or it may compare the thermodynamic requirements of different processes, products, or technologies. Exergy is a thermodynamic variable of state and is applicable to any thermodynamic system. When employed in LCA, exergy values are aggregated in a simple sum to compute the total thermodynamic requirements of any product or process in an aggregated measure called **cumulative exergy of consumption (CExC)**. *Waste exergy* should be accounted for separately from the exergy that remains in products. (See Chapter 4 for more on exergy.)

**Emergy** is a life-cycle accounting of the total quantity of exergetic inputs required to produce a product, animal, plant, or system. The word *emergy* comes from contraction of *embodied energy*. Unlike exergy or energy (or temperature or pressure), emergy is never a thermodynamic property of state; it is always an aggregated life-cycle accounting metric, like CExC. Emergy differs from CExC in that it specifically includes the exergy of ecosystem services such as solar radiation, wind, and rain. CExC typically *excludes* these from analysis and just focuses on the exergy of natural and economic resources without regard for the development of those resources. For example, CExC accounts for the exergy of 1 mol of methane gas as the maximum amount of work obtainable from the gas under ideal conditions, about 831 kJ, at the point at which it was consumed, and add to this the exergy required to locate, drill, pump, refine, and transport the methane to that point. Emergy, however, measures the solar exergy required to grow the plant material stored in sedimentary deposits tens (if not hundreds) of millions of years ago, in addition to everything else CExC accounts for. Production of natural gas (which is mostly methane) with an exergy of 1 J requires approximately 48,000 J of ancient solar energy (Odum, 1996). This broad definition of life-cycle boundaries is controversial because the solar radiation of the dinosaur age clearly cannot be captured for any chosen purpose (other than the current endowment of fossil fuels) and cannot be reversed or recovered. Consequently, it is not clear how this energy relates to decisions that must be made today. Nevertheless, emergy does make an interesting basis for comparison of different exergetic resources.

Figure 1.8 lists the net emergy yield ratio, or emjoules available for consumption per emjoule investments required to extract a variety of fuels. (An emjoule is a life-cycle metric that refers to the emergy approach of aggregating ecological with industrial exergetic resource inputs. An emjoule indicates that the energy referred to is expressed in equivalents of solar inputs and is derived from a life-cycle accounting of all exergetic sources, including those provided by ecosystem inputs such as solar insolation, tidal energy, or geothermal energy.) Note that fossil fuels generally have higher emergy yield ratios than renewable resources. Emergy yield calculations include only emergy expenditures *required of the economy* in the denominator—the life-cycle exergetic work of exploring, drilling, excavating, processing, and transporting fuels. In the numerator is the life-cycle accounting of all the economic and ecosystem inputs—including photosynthesis or evapotranspiration, for example. Rain forest logs, like fossil fuels, accumulate the exergy of sunlight over lengths of time that are long when compared to the typical time required to consume the exergy in our economy, and they have high emergy yields because they require little work investment per emjoule of emergy returned. However, the *exergy* embodied in rain forest logs is not the same as the *emergy*. Higher emergy does not necessarily mean that a material is capable of performing greater economically useful work. For example, rain forest logs are heavier, less easily transported, and do not burn as cleanly as an equivalent mass of petroleum. (On the other hand, furniture made from teakwood may be much more highly prized than that made from plastic. Not all high-exergy materials

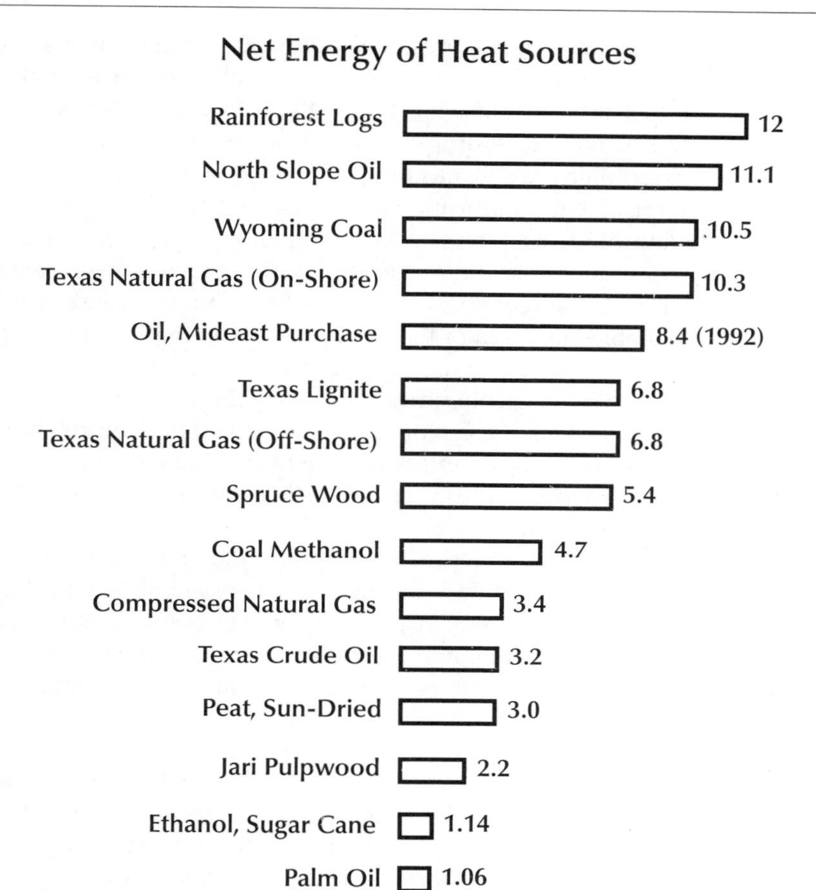


**Figure 1.8 Emergy yield ratios for renewable and exhaustible fuels.** Few renewable fuel sources are comparable in yield to exhaustible resources. Low yields are associated with fuels that require greater energetic inputs to process or extract, often because they come naturally in less concentrated forms. (From Odum, 1996. This material is used by permission of John Wiley & Sons, Inc.)

should be used for fuel.) Whether rain forest logs are considered a renewable resource may depend on the time frame under consideration and the method by which they are harvested. Fossil fuels represent the concentrated storage of eons of sunlight and are, therefore, superior energy sources; however, because they take such long periods to create, they are not renewable. Fossil fuel consumption is not sustainable indefinitely. Does this mean we should forgo the exploitation of fossil fuels to preserve them as a resource for future generations? Alternatively, should the highest quality resources be exploited first?

Another popular thermodynamic concept is that of the **ecological footprint**. It represents an estimate of the land area required to support a national economy or activity. Energy use is represented by the land area required to sequester carbon in photosynthetic products equivalent to the carbon released by industrial processes and dominates the analysis for developed countries. Land area used for buildings and roads, gardens, crops, pastures, and productive forests are also counted and aggregated in an unweighted single figure. The measure illustrates that industrial functions can create effects that reach far beyond local borders. Cities, for example, depend on resources extracted from a wide geographic area, just as they disperse pollutants that have continental, if not global, ramifications. Developed countries have ecological footprints of 3 to 5 ha/person, approximately 10 times that of less developed nations. It is possible, as a narrowly defined measure of sustainability, to compare the ecological footprint of the human economy with the total available land area of Earth. However, the implications of the comparison are difficult to assess without context because the land area calculated by the ecological footprint method is hypothetical rather than actual land. Is the ecological footprint better interpreted as an indicator of the unsustainability of human systems or as an expression of the productivity of natural systems? (See van den Bergh and Verbruggen, 1999, and Deutsch et al., 2000.)

Many different types of **environmental metrics** are available, but few have been designed with life-cycle assessment or systems analysis in mind. One of the most pressing problems of industrial ecology is to develop environmental metrics that are comprehensive, informative, and applicable to life-cycle assessment and systems analysis.

*Environmental Metrics*

Some examples of informative environmental metrics are presented here. **Global warming potential** (GWP) is an expression of the time-integrated radiative forcing effects of an atmospheric pollutant. It is a measure of how much energy a gas is capable of trapping but is not easily related to changes in global temperatures or population effects. GWP depends on the infrared radiation absorption spectra, atmospheric residence time, and molecular weight of a pollutant. It is expressed in comparison to an equivalent mass flux release of carbon dioxide over a particular time horizon in units of equivalent kilograms of $CO_2$ (eg. kg $CO_2$). Selecting an appropriate time horizon may have a profound effect on GWP because different gases have different atmospheric residence times. For example, methane is a more effective infrared radiation absorber than carbon dioxide on an equivalent mass basis. Over a period of 20 years, methane traps 64 times more energy in the atmosphere than an equivalent mass release of carbon dioxide and thus has a GWP of 64 eq. kg $CO_2$. Carbon dioxide, though, has a much longer atmospheric residence time than methane (approximately 80 years to about 12.2 years, respectively). Over 100 years, the GWP of methane is only 24 eq. kg $CO_2$; for 500 years the GWP is only 7.5 eq. kg $CO_2$. When working with GWP, it is essential to report the time horizon to which the data relate.

**Total equivalent warming impact (TEWI)** is an environmental metric that represents the combined effects of manufacture, use, and disposal of a product (e.g., a refrigerator) in units of mass of carbon dioxide release that would create an equivalent radiative climate forcing over a defined time horizon—typically several decades or more (McCulloch,

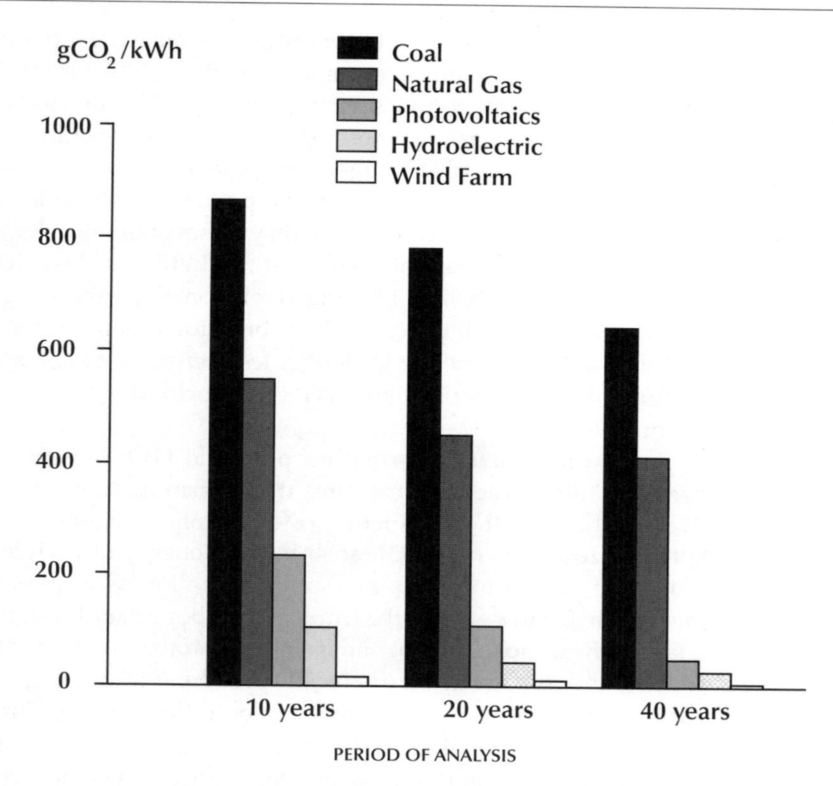


**Figure 1.9 TEWI of electricity generation technologies.** (Adapted from Pacca and Horvath, 2002. Used with permission of the American Chemical Society.)

1994). It is a GWP-weighted sum of climate-related emissions and may be aggregated along the entire life cycle. Figure 1.9 compares the TEWI of five different electricity generation technologies: coal, natural gas, photovoltaic, hydroelectric, and wind. The TEWI approach shows that even renewable resources can result in global warming effects during construction or production of the power-generating infrastructure. The global warming effects of hydropower are associated primarily with production of cement for the concrete that is used to construct large dams—a source that may have gone unrecognized had not a life-cycle approach been employed. Notice that the relative positions of the technologies do not change as the time horizon shifts from one to several decades. Over all lengths of time, wind power is preferred from a TEWI perspective.

**Ozone depletion potential** (ODP) expresses the capacity of a pollutant to destroy stratospheric ozone. It is usually applied to a class of chemicals called halogenated hydrocarbons that combine one or more carbon atoms with hydrogen, fluoride, chloride, bromine, or iodine. ODP is reported in comparison to CFC-11 ($CFCl_3$) like GWP is reported relative to carbon dioxide. A pollutant with an ODP of 0.1 is expected to result in one-tenth the reduction in stratospheric ozone as an equivalent mass release of CFC-11. ODP depends on several factors. The most important is the chloride or bromide content of a pollutant. Generally speaking, bromide is more reactive than chlorine; thus, the more bromine a pollutant contains the higher the ODP. Fluorine and hydrogen are not ozone active; therefore, substitution of fluoride for chloride results in lower ODP. Other factors include the molecular weight and the degradation properties of the pollutant. Molecules containing one or more carbon–hydrogen bonds will substantially degrade in the troposphere; therefore, pollutants with a lifetime of less than about 2 years contribute very little to ozone depletion. Once reaching the stratosphere, practically all the chloride or bromide in a molecule will eventually be released, meaning ODP is less sensitive to the atmospheric lifetime of the pollutant than GWP (see World Meteorological Organization, 1999.)

**Tropospheric ozone formation potential** (TOFP or photochemical ozone creation potential) captures the contribution of volatile organic hydrocarbons to the production of tropospheric ozone. In the stratosphere, ozone is beneficial. It absorbs high-energy ultraviolet light that would be deadly to plants and animals if allowed to pass to the surface of Earth. However, in the tropospheric boundary layer, the first 1 or 2 km of the atmosphere, ozone is an undesirable oxidant that can lead to lung and eye irritation or chronic breathing impairment. In combination with nitrous oxides, water, and pollution particles, tropospheric ozone forms smog. Like GWP and ODP, TOFP is reported in comparison to a reference pollutant. Its reference is ethylene, which is assigned a TOFP of 100. A photochemical trajectory model tracks the degradation and reactivity of the relevant atmospheric constituents and predicts increased levels of ozone created by marginal increases in emissions of a particular pollutant. Sunlight drives the reactions, providing sufficient energy to create ozone. Atmospheric systems often respond nonlinearly to changes in chemical constituents, however. TOFP models work best under conditions that are hydrocarbon limited; that is, all other chemicals are present in sufficient concentrations. TOFP data reported for one airshed may not apply to other airsheds with different chemical constituents. Halogenated hydrocarbons that contain at least one fluoride or chloride generally have low TOFP (Hayman and Derwent, 1997).

**Toxic release inventory (TRI)** is an aggregated environmental metric representing the mass-weighted sum of certain toxic pollutants. The TRI can also be aggregated using toxicity-weighted sums to obtain a better estimate of the relationship of the TRI to human health. Since 1987 TRI data have been compiled by polluters and reported to USEPA in accordance with regulations that were promulgated in response to the Bhopal disaster in December 1984 and a smaller but similar accident in Institute, West Virginia, in August 1985. During the last decade, reported toxic releases have declined significantly for many companies, partly because the TRI provides a way for the public to identify and pressure top polluters. The inventory data may represent nothing more than a best-guess estimate by plant managers, however, if the managers lack comprehensive environmental information management systems.

The data are not regularly checked, even by simple mass balance; it is, therefore, difficult to evaluate the inventory's accuracy (USEPA, 1998).

**Pollution potential** is a measure of the change in chemical composition of the environment that can be expressed as a singular environmental metric or aggregated in terms of total change in environmental exergy of mixing. Typically, several environmental metrics are employed in a single LCA, including global warming potential, ozone depletion potential, and a human health measure such as mass of hazardous waste generated or human toxicity potential (HTP). The purpose of pollution potential is to express total environmental effect in a single metric to help assess scenarios that involve trade-offs among incommensurate environmental criteria. (Chapter 4 further explains pollution potential and how it helps resolve ozone and global warming conflicts.)

**Human toxicity potential** is a toxicity- and exposure-weighted aggregation of toxic releases. Unlike the TRI, HTP incorporates additional factors that relate to human health: potency, persistence, and exposure pathways. Therefore, HTP combines environmental factors with ecological factors (such as the sensitivity of human beings to specific toxins). The advantage of using HTP is that the results are more closely associated with actual risk end points such as cancer. The disadvantage is that HTP requires more information and sophistication and may not be relevant to other species. Calculations depend on the mass loading and fate of pollutants released into the environment and the pathways and routes by which people may be exposed. Typical exposure pathways are contaminated air, water, foods, and soil; typical exposure routes include breathing, eating, or absorption through the skin. Multiple pathways and routes may be involved. For example, a pollutant with a high vapor pressure may be discharged in liquid form but vaporize into the atmosphere. Both effects are aggregated in a linear sum. Like many other environmental metrics, HTP is usually reported on a relative scale; that is, results are normalized in comparison to a reference pollutant. Although global warming potential is always reported relative to an equivalent mass release of carbon dioxide, HTP reference compounds can vary from study to study depending on the pollutants or technological choices under consideration. Also, HTP depends on local environmental conditions, population distributions, and behaviors (such as diet). Published results may vary, therefore, depending on the assumptions employed.

> An informative comparison of different toxicity-weighted metrics may be found in Hertwich et al. (1998). Estimated human toxicity potentials for over 300 compounds may be found in Hertwich et al. (2001).

*Ecological Indices*

Ecological indices express the abundance, health, and variety of species present in a specified system (Izsak and Papp, 2000). In general, the consensus among ecologists is that biodiversity contributes to ecosystem health and to the long-term viability of the human population. Ecological (also called biological) metrics may aggregate data on populations of several species, such as total ecosystem biomass, or may be focused solely on the health or abundance of a single species, such as breeding population. Biodiversity by itself may be computed from a species population inventory as a statistical measure (like variance), but more complex measures of ecosystem diversity have been devised that incorporate information regarding food webs or exergy flows.

For example, **ascendency** is defined as the product of total system throughput and average mutual information. Figure 1.10 depicts the flow of carbon in the mesohaline Chesapeake Bay (which is near the upper reaches above the Potomac River where fresh and ocean waters mix). Total system throughput is measured in units of mass carbon/hectare/time, whereas average mutual information is defined by distribution of carbon flows. A distribution in which all flows are equal is indistinguishable from random, but because carbon is more likely to flow from some areas of the diagram to others, these flows are constrained. The degree to which the distribution departs from a merely random distribution of carbon may be measured with the same statistical techniques as a population distribution. It is, therefore, analogous to a biodiversity index but is aggregated on a carbon-flow-weighted (or

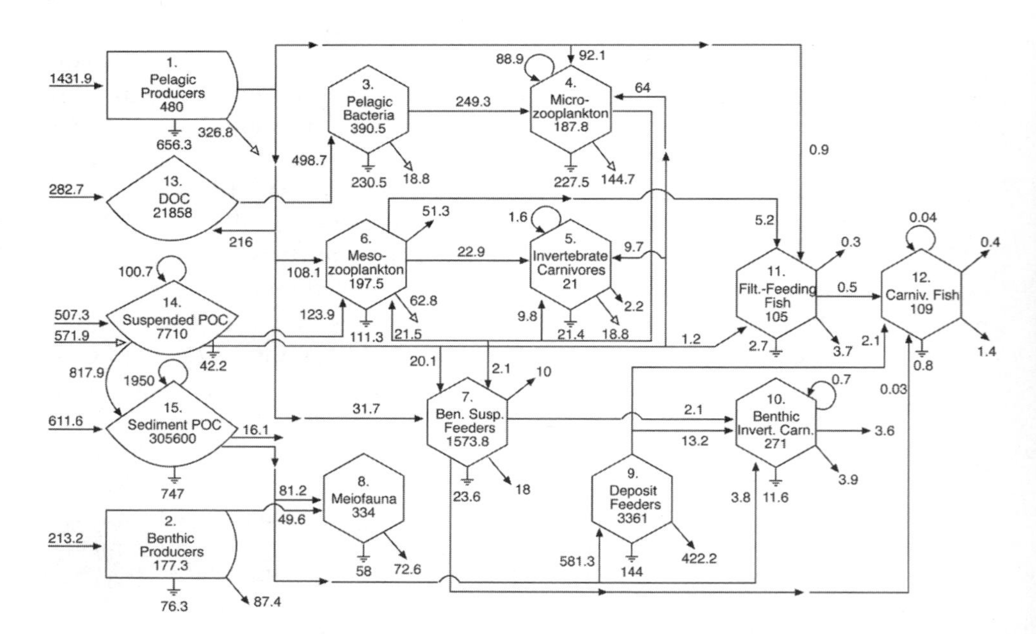


**Figure 1.10 Estimated carbon flows among principal components of mesohaline Chesapeake.** Flows are in units of milligram/square meter/day and standing stocks (inside boxes) are in milligrams/square meter. Electrical ground symbols indicate losses due to respiration; open arrows indicate returns to pelagic detritus, and filled arrows indicate returns to benthic detritus. DOC = dissolved organic carbon. POC = particulate organic carbon. (*Source:* Ulanowicz, 1997. Used with permission of Columbia University.)

other measure such as exergy-flow-weighted) basis rather than population weighted. As with biodiversity, greater average mutual information is generally regarded as consistent with improved ecosystem health (see sidebar below).

Human health metrics, such as life expectancy or infant mortality, may also be subsumed in the ecological category. Selecting an appropriate set of human health metrics can be a difficult or contentious process considering the importance of confounding factors such as personal risk-taking behaviors (such as smoking), access to modern medical treatment, individual variability in medical history, or environmental justice. The need for sublethal end points is particularly acute but problematic because of the difficulty of defining detection levels. Selection of any metric or aggregation scheme may codify value judgments regarding the time frame of analysis (acute or chronic) or age dependency of the population (old or young). Consequently, it may be necessary to maintain multiple measures that are not easily reducible or aggregated.

Because the potential health effects of any combination of environmental insults are myriad, representative symptoms, indicators, or diagnoses are selected under the assumption that they indicate general conditions. Typically, these are expressed in terms of increased incidence or prevalence of cases of disease or disability per capita per year. Potential metrics may include the severe (mortality or disease), to the serious (hospital admissions), or permanently or temporarily debilitating (neurocognitive impairment, lost work days), or irritating (loss of sleep). Collectively, one approach called *quality-adjusted life years* (or conversely, *disability-adjusted life years*) aggregates case incidences according to the severity (a quality-weighted factor) and duration (time-weighted factor) of the effect. The advantage of this approach is the expediency of expressing all health-related measures in a single unit. The disadvantage is the potential for cases that mildly affect large numbers of people to overwhelm cases that severely affect small numbers (especially older populations). Sometimes, such as in jury damage awards, it seems necessary to express human health metrics in terms of money, possibly confounding economic and ecological measures—a relation that some find ethically repugnant.

Hofstetter and Hammit (2002) explore the selection and aggregation of human health metrics for life-cycle assessment in detail, especially the sensitivity of impact assessment to the choice of metric.

**Ecosystem Health**

Ecosystems are made up of many plant and animal species that interact collectively to support functions that are analogous to those performed by individual cells in a single organism. The health of an ecosystem may be evaluated in a manner analogous to the health of a human being (or other organism) and has primarily three attributes:

Vigor is a measure of activity, such as metabolism or primary production. A vigorous ecosystem has long and complicated food webs. Solar insolation is converted by photosynthesis to plant material, which sustains animal life at multiple trophic levels. Vigor is lost when solar radiation (or other forms of exergy at lower trophic levels) is rapidly degraded into waste heat (such as in a parking lot) rather than employed to sustain life. Vigor might be measured in terms of total exergetic system throughput or biomass yields.

Organization is measured in the quantity, quality, and diversity of interactions between different components of an ecosystem. Organization is not synonymous with *order*, which is a measure of the predictability or reproducibility of patterns. An intensive monoculture (e.g., a corn field) is highly ordered. To the extent that it is designed around the single purpose of maximizing commodity yields, it may also seem *organized*, especially to the farmer. However, intensive monoculture is designed to limit the diversity of species interactions that characterize healthier ecosystems: grazing, predation, parasitism, and biodegradation.

Resilience indicates the system's response to stress (Figure 1.11). Healthy ecosystems are able to withstand or repair temporary insults to structure or function or even evolve to rely on these, such as floods or fires, to maintain some essential functions. Resilience can be measured in terms of time or rate of recovery after injury. When resilience is absent or diminished, an ecosystem may shift to an alternative state with dramatically different population distributions.

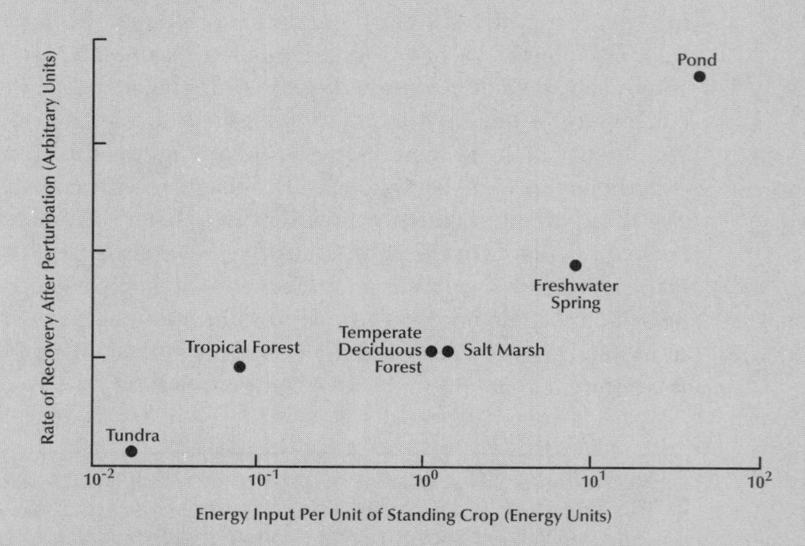


**Figure 1.11 Resilience vs. vigor.** Researchers removed 10% of the standing-crop vegetation in different ecosystems. The rate at which the removed biomass is replaced by new growth is a measure of resilience. Ecosystems with high-energy inputs per unit of standing crop (which is a measure of *vigor*) showed greatest resilience. (*Source:* Franzle, 1998.)

Human populations depend on healthy ecosystems to perform services that support human life or economic activities. For example, ecosystems provide waste attenuation, food, energy, and materials production as well as medicines. Loss of these services results in increased economic cost and greater dependence on exhaustible resources (to duplicate natural functions).

Adapted from Rapport et al. (1998) and Rapport (1999).

*Sociopolitical Metrics*

Metrics related to human society relate to sustainability in two ways. The first type of social metric evaluates whether industrial activities are consistent with political goals that relate to sustainability such as energy independence, ecojustice, or fair trade. This may be evaluated entirely subjectively or by consensus (such as an expert survey). The second type attempts to measure the extent or quality of collaborative relationships that can foster social solutions to shared problems typical of those found within the domain of sustainability. That is, the first measures progress toward certain goals while the second is a measure of the social resources (such as volunteer and nonprofit organizations or political participation) that may be available to achieve those goals.

A **social capital** index combines several measures of community interaction and support such as voter turnout, newspaper readership, membership in choral societies, confidence in public institutions, volunteerism, church attendance, or percentage of children with two-parent families (Rudd, 2000). Different studies have focused on different criteria; no one index is best. In theory, social capital may be substituted for monetary capital to the extent that cooperative systems (such as barter or volunteer work) can be substituted for cash. Social capital may make other forms of capital (financial or human, such as education) more productive. High levels of social capital may also reduce health care or government service costs by enhancing well-being or productivity.

Figure 1.12 shows that the well-being of high-risk children can be correlated with an index of social capital that includes five factors: two parent figures in the family, a perception of social support for maternal caregiver, no more than two children in the family, a perception of neighborhood support, and regular church attendance. Children with all five factors present were over five times as likely to be doing well as children with none (Runyan et al., 1998). Hypothetically, social capital may even be substituted for *thermodynamic* resources (giving new meaning to the term social *work*) as when people carpool. The sustainability of economic and thermodynamic measures must be related to *social* as well as technological systems; but even when a cooperative infrastructure is in place, such as in nonprofit organizations or government institutions, when economic capital and/or thermodynamic resources are plentiful and cheap, incentives for social cooperation are low. For example, people who live in suburbs often depend on cars to get to work. When interest rates are low and gasoline is cheap, newer and bigger cars are less expensive to buy and operate; carpooling seems less worthwhile, and public transportation seems unnecessary. Conditions in which social

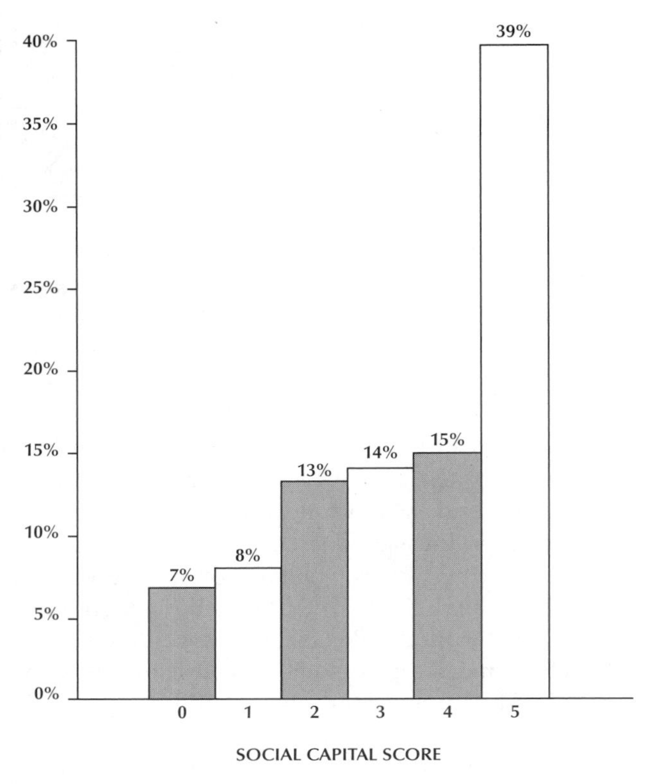


**Figure 1.12 Percentage of high-risk children doing well by social capital index score.** Children were scored by five factors: two parent figures in the family, a perception of social support for the maternal caregiver, no more than two children in the family, a perception of neighborhood support, and regular church attendance. The presence of any factor improves the chances that the child is doing well. The strongest predictor of child well-being was the linear combination of factors in a social capital index, which may be a better predictor than family income or maternal education level. (From Runyan et al., 1998. Reproduced with permission from *Pediatrics*, Vol. 101, Table 3, © 1998.)

capital, economic capital, human capital, and thermodynamic resources are all scarce are particularly impoverished.

The concept of **ecojustice** involves questions of whether the burden of pollution or environmental degradation is distributed fairly or equitably among different socioeconomic groups. For example, how would the costs or consequences of rising ocean levels affect different countries? Is it fair if rich countries that emit large quantities of greenhouse gases are nonetheless able to mitigate the effects of climate change by building dikes or coastline protection structures while poor, low-lying countries are dislocated? It may be especially difficult (if not impossible) to devise quantitative justice metrics that are scientifically reproducible and widely accepted. For example, racial minorities in the United States are, generally speaking, more likely to live near sources of pollution than whites. The correlation is understandably controversial, as are the possible reasons why (see Bowen, 2002, and Corburn, 2002, for some contrasting perspectives). Is it because hazardous waste dumps or industrial plants were intentionally located near communities of color such as American Indian reservations or black and Hispanic neighborhoods? Are regulatory or enforcement policies biased in favor of whites? Other studies suggest that energy price and distribution disparities correlate with greater environmental degradation, adverse public health outcomes, higher poverty, lower incomes, more pollution, and greater energy use at lower efficiency (Templet, 2001; Boyce et al., 1999). The possible consequences of such studies are far-reaching. It is likely that *fairness* is an essential aspect of achieving economic, thermodynamic, ecological, and environmental sustainability. All these criteria are interrelated.

## Design for Sustainability

It is important to understand the nature of any particular metric so that the underlying environmental management objectives can be clearly defined. A taxonomy of sustainability metrics helps to avoid confusing consequences of one type of metric with consequences of another type, and so that trade-offs may be made consciously rather than haphazardly. Although it sounds simple, confusion may be more often the norm than the exception. For example, thermodynamic, environmental, and ecological criteria are sometimes lumped together under a single environmental heading, such as in reference to the *triple bottom line*, which refers only to financial, social, and environmental assessment criteria. This is too simplistic; it may obscure the specific ramifications of various alternatives. Industrial ecology is really about discovering the trade-offs among these criteria and not about making the subjective judgments required to find the preferable technological alternatives when various criteria are not well aligned.

The catalytic converter example (Figure 1.2) illustrates how different metrics can provide conflicting signals. Consider that automobiles would certainly be more thermodynamically efficient (in terms of miles per gallon) without catalytic converters, which add weight and partially restrict exhaust flow. Catalytic converters may or may not be economically efficient, as the total health benefits of cleaner exhaust may outweigh the additional financial and fuel expense of making and operating the converters in many geographic areas—but not all. The environmental and ecological efficiency of catalytic converters has been questioned, however, on a life-cycle basis. Converters speed the oxidation of automobile exhaust containing carbon monoxide and incompletely combusted hydrocarbons into carbon dioxide and water, which are less harmful but not entirely environmentally benign. This environmental benefit may be outweighed by the increased carbon dioxide production that results from lower mileage and from the environmental consequences of mining the platinum group elements that make up the chemically active portion of the converter. While there is broad social and political support for converters, as evidenced by the regulations that require them, it must be understood that environmental and ecological improvement are not synonymous with energy efficiency. End-of-pipe treatments, such as converters and incinerators, may improve environmental qualities only at the expense of increased energy consumption. In some cases, environmental improvements may be dubious or difficult to quantify considering the complicated trade-offs involved.

## Environmental Limits of the Natural Analog

Incorporating natural principles into industrial systems may result in environmental and economic trade-offs that are difficult to assess objectively. For example, it is well understood that natural systems employ closed material loops of various time scales such that the remains or waste materials of one organism create the raw materials for others. Industrial systems are far less materially efficient. On the other hand, it is often overlooked that natural systems are highly energy intensive, compared to industrial systems. The quantity of solar exergy consumed by natural processes (e.g., photosynthesis, evaporation, and photochemical reaction) exceeds that which is consumed in the economy by a factor of approximately 7500 (Kåberger and Månsson, 2001). Very little solar exergy is ultimately stored in biochemical or physical reserves. For example, approximately 6600 kJ of sunlight are required to create 1 kJ of petroleum fuel, an efficiency of approximately 0.015% (Odum, 1996). The overall efficiency of the economy in the subsequent conversion of the petroleum into electricity (which is nearly completely convertible to useful work) may be estimated as 25 to 30%.

Empirical economic evidence may bear out the view that the material intensity of the economy has declined over time, but the energy intensity may not have improved significantly (Kaufmann, 1991). The trend in the economy of providing greater service or utility with fewer material requirements has been described as *dematerialization* and is sometimes cited as a hopeful sign of progress toward sustainability. The automobile industry exemplifies the concept of dematerialization. Modern designs substitute lighter high-technology materials (such as plastic, aluminum, and composites) for what was previously made of steel. As a result, modern automobiles are generally lighter, last longer, get better mileage, and are safer than previous models, but they are also more difficult to recycle because the larger variety of specialty materials require more separation and remanufacture. Conversely, the historical trend in patterns of energy use has been toward employing higher and higher quality fuels with concomitant increases in economic output. Electricity has been substituted for oil and gas, which were substituted for steam (from coal), which was substituted for wind and water, which was substituted (where possible) for animal labor, which was substituted for human labor. It may be difficult to determine whether the bulk of economic growth is attributable to technological progress (knowledge gains), economies of scale, or merely fuel switching. The result may be a generally increasing trend in the *energy* intensity of the economy, even as it has become less *materially* intensive over time (Peet, 1992). Although this is exactly what the natural analog might anticipate, it need not necessarily be perceived as preferable from the standpoint of sustainability because industrial energy sources are rarely environmentally benign (compared to natural energy sources). High-entropy wastes, in particular, require large exergy inputs to reconstitute or recycle (Ayres, 1999). Therefore, recycle of low-quality waste streams may *increase* thermodynamic resource consumption and/or total environmental burdens, which is counter to the first hypothesis of industrial ecology. It may be reasonable both economically and from the perspective of sustainability to employ the highest quality resources first (even if these are natural resources) rather than recycle poor-quality wastes. In the absence of some plentiful, inexpensive, environmentally benign energy source, however, it may be that to become sustainable, the human industrial metabolism must eventually *exceed* the material and energy efficiency of nature—not merely imitate it.

## Exercises

### Discussion Questions

1. Discuss the future of the environmental engineering profession in a zero-waste (or zero chemical pollution) economy. Is such an economy technical or hypothetically possible? Can you find evidence that our economy is moving in this direction?
2. Civil (and other) engineering works exist at the intersection of the economy with nature. They are subject to nature's forces (e.g., storms, earthquakes, pollution) and stress natural and social systems in many ways. What types of unintended or ancillary effects on sustainability do civil engineering works have? Consider dams, highways, tall buildings, water, and wastewater treatment plants, jetties, and bridges. List as many effects as possible. Categorize these by whether they are economic, thermodynamic, environmental, ecological, sociopolitical, or a combination of many factors.
3. Consider a genetically engineered crop that requires fewer pesticides or herbicides or produces an equivalent yield of food (or industrial products) with fewer irrigation or fertilizer inputs but may contaminate nonengineered organisms either by cross-pollination or by entering food webs. Contrast the potential economic, thermodynamic, environmental, ecological, and sociopolitical effects. Does your characterization change if the genetically modified organism (GMO) is engineered to produce medicine (e.g., human insulin) rather than food?

4. Use the economic input–output analysis tools available at http://www.eiolca.net/ to estimate the life-cycle inventories for the following product demands (in the United States): potable water, food, concrete, housing, photographs, computers. Try others. Evaluate the results against your intuition. Are there aspects that seem to be missing? How do the results surprise you?
5. Contrast the business operations of the following industries: health care, higher education, government, professional sports, software programming, computers. Are the companies that make up these industries profit maximizing? If not, by what criteria do they measure success?
6. Compare the diagram depicting carbon cycling in the Chesapeake (Figure 1.9) with that depicting ethanol production from sugar crops (Figure 1.3) and production of the catalytic converter (Figure 1.2). Do they lead to similar insights or understanding of the systems they are intended to portray?
7. Using the streamlined life-cycle impact assessment data depicted in Table 1.3 for alternative automotive technologies, what strategy would you employ to recommend an optimal alternative? (Check Table 1.6 for examples.) To what extent have available alternatives been implemented in your community? Why?
8. Research electricity prices and power provider options in your area. Do any advertise "green" power because the electricity is generated from renewable resources such as wind, solar, or hydro? How do prices for green power compare to conventional? Use the data presented in Figure 1.8 to estimate the marginal cost of reducing greenhouse gas emissions by switching electricity-generating technologies.

## Mathematical Exercises

1. Use the systems approach (see sidebar) to optimize the cylindrical dimensions of a 12-ounce aluminum soda can, assuming constant wall thickness and homogenous material. Repeat the exercise assuming the top and bottom of the can are twice as expensive (per area) as the material from which the rest of the can is made. How do your optimized dimensions compare with those of a real can?
2. How should a corral be constructed to contain the largest possible number of range animals if a finite amount of fencing is available? Can you arrive at your answer using a numerical example? Equations without any numbers? A scale drawing?

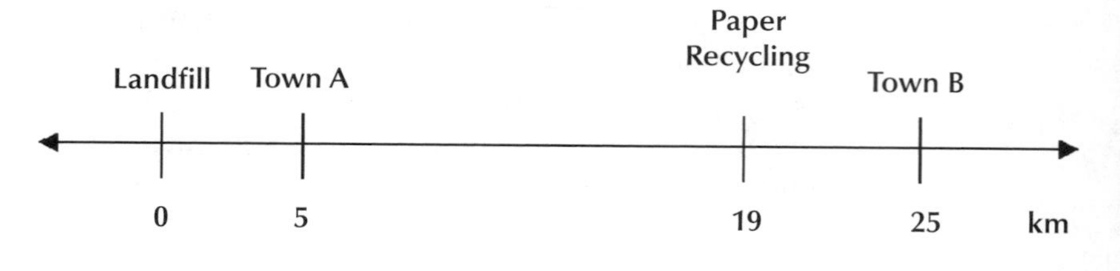

3. Town A generates 10 metric tons of solid waste per day, 20% of which is recyclable paper. Town B generates only 5 metric tons per day, but 30% is recyclable paper. The two towns, a public landfill, and a paper recycling plant are all located along a straight highway as shown in the accompanying figure. Transport costs are \$0.40/km/metric ton. Tipping fees at the landfill are \$100/metric ton. The paper plant pays \$5/ton for separated paper.
   - Compute the minimum daily waste disposal net costs to towns A and B.

   These towns are considering a waste-to-energy proposal in which tipping fees will be \$50/metric ton for regular waste but only \$5/metric ton for separated paper. The ash content of the regular waste is 15%; for the separated paper the ash content is only 5%. Each town must bear the cost of hauling and disposing of its ash in the landfill. The proposal is contingent on a guarantee of a minimum of 10 metric tons per day of total waste received at the waste-to-energy facility.
   - Where should the facility be located to minimize the total costs of operating the system? (Hint: Create a line graph of total cost versus position of waste-to-energy plant.)
   - Is the optimum location when *total* costs are considered the same as town A or town B would select if they were considering only their own costs?
   - Should the voters of town A approve the project? Should the voters of town B?
4. Use the emergy yield ratios listed in Figure 1.7 to estimate the factor by which consumption efficiency must be improved if we were to switch to ethanol from sugar cane as a fuel source instead of Middle Eastern oil (without increasing the total energy demands of the economy or reducing consumption).
5. Graph the exponential discount function versus time using interest rates of 3, 5, and 10% and a time horizon of 100 years. How does the function change with interest rate? At what time for each interest rate does the area under the curve equal 95% of the total area? Now select values of B in the hyperbolic model that roughly match each case. (Hint: Start by guessing twice the percentage interest rate for B.) How do the results differ? At what time does the area under the hyperbolic curve equal that under the corresponding exponential? (Hint: The area under the exponential decay curve is equal to $A$, the present value at $t = 0$, but the area under the hyperbolic curve is infinite.)

**Stepwise systems approach**

**Step 1:** Identify the Problem. The first step may be somewhat misnamed. Engineering is not always motivated by a problem; sometimes it is an opportunity. Nevertheless, the client's goals must be elucidated in detail such that design efforts may be focused toward accomplishing them with the fewest resources possible. A particularly important aspect of identifying the problem is deciding how to measure the effectiveness of the design. This is the first step toward drawing an objective function: deciding what resource or metric will be optimized. Shall it be least time? least cost? maximum profit? least labor? minimal adverse environmental effect? The client may be curious to understand how the system behaves or responds to different objective criteria and request that a broad range of design options be preserved. Where this is expensive, designers typically explore alternatives in a schematic (rather than detailed) design to inform the client sufficiently to constrain the design process and limit design costs.

**Step 2:** Assess Alternatives. It is possible at this step to overlook technological alternatives that seem disadvantageous at first glance or that the system analyst is unaware of. Clearly, optimization of a single technological alternative does not ensure the best possible design if superior alternatives have been overlooked. It is essential, therefore, to engage a wide range of experience and/or research in assessing alternatives. It may be that several must be investigated mathematically to determine which is the best available.

**Step 3:** Construct a (Mathematical) Model. Identify design variables. These are the factors that the engineer can exercise some control over within the technological constraints imposed by the alternative under selection. Design variables are the focus of engineering coursework (e.g., sizing pipes, pump rates, detention times, and foundation footings).

Identify constraints. Constraints may take the form of equations that describe technological, economic, chemical, or physical (or social) relationships. They may be equalities or inequalities; in any case, they are not subject to the engineer's control. (Self-imposed constraints, such as voluntary environmental objectives, may be considered as an exception.) For example, resource prices, physical laws, and budget limitations are typical constraints. Some confusion regarding the difference between constraints (which are inviolable) and the objective function is often experienced by students first introduced to the systems approach. Recall that the objective measures should be selected in identification of the problem, despite the fact that the exact relationship between the design variables and objective criteria has not yet been determined (see below).

Identify the objective function. The constraint relationships both limit the range of design variable values that may be feasible and govern the relationship between the objective function and the design variables. The objective function is the focus of mathematical analysis, and reflects the values of the client in that it is the sole measure of the effectiveness of a design recommendation. It is not possible to select a set of design variables that optimizes for more than one objective criterion. In identification of the problem, therefore, the client and engineer must select the criteria (which could be a weighted average of several different criteria appearing as different terms in a single objective function) that will guide the design recommendation. Sometimes, it is desirable to create several different objective functions and solve the system under alternative criteria to observe how the design recommendations might respond. In this case, the client may reconsider the problem (to the frustration of the engineer) as trade-offs inherent in the constraint relationships become readily (and quantitatively) apparent. To avoid confusion or design remorse, the engineer may best serve the client by discussing as many alternative objective criteria and the possible trade-offs involved in identification of the problem, rather than waiting until final design recommendations are imminent.

**Step 4:** Create a Design Recommendation. (Solve the set of mathematical equations for an optimal design.) Finding a set of design variables that satisfies the constraint equations at the optimum value of the objective function is sometimes a difficult and complicated mathematical problem, but it is the culmination of analytical efforts. Different methods are available, some of which involve sophisticated numerical guess, check, or search algorithms. Simpler problems with few design variables may best be solved by graphing the value of the objective (sometimes called merit function when nonmonetary units are employed) against different design values. Even after solution, however, design recommendations may be rejected by the client. This may be due to a failure of communication early in the systems approach (such as in identifying the problem) if the analyst has properly modeled the problem. Sometimes, clients may be frustrated by, or even in denial of, important constraint relationships, which may be due to an inadequate exploration of alternatives or a failure to involve important constituent parties early in the process.

## References

Amatayakul, W. and O. Ramnas. 2001. Lifecycle assessment of a catalytic converter for passenger cars. *Journal of Cleaner Production,* **9**(5):395–404.

Anex R. 2002. Public participation in lifecycle assessment and risk assessment: A shared need. *Risk Analysis,* **22**(5):861–877.

Ayres, R. U. 1999. The second law, the fourth law, recycling and limits to growth. *Ecological Economics,* **29**(3):473–484.

Ayres, R. U. 1995. Lifecycle analysis: A critique. *Resources, Conservation and Recycling,* **14**:199–223.

Ayres, R. U. 1989. Industrial metabolism. In *Technology and the Environment*. J. H. Ausubel and H. E. Sladovich, eds. National Academy Press: Washington, DC.

Ayres, R. U. and R. Axtell. 1996. Foresight as a survival characteristic: When (if ever) does the long view pay? *Technological Forecasting and Social Change,* **51**:209–235.

Belton, V. and T. Steward. 2002. *Multiple Criteria Decision Analysis: An Integrated Approach.* Kluwer Academic: Boston.

Bennet, D. H., M. D. Margni, and T. E. McKone. 2002. Intake fraction for multimedia pollutants: A tool for lifecycle analysis and comparative risk assessment. *Risk Analysis,* **22**(5):905–918.

Berthiaume, R., C. Bouchard, and M. A. Rosen. 2001. Exergetic evaluation of the renewability of a biofuel. *Exergy, an International Journal,* **1**(4):256–268.

Bowen, W. 2002. An analytical review of environmental justice research. *Environmental Management,* **29**(1)3–15.

Boyce, J. K., A. R. Klemer, P. H. Templet, and C. E. Willis. 1999. Power distribution, the environment, and public health: A state-level analysis. *Ecological Economics,* **29**(1):127–140.

Cairns, J., Jr. 1998. Endpoints and thresholds in ecotoxicology. In *Ecotoxicology: Ecological Fundamentals, Chemical Exposure and Biological Effects*. G. Schuurmann and B. Markert, eds. Wiley: New York.

Carnegie Mellon University Green Design Initiative. 2002. Economic input–output lifecycle assessment (EIO-LCA) model [Internet], http://www.eiolca.net/ [Accessed 15 Nov, 2002].

Cleveland, C. J., R. K. Kaufmann, and D. I. Stern. 2000. Aggregation and the role of energy in the economy. *Ecological Economics*, **32**(2):301–317.

Corburn, J. 2002. Environmental justice, local knowledge, and risk: The discourse of a community-based cumulative exposure assessment. *Environmental Management*, **29**(4):451–466.

Costanza, R. 1998. The value of the world's ecosystem services and natural capital. *Ecological Economics*, **25**(1):3–15.

CWRT 1999. *Total Cost Assessment Methodology*. Center for Waste Reduction Technologies, American Institute of Chemical Engineers: New York. (See also http://www.aiche.org/cwrt/.)

Deutsch, L., A. Jansson, M. Troell, P. Ronnback, C. Folke, and N. Kautsky. 2000. The "ecological footprint": communicating human dependence on nature's work. *Ecological Economics*, **32**(3):351–355.

Emblemsvag, J. and B. Bras. 2001. *Activity-Based Cost and Environmental Management: A Different Approach to the ISO 14000 Compliance*. Kluwer Academic: Boston.

Franzle, O. 1998. Sensitivity of ecosystems and ecotones. In *Ecotoxicology: Ecological Fundamentals, Chemical Exposure, and Biological Effects*. G. Schuurmann and B. Markert, eds. Wiley: New York.

Frosch, R. A. and N. Gallopoulos. 1989. Strategies for manufacturing. *Scientific American*, **261**(3):144–152.

Graedel, T. E. 1998. *Streamlined Lifecycle Assessment*. Prentice Hall: Upper Saddle River, NJ.

Green, L. and J. Myerson J. 1996. Exponential versus hyperbolic discounting of delayed outcomes: Risk and waiting time. *American Zoology*, **36**:496–505.

Gregory, R. 2002. Incorporating value trade-offs into community-based environmental risk decisions. *Environmental Values*, **11**:461–488.

Hayman, G. D. and R. G. Derwent. 1997. Atmospheric chemical reactivity and ozone-forming potentials of CFC replacements. *Environmental Science & Technology*, **31**(2):327–336.

Hertwich, E. G., S. F. Mateles, W. S. Pease, and T. E. McKone. 2001. Human toxicity potentials for lifecycle assessment and toxics release inventory risk screening. *Environmental Toxicology and Chemistry*, **20**(4):928–939.

Hertwich, E. G., W. S. Pease, and T. E. McKone. 1998. Evaluating toxic impact assessment methods: What works best? *Environmental Science & Technology*, **32**(5):138A–144A.

Hofstetter, P. and J. K. Hammit. 2002. Selecting human health metrics for environmental decision-support tools. *Risk Analysis*, **22**(5):965–982.

Izsak, J. and L. Papp. 2000. A link between ecological diversity indices and measures of biodiversity. *Ecological Modeling*, **130**(1–3):151–156.

Kåberger, T. and B. Månsson. 2001. Entropy and economic processes—physics perspectives. *Ecological Economics*, **36**(1):165–180.

Kennedy, M. L. 1998. *Total Cost Assessment for Environmental Managers*. Wiley: New York.

Kaufmann, R. K. 1991. A biophysical analysis of the energy/real GDP ratio: Implications for substitution and technical change. *Ecological Economics*, **6**(1992): 35–56.

Keeney, R. L. 1992. *Value Focused Thinking: A Path to Creative Decision Making*. Harvard University Press: Cambridge, MA.

Lave, L. B., H. MacClean, C. T. Hendrickson, and R. Lankey. 2000. Lifecycle analysis of alternative fuel/propulsion technologies. *Environmental Science & Technology*, **34**:3598–3605.

Lave, L. B., E. Cobas-Flores, E. T. Hendrickson, and F. C. McMichael. 1995. Using input–output analysis to estimate economy-wide discharges. *Environmental Science & Technology*, **29**(9):420A–426A.

McCulloch, A. 1994. Lifecycle analysis to minimise global warming impact. *Renewable Energy*, **5**(Part 2):1262–1269.

Nishioka, Y., J. I. Levy, G. A. Norris, A. Wilson, P. Hofstetter, and J. D. Spengler. 2002. Integrating risk assessment and lifecycle assessment: A case study of insulation. *Risk Analysis*, **22**(5):1003–1017.

Norris, G. A. 2001. Integrating lifecycle cost analysis and LCA. *International Journal of Lifecycle Assessment*, **6**(2):118–121.

Odum, H.T. 1996. *Environmental Accounting: Emergy and Environmental Decision-Making*. Wiley: New York.

Pacca, S. and A. Horvath. 2002. Greenhouse gas emissions for building and operating electric power plants in the upper Colorado River basin. *Environmental Science & Technology*, **36**(14):3194–3200.

Paton, B. 2001. Efficiency gains within firms under voluntary environmental initiatives. *Journal of Cleaner Production*, **9**(2):167–178.

Peet, J. 1992. *Energy and the Ecological Economics of Sustainability*. Island Press: Washington, DC.

Rapport, D. 1999. Gaining respectability: Development of quantitative methods in ecosystem health. *Ecosystem Health*, **5**(1):1–2.

Rapport, D. J., R. Costanza, and A. J. McMichael. 1998. Assessing ecosystem health. *Trends in Ecology and Evolution*, **13**(10):397–402.

Rudd, M. A. 2000. Live long and prosper: Collective action, social capital and social vision. *Ecological Economics*, **34**(1):131–144.

Runyan, D. K., W. M. Hunter, R. R. S. Socolar, L. Amaya-Jackson, E. English, J. Landsverk, H. Dubowitz, E. H. Browne, S. I. Bangdiwala, and R. M. Mathew. 1998. Children who prosper in unfavorable environments: The relationship to social capital. *Pediatrics*, **101**(1):12–18.

Seager, T. P. and T. L. Theis. 2002. A uniform definition and quantitative basis for industrial ecology. *Journal of Cleaner Production*, **10**(3):225–235.

Seppälä, J., L Basson, and G.A. Norris. 2002. Decision analysis frameworks for life-cycle assessment. *Journal of Industrial Ecology*, **5**(4):45–68.

SETAC. 1993. *Guidelines for Lifecycle Assessment: A "Code of Practice."* F. Consoli, D. Allen, I. Boustead, J. Fava, W. Franklin, A.A. Jensen , N. de Oude, R. Parrish, R. Perriman, D. Postlethwaite, B. Quay, J. Seguin, and B. Vignon, eds. Society of Toxicology and Chemistry Press: Pensacola, FL.

Templet, P. H. 2001. Energy price disparity and public welfare. *Ecological Economics*, **36**(3):443–460.

Ulanowicz, R. E. 1997. *Ecology, The Ascendant Perspective*. Columbia University Press: New York.

USEPA. 1998. *1996 Toxic Release Inventory Data Quality Report*. EPA 754-R-98-016.

USEPA. 1993. *Lifecycle Assessment: Inventory Guidelines and Principles.* Office of Research and Development: Cincinnati, OH. EPA/600/R-92/245.

van den Bergh, J. C. J. M. and H. Verbruggen. 1999. Spatial sustainability, trade and indicators: An evaluation of the "ecological footprint." *Ecological Economics,* **29**(1):61–72.

World Meteorological Organization. 1999. *Scientific Assessment of Stratospheric Ozone Depletion: 1998,* Vols. 1 and 2. United Nations Environment Programme: Geneva, Switzerland. WMO No. 44.

# CHAPTER 2

# Life-Cycle Management

SUSAN E. POWERS AND ANAHITA A. WILLIAMSON

Since the 1960s, environmental issues have become a stalwart part of public awareness. The public recognizes that the manufacturing of products is one of the main sources of environmental deterioration. Products are regarded as a cause of resource depletion, energy consumption, and emissions during their life (Frankl and Rubik, 2000). These environmental effects occur throughout the life cycle of a product, beginning with the extraction of raw materials and ending with disposal. The effects also occur during product consumption and continue with a variety of waste management options such as incineration, landfilling, and recycling.

Because of the increased public concern, industry and government are continuing to develop methods to understand better and to reduce the adverse environmental effects of these activities (Vigon et al., 1993). The trend is away from end-of-pipe rules and regulations and toward shared responsibility among different organizations to minimize the effects of materials extraction, manufacturing, and product use. Business corporations have the responsibility to promote sustainable development. The government's task is to provide the policy framework to foster the greening of product management by business (Frankl and Rubik, 2000).

In the last few decades, tools, concepts, and ideas have been developed to "green" business. This chapter examines one tool, life-cycle assessment (LCA). It is used to assess the environmental effects of the overall system from the extraction of raw materials through manufacturing processes, product use, and ultimate disposal. LCA provides a family of methods for evaluating materials, services, products, processes, and technologies over their entire life. The main goal of LCA is to evaluate the environmental effects of a particular process or product from the point where raw materials are extracted from the earth, manufactured into the product, used, and disposed.

Many corporations have performed environmental audits or assessments to measure their environmental performance (see Chapter 5). Such assessments, however, often focus on a specific site or facility and do not consider the overall life cycle of which they are a part. Industrial processes do not function in a vacuum; they are linked through suppliers and customers with other processes, services, and materials. LCA broadens the decision-making process of corporations by providing a systematic approach that considers the energy and material use, transportation, postconsumer use, disposal, and environmental releases associated with a process or product. LCA can also enhance understanding of the benefits and risks associated with specific change in a process or product.

Life-cycle assessment emerged in the 1960s (Vigon et al., 1993) with an initial focus on calculating energy requirements, although some studies also included limited estimates of emission releases (Curran, 1996). In 1969, a study for The Coca-Cola Company became the foundation for current life-cycle analysis methods in the United States (Ciambrone, 1997). The main objective of the study was to determine whether glass or plastic containers would least affect the supply of natural resources and would have the smallest releases to the environment. Contrary to all expectations, the study revealed that the plastic container was the better choice. Other U.S. and European companies performed similar life-cycle analyses during this time (Ciambrone, 1997), but many were subject to biases. Because of the oil shortages in the early 1970s, both the U.S. and British governments commissioned extensive reviews of industrial studies to conduct detailed energy analyses (Curran, 1996). Once the oil crisis faded, however, so did interest in performing LCAs to evaluate energy use. A renewed interest was found in LCA in the 1980s as the "green movement" in Europe focused public attention on issues related to solid waste disposal and recycling. Environmental releases then began to be added to assessments of the effect of raw materials, solid waste, and energy consumption (Curran, 1996).

Industry leaders in product manufacturing sectors initially expressed interest in LCA, particularly when they tried to demonstrate their product's environmental superiority over that of their competitors' products. Consumer groups also wanted to use LCA to compare products to distinguish which were environmentally preferable (Curran, 1996). Because early LCAs lacked a standard approach and their results were based on diverse assumptions, it was difficult to justify many of their conclusions and the value of LCAs. These complications and the extensive data required for detailed LCAs pointed to the need for standards for methods to perform reliable and comparable assessments, and international standards were established. The ISO 14040-14043 series covers the primary steps in an LCA, yet a full LCA performed following International Standardization Organization (ISO) 14040-14043 can be an expensive and time-consuming task. Other options streamline this analysis. This chapter introduces the comprehensive approach to LCA. It then discusses simplifying methods.

Life-cycle assessment is only one of several environmental management tools. It may not always be the most appropriate one and will sometimes be used in conjunction with other tools such as those discussed in other chapters of this book.

## Standard Implementation of LCA

Life-cycle management is a philosophy and method for understanding the environmental effects of industrial systems or products in a "cradle-

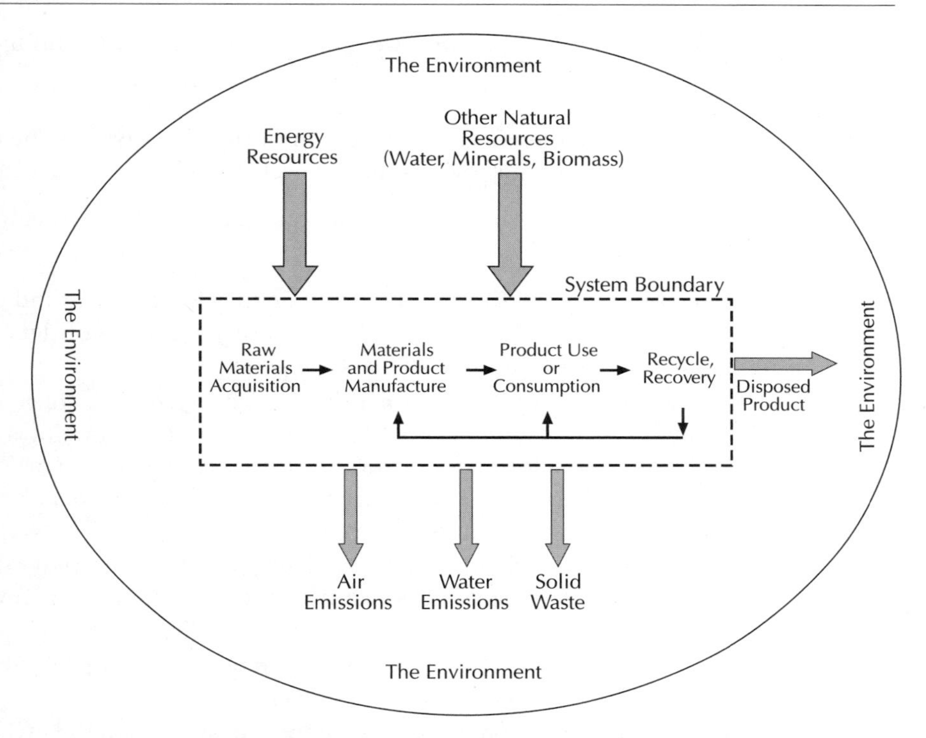


**Figure 2.1 Life-cycle concept.** (Adapted from Curran, 1996)

to-grave" approach. The system includes all activities to create a product or service; it begins with gathering raw materials from the earth and ends when materials are returned to the earth (Figure 2.1). Specific tools and methodologies have been developed to conduct LCAs for various industries. The systems approach that an LCA offers ensures that the interdependencies and cumulative effects of various steps in the product's creation, marketing, use, and disposal are integrated into the analysis and interpretation of environmental effects. The objective is to understand better the trade-offs between environmental effects and disparate products or processes. The life-cycle approach provides a comprehensive picture of the overall effect, as compared with traditional approaches that might assess only one stage in the life cycle or one particular environmental effect (e.g., air emissions).

An LCA, however, generally does not provide information about the life-cycle costs of the process or improvements defined to reduce environmental effects. Thus, the LCA is only one tool used in making an informed decision. **An accounting of the costs is required for a final assessment.**

In the United States, most tools available to conduct LCAs have evolved from an early approach developed and published by the U.S. Environmental Protection Agency (USEPA) (Vigon et al., 1993) and the Society of Environmental Toxicologist and Chemists (SETAC, 1993). The concept requires:

- Defining and understanding mass and energy flows throughout the life cycle
- Relating these flows to environmental effects
- Interpreting the net effects so that informed decisions can be made to change any processes in the life cycle to reduce these effects

The basic steps for most LCA techniques include the following (Boguski et al., 1996):

1. **Define *Scope* of Study** This step defines the purpose and goals, boundaries and environmental concerns, and level of detail and data accuracy required to meet the goals.
2. **Develop life-cycle *Inventory* (LCI)** The inventory collects and analyzes information related to the mass and energy flows of all phases of the life cycle, including emissions. The LCIs particulars can range from a list of materials used and emitted to a quantitative account of all mass and energy flows.
3. **Create Life-Cycle *Impact* Assessment (LCIA)** It employs the inventory data in a perspective of environmental effects, not just mass and energy units. Effects can include human health, global warming, eutrophication, and so forth. Many approaches are available to convert from inventory data to impacts.
4. **Interpret Data and Make Decisions** Results from the LCI and/or LCIA are analyzed to provide informed decisions about reducing the overall effect of the product on the environment.

Figure 2.2 shows that these four steps must be completed in an iterative manner. They cannot be completed in just one pass.

Many corporations choose to conduct LCAs under the guidelines of ISO 14040 (ISO, 1996). Others use streamlined approaches that reduce the time and expense associated with the LCA. In either case, the quality of the analysis and interpretation can vary widely depending on the availability of suitable data and assumptions made during the process. Defining the availability of data and the level of depth that will be used in the LCA is a critical component of the scoping stage. This sets the stage for the level of analysis that can be completed in the other steps.

Although LCA methods are being employed in corporations internationally, several barriers limit their use:

- Lack of awareness of importance of life-cycle concept
- Inaccessibility of LCA data
- Assumptions and subjective valuation procedures
- Lack of an objective impact assessment method
- Lack of advocacy for LCA within the corporation

## Life-Cycle Scoping Stage

All interested parties (stakeholders) must agree to the scope and objectives of the LCA. The different parties interested in the LCA will have

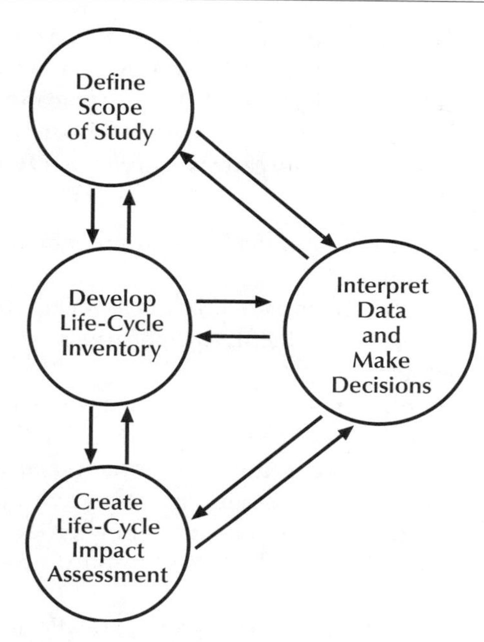


**Figure 2.2 Primary steps in an LCA—an iterative process.** (*Source:* Ahmadi et al., 2003.)

different opinions about the purpose of the study and the approach used. Many LCAs are conducted for products, but the product itself should not be the focus. The function or service that that product serves is more important. This requires that the product be viewed as part of a *system* rather than in isolation. This systems perspective is the primary basis for an LCA, and it raises the critical questions that must be addressed in the LCA scoping phase (ISO, 1997):

- What is the function of the product or process within a system?
- Do we want the LCA to consider a single product, or will we compare two or more products (or processes) with our assessment?
- What are the boundaries of the system of interest?
- What do we want to accomplish with this study?
- If more than one product is generated within this system, how do we allocate environmental effects among the products?
- What LCA method should we use to meet our goals (and our financial resources)?
- How much data do we need?
- What will the limitations of the study be?

These scoping questions should be revisited during the later iterations of the LCA to ensure that the answers remain appropriate and the goals realistic.

*Defining the Goals*

For every LCA, a statement of the goals should unambiguously define the reasons for completing the study and the intended audience (ISO,

1997). The goals should also be revisited as the LCA is developed to ensure that they are realistic and that the work being conducted is relevant to the stated goals. LCAs are typically used to identify the best product, process, or service with the least effect on human health and the environment. They can also serve other functions (USEPA LCA 101):

- Marketing
  - Prove one product is environmentally superior to a competitive product.
  - Justify labeling the product as an environment-friendly product.
  - Meet green labeling specifications.
- Engineering
  - Identify stages in the life cycle where resource use and emissions might be reduced.
  - Establish a baseline of information on a system's overall resource use, energy consumption, and environmental loadings.
  - Guide the development of new products, processes, or activities toward a net reduction of resource requirements and emissions.
- Management
  - Determine the consequences to affected parties.
  - Determine if progress is being made toward corporate environmental or sustainability goals.

Specifying the goals is necessary before the data required and depth of the LCA can be defined. For example, establishing a baseline of information on a system's overall environmental burden would require a substantial amount of quantitative data and analysis, whereas an LCA comparing two products for marketing purposes could focus only on aspects that differ between the two products.

Two examples of LCAs are used in many problems and examples throughout this chapter: paper products and toner (see sidebars). In both cases, the analysis was limited to a life-cycle inventory, yet because the goals of these two studies were very different, the nature of the assessments also varied greatly. The first LCA provides an industry-average look at the environmental effects of general paper products to develop recommendations about the purchase of environment-preferable products. In contrast, the toner LCA is much more specific. Its goal was to inform process engineers and product developers how to reduce adverse environmental effects associated with toner use in a copier.

**Paper LCA**

The average American consumes 700 pounds of paper annually, and Americans together discard an amount equivalent to 20% of all of the paper manufactured in the world (Environmental Defense, 1995). With this level of consumption, the paper manufacturing process is the fourth largest user of energy, largest generator of wastes (by weight), and the largest consumer of water among the primary industries in the United States. The significant environmental effects associated with these startling statistics prompted the Environmental Defense Fund (now called Environmental Defense) to organize a voluntary private-sector task force to identify ways to reduce these effects while still meeting industry's paper needs in terms of functionality and economics.

Duke University, Johnson & Johnson, McDonalds, The Prudential Insurance Company of America, and Time Inc. joined the task force. They completed an extensive life-cycle inventory for several grades of paper and cardboard products. Because the scope of this study was to look at industry-wide activities and effects, average or typical values of data were of interest—not the industry's best or worst examples. Data were collected from published sources, site visits to facilities at all stages of the life cycle, and discussions with industry experts.

The results clearly support the need for paper consumers not only to recycle their own discarded paper but also to purchase paper with recycled content to assure the economic viability of this paper source. More detailed recommendations were made for processes associated with each stage in the paper system.

## Xerographic Toner LCA

The xerographic industry has seen remarkable advances in speed and quality of image production in the past few decades. This technology is still improving today with many companies competing for the future in copy and print technology. One of the main components in the xerographic process is the toner, the dry ink that creates the image on paper during the xerographic process. Toner is used in most copiers and some large printers. The system boundaries for the toner LCA are shown in the accompanying figure.

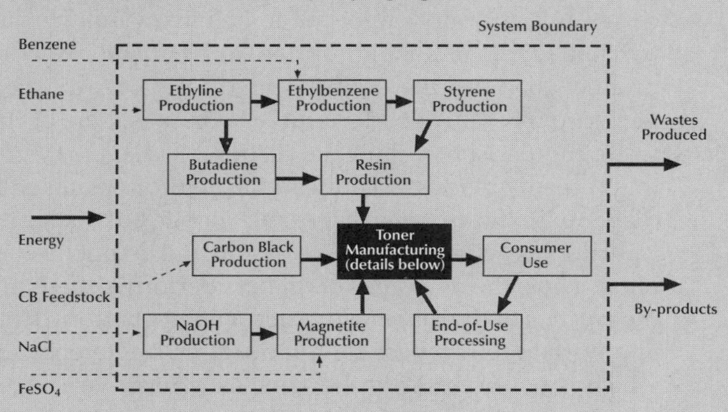


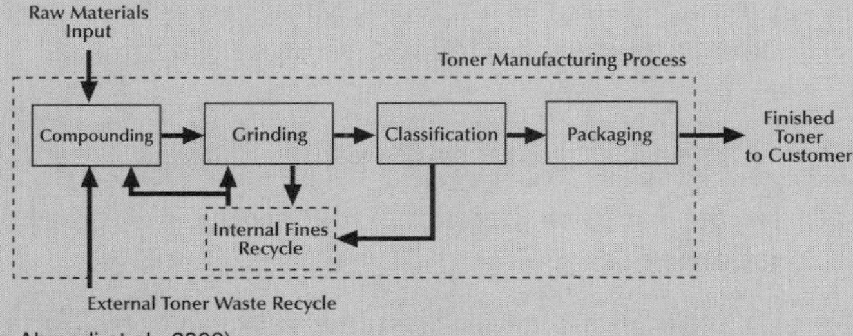


(*Source:* Ahmadi et al., 2003)

In conventional method of toner production, particles are mechanically fractured to the desired size distribution. (See lower part of figure.) In this process, the raw materials (resin, magnetite, and carbon black) are mixed and heated in the compounder to a homogeneous polymer melt. The cooled solid pellets are ground and screened to obtain the proper size distribution and then packaged for distribution.

The industry has taken steps to incorporate recycling where possible. One recycling loop is internal, where toner waste is recycled from the grinder and classifier/screener. The other loop is external, where toner waste from the consumer is returned to the manufacturing facility for recycling.

Working in conjunction with an imaging technology corporation, toner plant engineers and upper management requested that the conventional manufacturing process of toner be evaluated to identify the greatest environmental concerns. Quantifying the improvements associated with recent recycling initiatives was also desired. A life-cycle inventory was chosen to carry out the baseline evaluation, specifically focusing on the material and energy inputs into the system along with the wastes produced and by-products leaving the system. The inventory categories included energy use (fossil fuel use and electricity); $CO_2$, $NO_x$, $SO_2$, volatile organic compunds (VOCs), and particulate emissions; wastewater volume; and solid waste. The approach and results of this analysis were described by Ahmadi et al. (2003).

*Defining the Functional Unit*

Because many LCAs are conducted to compare two products or processes, it is critically important to identify a suitable basis for the comparison. Often, it is inappropriate to compare individual units of two products because the products might perform differently in their intended purpose. A good example is diapers. One of the earliest LCAs was a comparison of cloth and disposable diapers (Vigon et al., 1993). It might seem obvious to compare the environmental effects of one cloth diaper with those of one disposable diaper, but disposable diapers have a greater capacity for absorbing liquid and do not need to be changed as often. Thus, the two types of diapers perform differently. In both cases, the *function* of the diaper is to keep a baby's bottom dry. Thus, it is more appropriate to compare the products with reference to their function, which is not necessarily the same as a mass, volume, or unit basis. A *functional unit* is the formal term used to quantify the performance of a product system as a reference unit in a life-cycle assessment. The functional unit must be defined in the scoping phase.

**Example 2.1 Functional Units**

Define a suitable functional unit for the life-cycle system for each of the following:

1. Ethanol-blended gasoline versus a petroleum gasoline

2. Wool versus fleece fabric for winter garments
3. Spray versus roller-applied paint

### *Solutions*

Ethanol-blended gasoline versus a petroleum gasoline:

A volume basis might be the easiest to use since that is the unit measure used to buy gasoline. Since many cars get a lower mileage with ethanol-blended gasoline, however, it would be more appropriate to consider the function—to propel a car. Thus, distance driven would be a better functional unit.

Wool versus fleece fabric for winter garments:

The function is to keep a person warm and dry. Since wool garments are often heavier than a comparable fleece garment, it would be inappropriate to compare the fabrics on a mass basis. Thus, the amount of fabric for a garment designed for comfort at a certain temperature should be the basis for comparing the function of these two fabrics.

Spray versus roller-applied paint:

Both the paint and the application method are different, but the function—to cover a surface for aesthetics and surface protection—is the same. The two processes could differ in the amount of paint required to cover a specified area and also in the durability of the paint. The functional unit should incorporate both surface area covered as well as a specified time frame.

#### *Defining the System*

The "system" studied in an LCA often incorporates all raw materials, other inputs, products, wastes, and emissions associated with the product from cradle to grave. For some studies, especially comparative studies for products that have the same raw materials, some steps of the life cycle may be omitted if they are very similar. Thus, some LCAs only cover cradle-to-gate or gate-to-grave approaches, where the term *gate* refers to the end of the manufacturing process, thus implying that the use and disposal aspects or raw material extraction and processing stages are not considered.

The USEPA diagrammed the range of activities that could be included within a system boundary (Figure 2.3). It is essential to understand thoroughly the process and materials required and generated at all life-cycle stages to define the system for a given product or process. Ancillary processes, including packaging and transportation of all materials, should be included as part of the system. These considerations can make the system quite extensive!

The manufacture of aluminum is a good example of a product that would be best studied in a cradle-to-gate approach. An aluminum manufacturer has significant control over the extraction and processing of

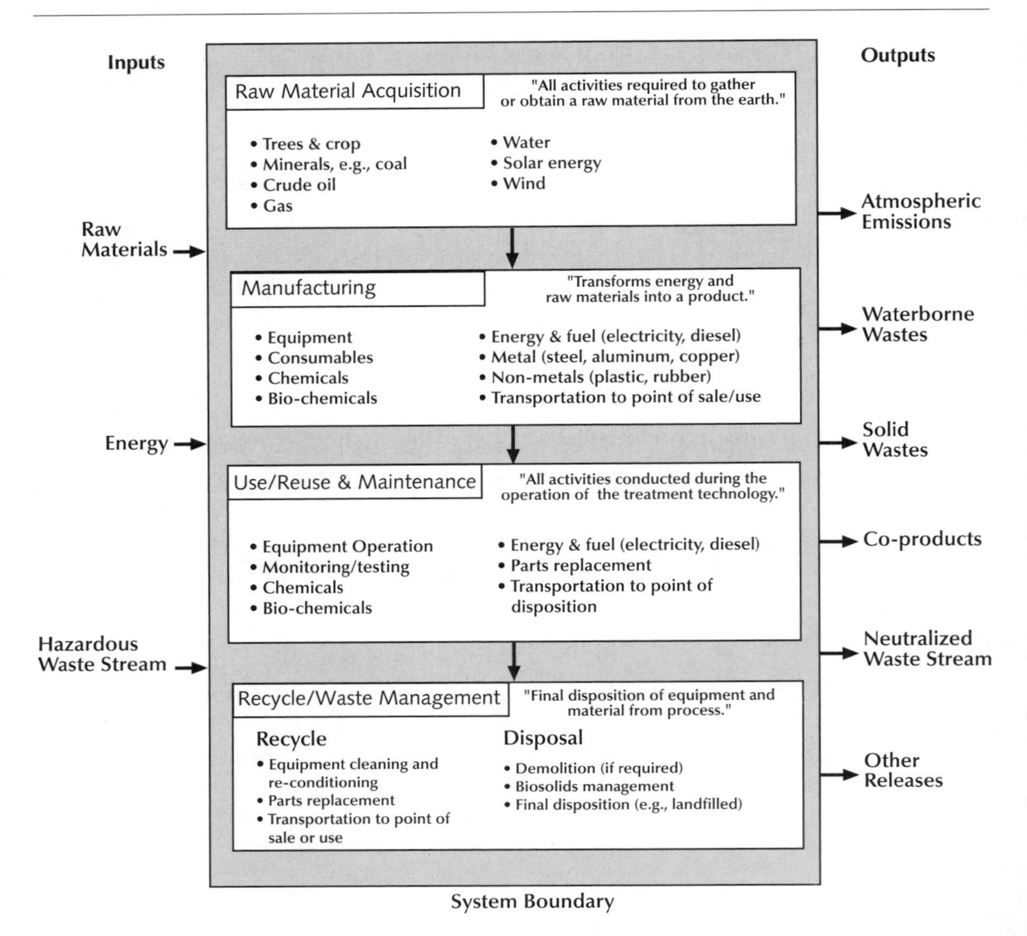


**Figure 2.3 Range of activities and flows that could be included within a system boundary.**

bauxite and the production of aluminum metal and is able to use the results of an LCA to reduce environmental effects associated with these life-cycle stages. The aluminum products are used for a wide variety of applications, and it would be nearly impossible to include all of the uses and recycle/disposal options in an LCA. It is, therefore, appropriate to limit the boundaries of this type of LCA to the end of the aluminum manufacturing stage—a cradle-to-gate assessment.

Once the general life-cycle stages included in the study are defined, the specific processes used in each of these steps must be understood and identified. A systems flowchart that shows the material and energy flow streams is necessary at the scoping stage. The flow diagrams help stakeholders to visualize all of the input and emissions streams associated with the product or process. More detailed process flow diagrams for each stage of the life cycle are required for completing the life-cycle inventory. (See sidebar for help with resources for gathering information to construct process flow diagrams.)

**Finding Information to Draw Life-Cycle and Process Flow Diagrams**

*Kirk–Othmer Encyclopedia of Chemical Technology* (Kroschwitz and Howe-Grant, 1991) provides an introduction to most major chemical processes and products

Chemical or mechanical engineering textbooks

Internet-based search (using results only from legitimate and authoritative sources)—for example, a search on "aluminum" provided a link to the World Aluminum Organization, which provides details on the aluminum life cycle from bauxite mining through use in final products (http://www.world-aluminium.org)

Figures 2.4 through 2.6 illustrate the flow diagrams necessary to understand and define the system associated with paper manufacture and use. Figure 2.4 provides a general understanding of the flow of paper through its life cycle. As shown in the more detailed flow diagram of the pulping process (Figure 2.5), a significant number of other raw materials, energy inputs, and emissions to water and air should be considered in an LCA. These secondary systems introduce the need to incorporate the life cycle of additional materials. Figure 2.6 shows the manufacturing process for the sodium hydroxide and chlorine gas that are used in the pulping and bleaching process. The raw materials for this process come directly from the earth.

The paper-manufacturing example illustrates the complexity of defining the *system* and the system's boundaries. The big question is: *How many materials and processes in advance of the primary product of interest should I include?* For any particular product or process, there is no one right answer; system boundaries of some products could comprise 40 to 50 different processes. Significant environmental effects are less likely to be omitted if the system boundaries incorporate all of the life-cycle effects of all raw materials, energy sources, and products. The trade-off is that the broader the boundaries, the more difficult it is to find the necessary data and to complete the LCA in a timely manner. The breadth of the LCA, therefore, must depend on the particular goals of the study and the availability of time, data, and financial resources.

Some rules of thumb are (Boguski et al., 1996):

- A step may be excluded only if doing so does not change the conclusions of the study.
- Inputs and outputs from the manufacture of capital equipment and buildings can generally be excluded.
- In a comparative study, steps that are identical between the two products that will be compared can be excluded.

Other LCA practitioners choose to limit the scope of their studies by only considering a few environmental categories. For example, LCAs

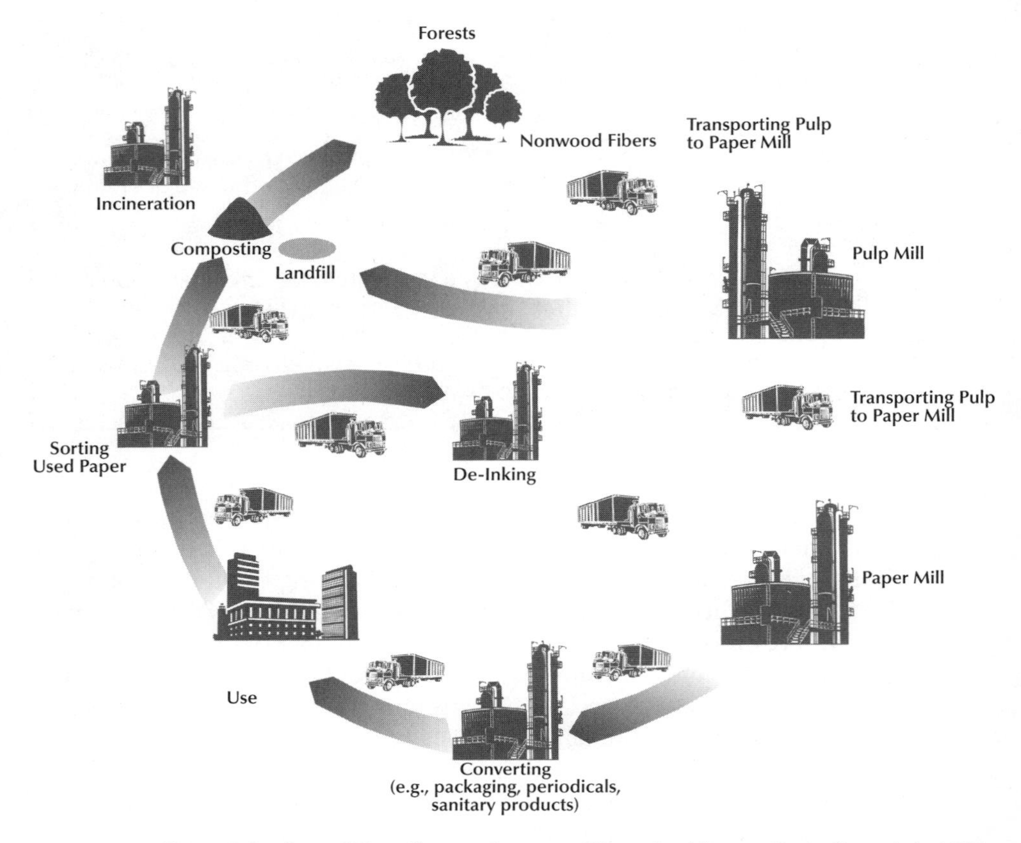


**Figure 2.4 Overall flow diagram for paper life cycle.** (*Source:* Greig-Gran et al., 1997. Reprinted by permission of MIT Press journals.)

related to energy use often focus on the consumption of nonrenewable resources and air emissions, including greenhouse gases, ozone, $SO_x$, $NO_x$, and air toxics [e.g., see Winebrake et al. (2000) and Kadam et al. (1999)]. In such a study, these factors are assumed to be the most important environmental effects.

The simplification of an LCA by neglecting some environmental effects, however, can lead to serious omissions and erroneous conclusions. A good example is the widespread use of MTBE (methyl *tert*-butyl ether) as an oxygenate component in gasoline during the late 1990s. This additive was purported to decrease atmospheric emissions—a key objective of reformulated gasoline during the use phase of the gasoline life cycle. However, the narrow focus on atmospheric effects resulted in the lack of adequate consideration of the fate of MTBE in aqueous systems. Because of its high solubility and very low biodegradation rates, MTBE is now found in many U.S. groundwater aquifers and drinking water supplies (Gullick and LeChevallier, 2000). A broader LCA before the decision was made to increase our use of MTBE could have averted this significant environmental outcome (Franklin et al., 2000).

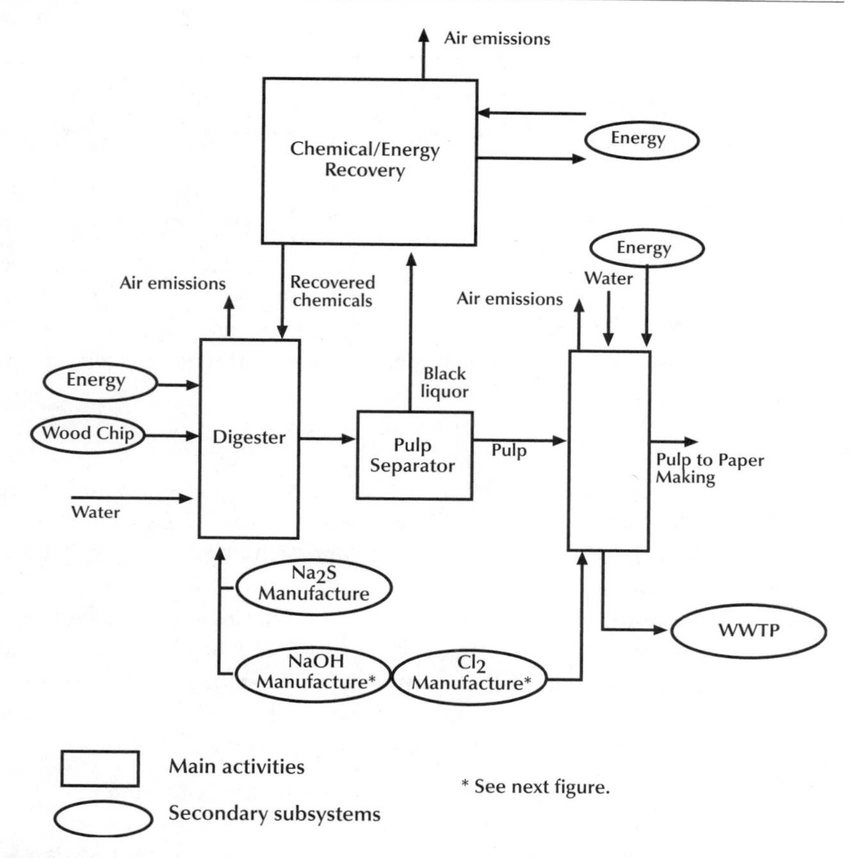


**Figure 2.5 Detailed process flow diagram for Kraft pulping process at the paper mill.**

## Life-Cycle Inventory

The life-cycle inventory (LCI) provides quantitative information on energy and raw material requirements, environmental emissions, and wastes for each process included in the system boundaries (Vigon et al., 1993). The life-cycle and process flow diagrams developed in the scoping stage to help define the system's boundaries are the starting point for the LCI. When completed, the LCI provides mass flow rate and energy consumption data for each of the flow streams previously identified. The process of completing this inventory requires the following:

- Collection of a significant amount of data
- Mass and energy balance calculations
- Organization of the data by type of material and emission so that similar emissions (e.g., $CO_2$) can be aggregated by life-cycle stage, by media (air, water, land), and/or by specific processes
- Presentation of the data in a fashion that is useful for further analysis in the life-cycle impact or interpretation stages

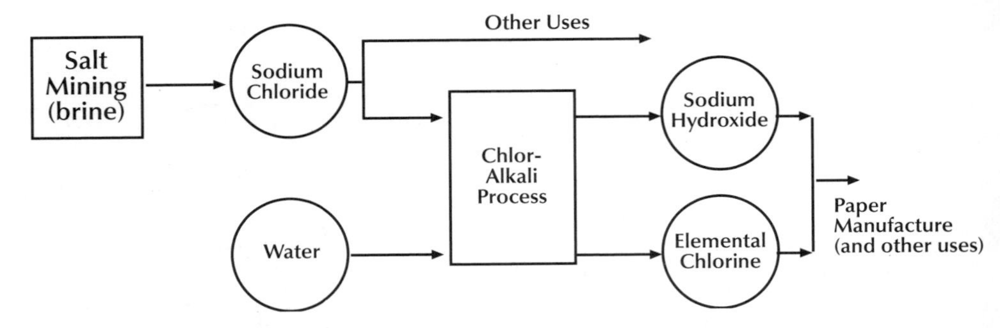


**Figure 2.6 Process flow diagram for the manufacture of sodium chloride and chlorine gas.** Materials used in pulping process.

With the need for process flow diagrams and mass and energy balance calculations, Schenck (2000) refers to the LCI stage as an "engineers dream." Process engineers are particularly well prepared to develop LCI studies.

Guidelines for completing an LCI are available from the USEPA (Vigon et al., 1993) and the ISO 14040 (1998) series of standards. These documents provide a framework for performing an inventory analysis and evaluating the quality of the results based on the quality of the data used. The key components of an inventory include:

- Flow diagram of the processes being evaluated
- Data collection plan
- Data collection to quantify input and output streams from processes included in the flow diagram
- Modeling and calculation to complete mass and energy flow estimates
- Analysis of the quality of the data and LCI results
- Report to communicate results.

In each of these steps, the level of detail and time requirements depend on the goals and scope of the assessment project that were defined in the scoping phase. For example, significant differences in the nature of the data gathered would occur if the LCI were being completed as an industry-wide assessment of a product. In that case, average values for many of the mass and energy flows would be used. On the other hand, such data would be unusable for a site-specific assessment aimed at reducing the environmental effects associated with a specialized product or a particular manufacturing facility.

Figure 2.7 presents some of the final aggregated LCI data collected by the EDF Paper Task Force. In this example, inventory data from virgin paper was compared to postconsumer recycled paper. This example showed the inventory categories that were chosen to represent the individual mass and energy flows through the system. The inventory data were divided into several primary categories: solid waste, energy usage, atmospheric emissions, waterborne wastes, and effluent flow. All mass

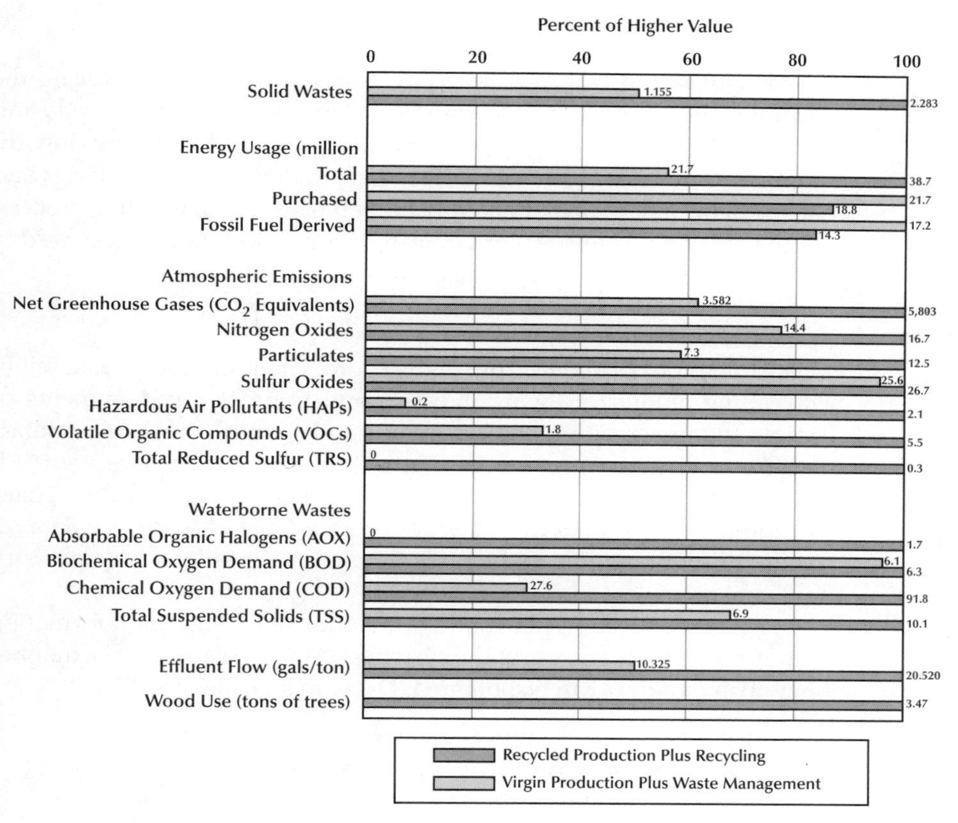


**Figure 2.7 Example results of LCI.** LCI compared mass and energy flows for virgin versus postconsumer recycled office paper systems. (*Source:* Environmental Defense Fund, 1995. Figure revised 2002.)

or energy flows in these categories were summed across all life-cycle stages for this particular report.

Several of the primary inventory categories—especially air emissions —were further divided into subcategories. This is critically important because different air emissions have substantially different environmental effects. For example, the paper life-cycle system releases a much larger mass of carbon dioxide than hazardous air pollutants (HAPs). The contributions of $CO_2$ to global warming and HAPs to human health effects are certainly not comparable. The magnitude of the releases are several orders of magnitude different—another reason not to add the emissions—the HAPs would be insignificant if they were summed with $CO_2$. In presenting these inventory data, the Paper Task Force chose to use bar graphs representing the percentage of the higher value. Thus, for each comparison between virgin and recycled paper, it is easy to see that the virgin paper had higher values in almost all inventory categories. These results would help a manager to substantiate a decision to buy more recycled paper. In this final form, however, they would not be useful to a plant engineer who wants to know how to reduce HAPs.

More detailed data for each stage of the life cycle are required if that is the intended purpose of the LCI.

Several of the LCI steps overlap with work completed during the scoping phase. For example, an understanding of the data needs and availability and the generation of at least preliminary process flow diagrams were required in the earlier stage to define the boundaries and scope of the assessment. Remember that the LCA is an iterative process! Some of the preliminary work done for the scoping stage might need to get revised during work on the inventory.

*Data Collection*

Data for an LCI can come from many sources. Finding useful data might take time online reviewing industry and trade group documents or might require crawling around pumps and compressors in a manufacturing facility to find the tags that indicate energy consumption. Although the sources of data might vary depending on the LCA's goals, the quality of the data must always be known. Highly precise data often requires more time and money to collect than estimates or industry average data.

The data collection plan should include a list of all the information required for each process of the system and identification of the manner in which that data can be obtained. Data are available from:

- Records and inspection at a manufacturing facility
  - Purchasing records (e.g., raw materials, utilities)
  - Environmental compliance reports
  - Production records
  - Equipment design and specifications
  - Meter readings from equipment
  - Equipment operating logs and journals
- Government reports, database records (e.g., Department of Commerce)
  - Technical reports from industry trade groups
  - Published life-cycle databases
  - Reference textbooks and journals

The data collected from these sources should be precise, complete, representative, consistent, and reproducible as required to meet the goals of a particular LCA. This is difficult when data are collected from a variety of sources. The reference materials often do not identify the manner in which the data were generated nor the assumptions involved. This lack of *transparency* in published data continues to be a problem in the implementation of LCAs. A good LCA should state all assumptions, sources of data, and uncertainties in the quality of the data.

The need to keep the data and LCA transparent is a major topic discussed in the ISO 14040 standards. A good LCA should state all assumptions, sources of data, and uncertainties in the quality of the data. Table 2.1 illustrates a data collection plan for the Kraft pulping

**Table 2.1** Data Collection Plan for the Kraft Pulping and Bleaching Process

| | |
|---|---|
| Completed By: | A. Smith |
| Process: | Kraft pulping operation |
| Date: | January 2005 |
| Description of the unit operation: | The Kraft pulping operation uses high temperatures, pressures, and pH to break wood into lignin and cellulous. The lignin, which is separated from water as black liquor, is combusted to recover its heating value. Process chemicals are also recycled from the energy recovery operation and returned to the digester. The cellulous is rinsed with copious amounts of water and bleached to make white office paper. |

| | Information required | Source of Information |
|---|---|---|
| **Material Input** | | |
| Wood chips<br>Sodium sulfide<br>Sodium hydroxide<br>Water<br>Chlorine | Input flow rates<br>Environmental emissions, energy consumed, waste generated with raw material life cycle<br>Source of material and transportation distances, methods | Technical engineering data sheets from trade group (e.g., Pulp and Paper Technical Association of Canada, TAPPI)<br>Published LC database for information on raw material processes |
| **Energy Inputs** | | |
| Electricity<br>Fossil fuel inputs<br>Recovered energy | Usage rates from each unit operation<br>Regional source of electricity<br>Source and transportation information on fossil fuels<br>Environmental emissions, energy consumed, waste generated with energy life cycle | Technical engineering data sheets from trade group<br>Electric power/petroleum trade groups<br>Published LC database for information on energy source LCA |
| **Material Outputs** | | |
| Paper pulp | Output flow rate<br>Conversion factor—quantity of pulp to make 1 ton of paper | Technical engineering data sheets from trade group |
| Air emissions | Flow rates from each unit operation in the process<br>Concentrations ($CO_2$, $SO_x$, $NO_x$, VOCs, HAP, particulates) | Review of TRI (www.epa.gov/tri) and air emission reports from three specific manufacturing plants to obtain averages |

(*continued*)

**Table 2.1** (*Continued*)

| | Information required | Source of Information |
|---|---|---|
| **Material Outputs** (*continued*) | | |
| Water emissions | Flow rates to wastewater treatment plant<br>Concentrations of chlorinated organic chemicals, biochemical oxygen demand, chemical oxygen demand, suspended solids, particulates | Review of SPDES reports from three specific manufacturing plants to obtain averages |
| Solid waste | Mass produced in each unit operation | Technical engineering data sheets from trade group |

The goal of this LCA was to assess differences in the use of recycled versus virgin paper. Industry-wide averages were appropriate for this analysis. Note that data requirements are consistent with defined inventory categories (Figures 2.4 through 2.6).

and bleaching process (Figure 2.5). A clear understanding of the process is required for identifying the data needs. The specific data collected for each process of the product system must be consistent with the inventory categories specified during the scoping stage. Data required to relate the material and energy flows to the specified functional unit are also required for each process in the system.

Sources for data collection vary widely depending on the scope and goals of the study. The Kraft pulp example shown in Table 2.1 illustrates a relatively easy collection of much of the data required to do an LCI—at least the material and energy requirements are well defined and published for this old and well-established process. The team completing the LCI must gather and organize the data for analysis.

An LCI on a particular product that does not have industry standards and documentation requires much different sources of data. The data collection plan for the xerographic toner required extensive searching through a manufacturing facility's records and examination of the manufacturing equipment. For the entire system (downstream, toner manufacturing, and upstream processes), the information required included material inputs, flow diagrams, and flow rates for each process, energy inputs, emissions, wastes, and by-products. The sources of information for the system varied. The magnetite and carbon black are standard processes, and information was obtained on their manufacturing processes by referring to technical references. The resin process was unique, and the supplier had to be contacted to obtain the flow sheet and details on the manufacturing process. Collecting data about the toner manufacturing process required the most intensive effort. To obtain material flow rates, collaboration with toner plant engineers led to the acquisition of process yields for the individual steps of the manufacturing process. The energy consumption data were not readily available for the process. Therefore, the energy requirements were estimated by inspection of the individual motors in the process. This meant walking through and reading the nameplate information directly from the motors in the toner manufacturing plant. The most energy intensive step, it became apparent, was with the grinding process. Plant engineers monitored this step directly, and they provided the data for the LCI. The information for postconsumer processes was obtained by using resources related to paper use and recycling rates, personal communication with the toner manufacturing facility, and technical reference sheets that showed the energy consumed by a specific copier during use. The collection of this type of detailed data is time consuming and requires a level of openness between the LCA practitioner and the industry. Lack of access to proprietary data can stymie an LCA.

During the maturation of LCA methods over the past decade, the body of resources that can help with the collection of data has grown. Most helpful are the inventory databases available for raw materials and energy used in production. The LCI practitioner can now rely on these databases to provide inventory information for these secondary inputs. Most LCI databases are proprietary and require purchase; others are readily available. (See the end of this chapter for examples of online sources.)

Not all data required for an LCI are available in records and published information. Sometimes, it is necessary to design experiments to measure flow rates or concentrations or to estimate data with standard engineering calculations and models. Estimated data are also important to verify the quality of data. Basic laws of nature require that mass and energy both be conserved. The flowchart becomes an important tool for these calculations. It is essential to quantify pertinent flows that were identified in the scoping stage.

#### *Estimating Mass Flow Streams and Verifying Data*

Mass and energy balances are powerful tools for estimating missing values and for verifying that the collected data make sense. These balances can be completed at various spatial scales ranging from an individual unit operation to the entire life cycle of the primary material component (e.g., at the scale shown in Figure 2.4). Balances for LCAs are generally done at a time scale that permits an assumption that the system is at a steady state. A balance for the total mass flows of all component streams simply becomes

$$\sum_{i} \text{mass flows in} = \sum_{j} \text{mass flows out} \tag{2.1}$$

where $i$ and $j$ are the number of input and output streams, respectively. A mass balance on an individual component within the unit operation or system of unit operations becomes

$$\sum_i \text{mass flows in} + \text{mass generated by reaction}$$
$$= \sum_j \text{mass flows out} + \text{mass consumed by reaction} \quad (2.2)$$

The mass balance approach works well for major components in a process. Unfortunately, some minor emissions are insignificant relative to the total mass flow rates but are critical for environmental emissions. A mass balance approach generally does not work effectively for estimating or verifying the data collected for these streams.

### Example 2.2 Use of Mass Balances to Verify Data

Three processes at a semiconductor manufacturing facility use ultra-clean water to rinse an aqueous surfactant from the chips. The chips are dipped into the surfactant solution in an ultrasonicator to remove debris. Racks of the chips are then transferred to a countercurrent rinsing tank to remove the surfactant solution. The amount of surfactant solution carried by the rack into the rinse tank varies based on the number and size of chips on the rack. The rinse water, which has a discharge limit of 5 mg/L for the surfactant, is sent to a publicly owned treatment works (POTW)—a conventional municipal wastewater treatment plant.

Water and surfactant inflow data for each of the processes are shown in the accompanying figure. A review of the company's discharge records indicates that 70 L/min rinse water with an average concentration of 2 mg/L flows to the POTW. Do these numbers seem reasonable? Verify your answer with mass balance calculations.

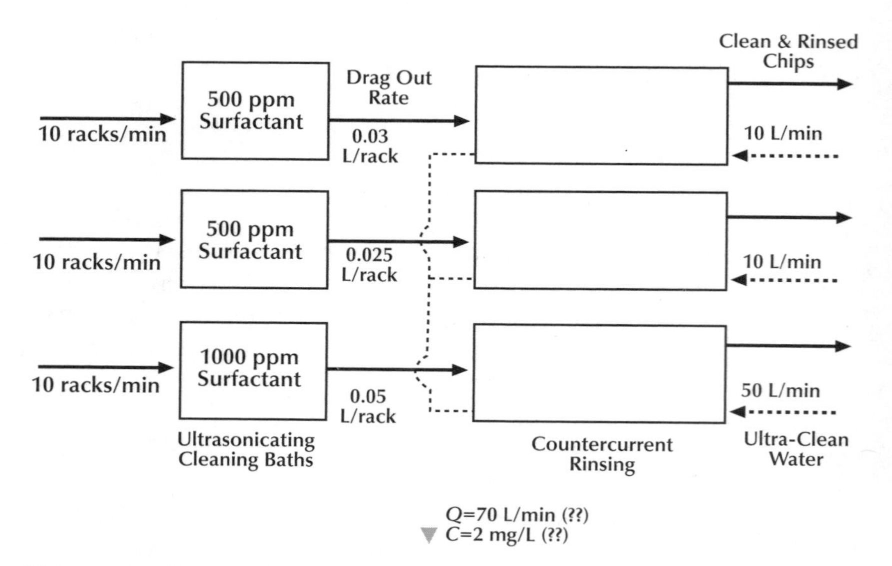


**Water and surfactant inflow data for each process.**

***Solution***

1. Complete a mass balance on the total flow first to verify the effluent flow rate. Assume that the low concentrations of surfactant do not affect the solution density so that we can assume that volumetric flow rates are conserved. Also assume that the volume contributed by the drag out is negligible relative to the rinse water flow rate:

$$Q_{\text{out}} = \sum Q_{\text{in}} = 10 + 10 + 50 \text{ L/ min} = 70 \text{ L/ min}$$

2. Complete a mass balance on the surfactant to estimate the effluent concentration. Mass flows for a minor component are generally estimated as $M = QC$. Assume that a negligible mass of surfactant remains on the cleaned and rinsed chips. Thus, the mass balance on the surfactant can be written as:

$$M_{\text{out}} = Q_{\text{out}} C_{\text{out}} = M_1 + M_2 + M_3$$

where $M_i$ represent the mass of surfactant entering rinse tank $i$.

An example calculation of the mass flow of surfactant into tank 1:

$$M_1 = 10 \frac{\text{racks}}{\text{min}} \cdot 500 \frac{\text{mg}}{L} \cdot 0.03 \frac{L}{\text{rack}} = 150 \frac{\text{mg}}{\text{min}}$$

Based on this mass balance, the effluent concentration can be verified:

$$C_{\text{out}} = \frac{150 + (10 \cdot 0.025 \cdot 500) + (10 \cdot 0.05 \cdot 1000)\,{}^{\text{mg}}/_{\text{min}}}{70\,{}^{\text{L}}/_{\text{min}}} = 11\,{}^{\text{mg}}/_{\text{L}}$$

3. Analyze results: Based on the mass balance analysis, it appears that the flow rate data found in the records are accurate but that the concentration of surfactant in the wastewater was underestimated by approximately a factor of 5. Because of this discrepancy, several samples of the wastewater should be collected when the rinsing operations are in process to provide the most accurate information.

*Incorporating Energy Use and Balances*

Energy use is a critical component of most life-cycle assessments. Depending on the goals of the assessment and the environmental effects of concern, energy use could be important in terms of the depletion of a nonrenewable resource and/or the environmental effects that arise as the energy is used. These could include global warming because of carbon dioxide and other greenhouse gases, acid rain from $SO_x$ and $NO_x$, and oil spills to aquatic systems. Including energy in the life cycle of a product system requires that the life cycle of the entire energy system also be understood.

Many of the environmental impacts associated with energy use arise from the combustion of fossil fuels. The release of heat energy from the

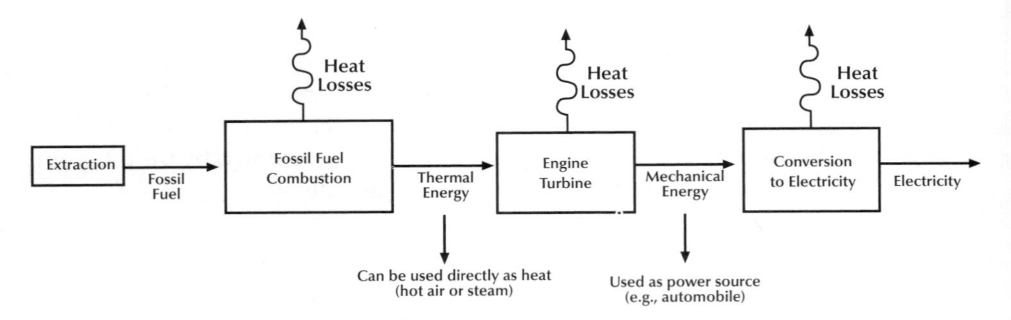


**Figure 2.8 Simplified schematic of energy system.**

combustion process can be used directly as a heat source, or the heat can be converted to mechanical energy to do work (Figure 2.8). Mechanical energy can also be converted to readily usable electricity.

Energy is expressed in units of joules (or British thermal units or kilocalories). *Power* is also often used in relation to energy, especially as related to electricity; it is a measure of the *rate* of energy usage. Some useful conversions:

$$1 \text{ joule} = 1 \text{ N/m} = 9.486 \times 10^{-4} \text{ Btu} = 2.39 \times 10^{-4} \text{ kcal}$$

$$1 \text{ W} = 1 \text{ J/s} = 1.341 \times 10^{-3} \text{ hp}$$

The source of energy for electric generation has a significant influence on the environmental effects. During the first 6 months of 2002, total U.S. net generation of electricity was 1,836 billion kWh [U.S. Department of Energy (USDOE)]. Approximately half of this generation was accounted for by coal-fired power plants, followed by 21% from nuclear, 17% percent from natural gas, 8% from hydroelectricity, 3% from renewable sources, and 2% from petroleum. As shown in Figure 2.9, however, the sources of electricity among regions of the country vary greatly. Thus, the environmental effects also vary greatly among

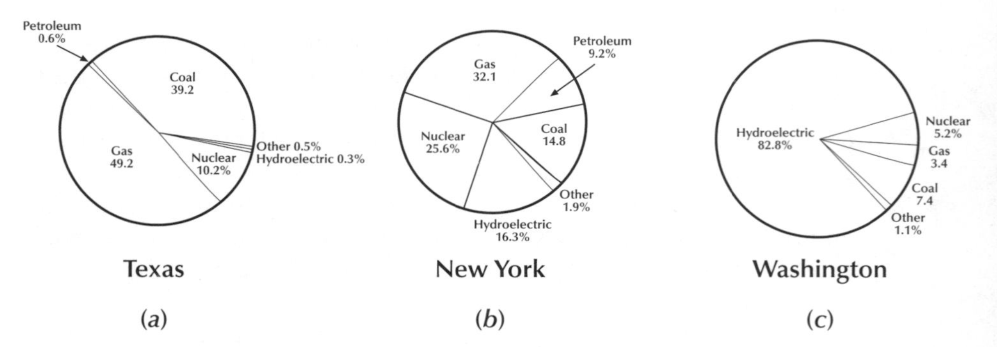


**Figure 2.9 Energy sources for electricity generation in various U.S. states in 1999.** (*Source:* USDOE.)

different states. Washington State has significant concern about habitat and ecosystem effects of hydroelectric dams, whereas the fossil fuel combustion in Texas gives rise to substantial greenhouse gas and other atmospheric emissions. An LCA should adequately reflect the regional electricity generation and the associated environmental effects.

These sources of electricity also vary in terms of their efficiencies. Because of the losses at each stage in the energy conversion processes, the energy that is received in a usable form by consumers is far less than the original energy content of the source. From a life-cycle perspective, the loss of energy at each step represents additional environmental effects throughout the system's life cycle without added value to the product. This certainly has economic as well as environmental repercussions! The overall efficiency of an energy system is calculated as:

$$\eta = \frac{\text{energy output}}{\text{energy input}} \tag{2.3}$$

where $\eta = 1.0$ means perfect efficiency. For several processes in series, the overall efficiency is defined as the product of the efficiencies of each of the steps.

In general, the conversions of chemical, mechanical, or solar forms of energy into electricity are among the least energy-efficient processes. Example 2.3 illustrates this for a coal-to-electricity system. The low efficiency for the conversion process (combustion, steam turbine, and generator) suggests that technological improvements would be most beneficial in this part of the life cycle. The system presented in this example does not include the "use" stage of the energy system. Clearly, inefficiencies in the manufacturing processes and consumer products powered by the electricity add to the low overall system efficiency. A choice of lighting is a good example. Fluorescent bulbs have an efficiency of 20%, whereas incandescent bulbs are only 5% efficient (Dorf, 2001). When this use stage is included as part of the life-cycle system, the efficiency is reduced from 0.25 (shown in the example) to 0.05 for fluorescent and 0.0125 for incandescent bulbs.

### Example 2.3 Calculate Life-Cycle Efficiency of Electricity Generated from Combustion of Coal

#### *Solution*

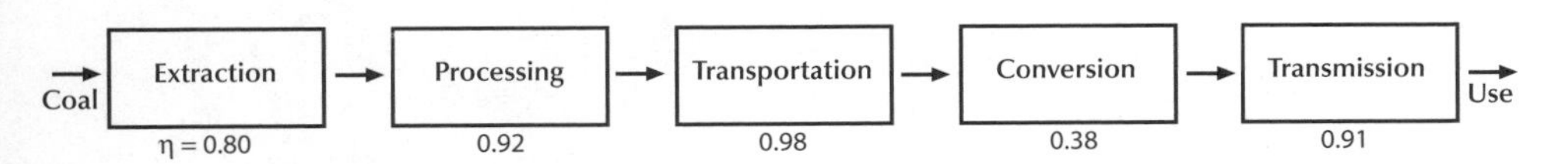


Overall efficiency: $\eta_{\text{tot}} = \Pi\eta_i = 0.80 \times 0.92 \times 0.98 \times 0.38 \times 0.91 = 0.25$

Most of the efficiency is lost in conversion of coal to electricity (combustion, steam generation, steam turbine). [Based on an example in Dorf (2001).]

*Energy Balances*

Energy balances can help to verify data or provide missing information. As with the mass analysis, the conservation of energy is a useful tool. The first law of thermodynamics—*energy is conserved; it can neither be created nor destroyed*—is the basis for an energy balance. Processes in which energy is used generally convert energy sources into work and/or heat:

$$\Delta U + \Delta E_k + \Delta E_p = Q + W \tag{2.4}$$

where $U$ = internal energy, because of motion of atoms and molecules
$E_k$ = kinetic energy, because of motion of system as a whole (velocity)
$E_p$ = potential energy, because of elevation of system relative to some datum
$W$ = work done on system by its surroundings (pistons, pumps, etc.)
$Q$ = heat transferred to the system by its surroundings (change in temperature)

Often, it is the *heat* term that represents a loss in efficiency. In many mechanical processes, the goal is to minimize $Q$ and maximize $W$. In other cases, for example, the burning of a fossil fuel for heat, the goal is to maximize $Q$.

For combustion and other chemical reactions (e.g., Figure 2.10), the energy conservation equation can be simplified. In this case, assume:

- Constant pressure (internal energy is described by enthalpy, $\Delta U = \Delta H$)
- No energy transformed into work
- No significant changes in elevation ($\Delta E_p \sim 0$)
- No significant velocities of materials ($\Delta E_k \sim 0$)

With these changes, the energy balance becomes

$$Q = \Delta U = \Delta H \tag{2.5}$$

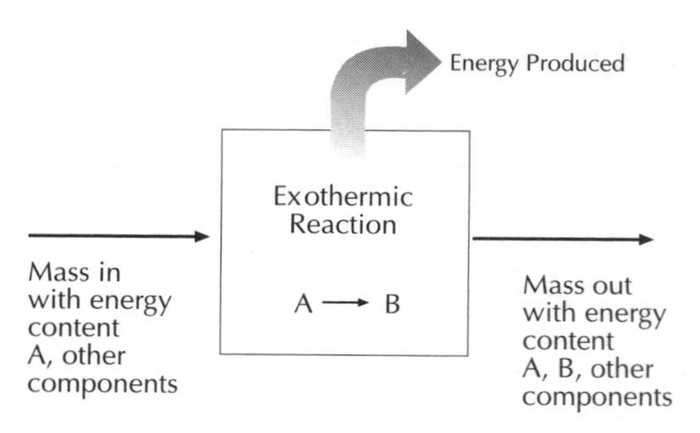


**Figure 2.10 Example application of energy balance.**

where the change in enthalpy in the system is defined by the difference in enthalpies of the materials entering the reactor and those leaving plus the heat released in the chemical combustion reaction:

$$Q = \Delta H = \frac{n_{AR} \Delta \hat{H}_r^o}{v_A} + \sum_{\text{outlet}} n_i \hat{H}_i - \sum_{\text{inlet}} n_i \hat{H}_i \tag{2.6}$$

where $A, i$ = any reactant or product
$n_{AR}$ = moles $A$ produced or consumed in the process
$v_A$ = stoichiometric coefficient of $A$
$\Delta \hat{H}_r^o$ = specific heat of reaction (energy released per mole reactant)
$\hat{H}_i$ = specific enthalpy (per unit mass or mole)

If water is present in the fuel that is combusted, the energy required to heat and evaporate the water must also be included in the energy balance.

Applying these energy balance concepts is easy, once all the necessary data are in hand. Specific enthalpies can be found or estimated:

**Directly from Tables** These data are available especially for gases involved in combustion reactions. In all cases, the specific enthalpy is presented relative to some reference state (LaGrega et al., 1994). For example, tabulations of the enthalpy of combustion gases in Btu/lb with a reference temperature of 60°F. As long as the same reference temperature is used for all flow streams, this approach works well and is easy.

**Heat Capacity of Material** This value can also be used to find the enthalpy of a material relative to a reference temperature:

$$\Delta \hat{H} = c_p(T_2 - T_l) \quad \text{or} \quad \Delta \hat{H} = \int_{T_1}^{T_2} C_p(T)\, dT \tag{2.7}$$

where $C_p$ is the heat capacity (often dependent on phase and temperature). The heat capacity of materials can be found in handbooks (e.g., Perry and Chilton's *Chemical Engineering Handbook,* 1973) or textbooks (Felder and Rousseau, 1986).

Heat of reaction data are also readily available in reference texts. In most cases, heat of reaction data are presented in units of kJ/mol. For combustion processes, however, data are often presented in a variety of units (Btu/lb, kJ/kg, etc.) and with a variety of interpretations. Two important differences should be understood:

- Organic materials that contain some moisture often have a notation that the heat of reaction data is "as fired." This implies that the values incorporate the energy losses because of the heating and evaporation of water in the materials.
- The complete combustion of organic materials ideally produces carbon dioxide and water. The energy released in this process

**Table 2.2** Heating Values of Various Fuels

| Fuel | Heat Content (kJ/kg) |
|---|---|
| Hydrogen (pure) | 110,000 |
| Natural gas (mostly methane) | 35,000 |
| No. 2 heating oil | 45,600 |
| Coal, bituminous | 24,000 |
| Polyurethane (foamed) | 30,300 |
| Wood sawdust (10% moisture) | 22,000 (as fired) |
| Newspaper (6% moisture) | 18,600 (as fired) |
| Citrus rinds (75% moisture) | 4,000 (as fired) |

*Sources:* Dorf, 2001, LaGrega et al. 1994.

because of the C → $CO_2$ reaction is defined as the *low heating value* (LHV) of the reaction. The heated water vapor also contains energy. If this steam is condensed and the energy is recovered, then the overall combustion process has a higher energy release. The combination of energy because of combustion (LHV) and the condensation of water vapor is termed the *high heating value* (HHV). Processes such as an internal combustion engine do not recover the energy content of the water vapor; so, calculations involving the energy value of gasoline in an automobile should use the low heating value.

Table 2.2 presents some heat of combustion data to illustrate these differences. In general, organic materials that have high carbon content have higher heating values, and the presence of moisture significantly decreases the heating value. Example 2.5 illustrates the use of an energy balance with a combustion component to provide missing data for a manufacturing process.

### Example 2.4 Estimate $\hat{H}$ of Air at 1000°C in j/mol with Reference Temperature of 15°C.

***Solution***

*Method 1*: Find tabulated data.

The following table includes enthalpy data for air with a reference temperature of 60°F [extracted from LaGrega et al. (1994, Table 12.6)]. Conversions between °F and °C, Btu to joules, and pounds to moles are all required to use the tabulated data.

| Temp (°F) | Enthalpy of Air (Btu/lb) |
|---|---|
| 1400 | 343.0 |
| 1600 | 398.0 |
| 1800 | 455.0 |
| 2000 | 513.0 |
| 2200 | 570.7 |

Determine required temperature range in Celsius:

$$(60°\text{F} - 32)/1.8 = 15.5°\text{C}$$

$$(1000°\text{F} - 32)/1.8 = 1832°\text{C}$$

The 15.5°C is close enough to the required reference temperature to use these data directly. The enthalpy of air at 1832°C can be determined by interpolation

$$\hat{H}_{\text{air},1000} = 455 + \frac{1832 - 1800}{2000 - 1800}(513 - 455) = 464\ \text{Btu/lb}$$

Conversions are required to put this into the required units. *Note:* The molecular weight for air (28.8 g/mol) is calculated as a weighted average between the $N_2$ (79%) and $O_2$ (21%) components:

$$\hat{H}_{\text{air},1000} = 464\frac{\text{Btu}}{\text{lb}} \cdot 1055\frac{\text{J}}{\text{Btu}} \cdot \frac{\text{lb}}{454\text{g}} \cdot 28.8\frac{\text{g}}{\text{mol}} = 3.11\ \text{J/mol}$$

*Method* 2: Use the heat capacity.

Data from Perry and Chilton (1973):

$$C_{p(\text{air})} = 28.94 + 0.0041447(T) + 0.3191 \times 10^{-5}\left(T^2\right) - 1.965 \times 10^{-9}\left(T^3\right)\ (\text{J/mol°C})$$

Integrate this from the reference state (15°C) to the desired temperature:

$$\begin{aligned}\hat{H}_{\text{air},1000} &= \int_{15}^{1000} c_p\, dT \\ &= 28.94(1000 - 15) + 0.0041447\left(\frac{1000^2 - 15^2}{2}\right) \\ &\quad + 0.319 \times 10^{-5}\left(\frac{1000^3 - 15^3}{3}\right) - 1.965 \times 10^{-9}\left(\frac{T^4}{4}\right) \\ &= 3.12 \times 10^{-4}\ \text{J/mol}\end{aligned}$$

Note that the estimates are nearly identical. Depending on the desired units, the heat capacity method is not much more difficult than finding information directly in a tabulated form.

### Example 2.5 Use an Energy Balance to Estimate Missing Data

The Kraft pulping process (Figure 2.5) has been used for many years because of the substantial energy that can be recovered from it. The lignin from the wood chips is considered a waste product in the paper pulping operation, but it has a relatively high heating value that makes it valuable for energy recovery. The lignin in the black liquor from the pulping digester is first concentrated by evaporating the water. The remaining material is then combusted for energy recovery. Other chemicals are also recovered in this process. Energy data acquired for a life-cycle assessment that includes this process is shown below.

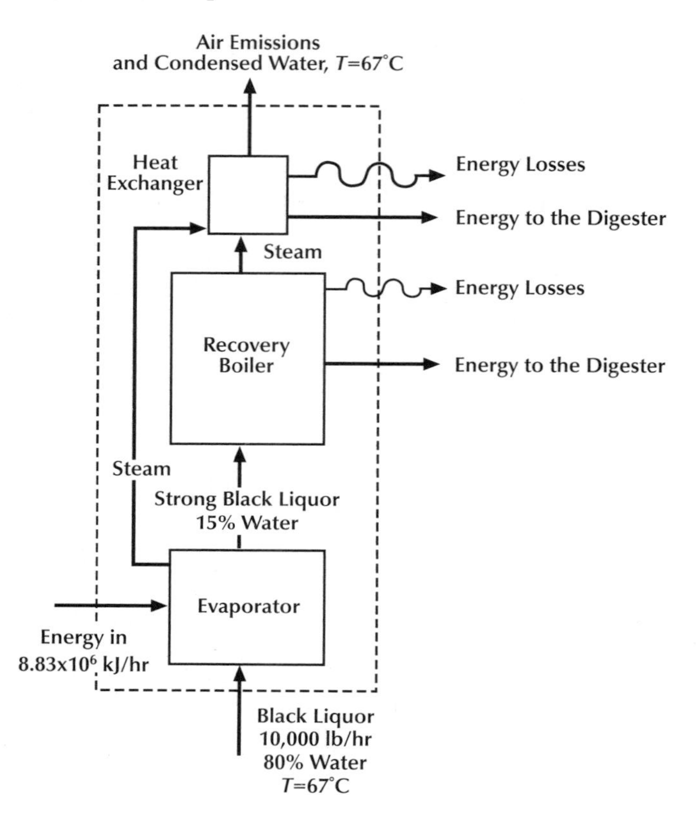


## Critical Values

73.3 MMT black liquor, 15% water (in USA 1988) (Ayers and Ayers 1999)

Heat value: 6000 Btu/lb dry solids

Efficiency of boiler: 65%

Efficiency of heat exchanger: 78%

The energy returned to the digester is required for the LCI so that the reduced need for coal combustion can be estimated. Calculate the net energy available from this system and the energy efficiency of the process.

### *Solution*

1. Define the system: The dashed line on the illustration defines a system that is suitable for this problem. No energy balance on each individual unit operation need be calculated because the data required for the LCI only involve the net energy (energy out − energy in) for the overall process.

2. Write an energy balance in conceptual terms:
   a. Energy flows in include:
      - Energy value of the influent black liquor
      - Energy added to heat and evaporate the black liquor (unknown)
   b. Energy is generated and recovered by:
      - Condensation and cooling of the water vapor in the heat exchanger
      - Boiler during the combustion of the black liquor solids—given the constraints of the efficiency of the boiler
   c. Energy flows out include:
      - Energy value of the air emissions (assumed to be mostly water)
      - Heat energy returned to the digester from the boiler and heat exchanger
      - Heat lost in the inefficiencies of the boiler and heat exchangers
3. Define assumptions:
   a. Heat capacity of the black liquor is approximately equal to pure water.
   b. For lack of other data, assume that the heating and evaporation of water occurs at atmospheric pressure.
   c. Define the reference state for energy content as $T = 67°C$ so that the enthalpy of the influent black liquor and effluent air emissions are zero.
4. Quantify energy associated with each of the above components:
   a. Energy flows in include:
      - Energy value of the influent black liquor (equal to zero based relative to the defined reference point)
      - Energy added to heat and evaporate the black liquor

   Use the heat capacity to estimate energy required to raise the temperature of the black liquor [find data from Felder and Rousseau (1986) → $C_{p(\text{water})} = 75.4$ J/mol/°C]:

$$m_w \, \Delta\hat{H}_w = 75.4 \frac{J}{\text{mol °C}} \cdot 10{,}000 \frac{\text{lb}}{\text{hr}} \cdot 454 \frac{\text{g}}{\text{lb}} \cdot \frac{\text{mol}}{18\text{ g}} \cdot (100 - 67)°\text{C}$$

$$= 6.27 \times 10^5 \frac{\text{kJ}}{\text{hr}}$$

   Use the heat of evaporation required to evaporate water [find data from Felder and Rousseau (1986) → $\Delta\hat{H}_{\text{vap}} = 40.656$ kJ/mol]:
      - Total energy to evaporate water:

$$m_w \, \Delta\hat{H}_{\text{vap}} = 40.656 \frac{\text{kj}}{\text{mol}} \cdot 10{,}000 \frac{\text{lb liquor}}{\text{hr}} \cdot 0.8 \frac{\text{lb water}}{\text{lb liquor}} \cdot \frac{454\text{ g}}{\text{lb}} \cdot \frac{\text{mol}}{18\text{ g}}$$

$$= 8.20 \times 10^6 \frac{\text{kJ}}{\text{hr}}$$

b. Energy is generated and recovered by:
   - Condensation and cooling of the water vapor in the heat exchanger. The calculation is the same as above for raising temperature and evaporating (since the final temperature is the same) for cooling – total energy released = $8.83 \times 10^6$ kJ/hr. However, only 78% of this is recovered:

$$\text{Energy recovered} = 0.78 \times 8.83 \times 10^6 \text{ kJ/hr} = \boxed{6.89 \times 10^6 \text{ kJ/hr}}$$

   - The boiler during the combustion of the black liquor solids—given the constraints of the efficiency of the boiler:

Use the heat of combustion to determine the energy released during the combustion of the black liquor solids:

$$m_s \, \Delta \hat{H}_{\text{comb}} = 6000 \frac{\text{Btu}}{\text{lb}} \cdot 10{,}000 \frac{\text{lb liquor}}{\text{hr}} \cdot 0.2 \frac{\text{lb solids}}{\text{lb liquor}} \cdot \frac{J}{9.486 \times 10^{-4} \text{ Btu}} = 1.26 \times 10^7 \frac{\text{kJ}}{\text{hr}}$$

But this process is only 65% efficient: Energy recovered = $0.65 \cdot 1.26 \times 10^7 \text{kJ/hr} = \boxed{8.19 \times 10^6 \text{ kJ/hr}}$.

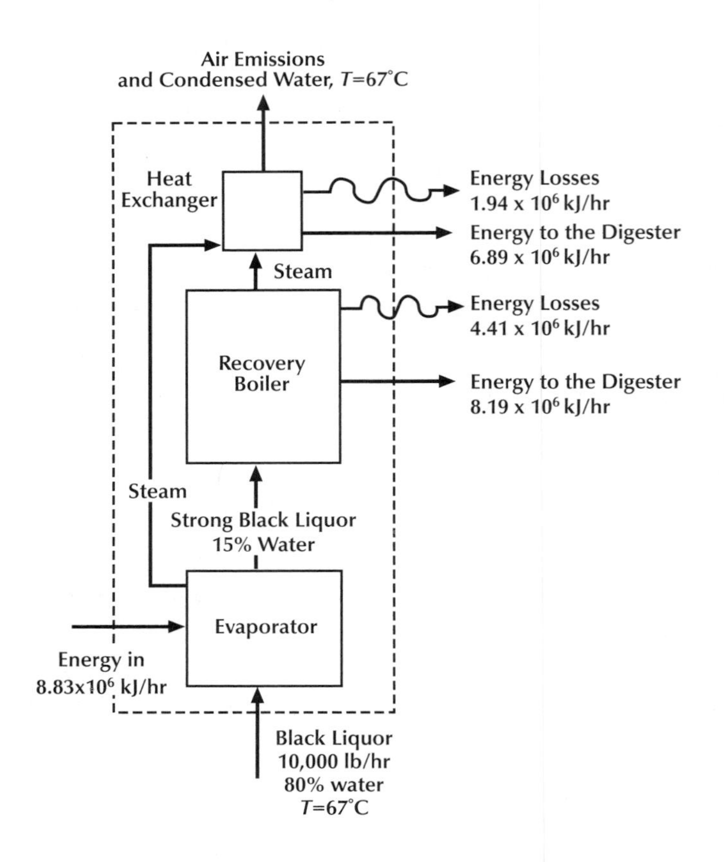

5. Calculate the required energy inflow and recovery:
   a. Total energy required for heating and evaporation $= 8.20 \times 10^6 + 6.27 \times 10^5 = \boxed{8.83 \times 10^6 \text{ kJ/hr}}$.
   b. Total energy recovered $= 6.88 \times 10^6 + 8.19 \times 10^6 = \boxed{1.51 \times 10^7 \text{ kJ/hr}}$.

6. Determine net energy recovered in this process:

$$\text{Net energy} = \text{energy recovered} - \text{energy input}$$
$$= 1.51 \times 10^7 - 8.83 \times 10^6 = \boxed{6.37 \times 10^6 \text{ kJ/hr}}$$

   This recovery is excellent. It is only possible, however, with recovery of the energy value of the steam released during the evaporation process.

7. Reporting and using results: The unknown values can now be added to the process flow diagram and used to define better the reduced need for coal combustion for heating the digester in the Kraft pulping process. Note that the energy losses are calculated as 35% of the possible heat recovery in the boiler and 22% of the heat exchanger.

A quick energy balance check confirms that energy is conserved in this example (all numbers in millions of kilojoules per hour):

$$\text{Energy in flows} + \text{energy generation} = \text{energy recovered} + \text{energy losses}$$
$$8.83 + 12.6 = (6.89 + 8.19) + (1.94 + 4.41)$$
$$21.43 = 21.43$$

*Allocating Mass and Energy Flow Streams*

The mass and energy flow analyses provide the necessary data to quantify all streams on the process flow diagrams. This is still not enough to complete the LCI. At times, the primary product or secondary inputs to the process are among many products being produced in one unit operation. In such a case, it is unreasonable to attribute all of the raw material consumption, energy use, and associated environmental effects with the single material of interest. It is necessary to determine what fraction of the total flows should be allocated to the life-cycle system being assessed. Figure 2.11 illustrates this concept for the chlor-alkali process, which contributes process chemicals to the Kraft pulping process. Clearly, the environmental effects associated with all of the salt mining activities should be split among the products: NaCl, NaOH, and $Cl_2$.

Different procedures are available to determine which fraction of a process should be incorporated into an LCA when multiple products are associated with a particular process. The goal is to partition the mass and energy flows in a way that reflects the physical behavior of

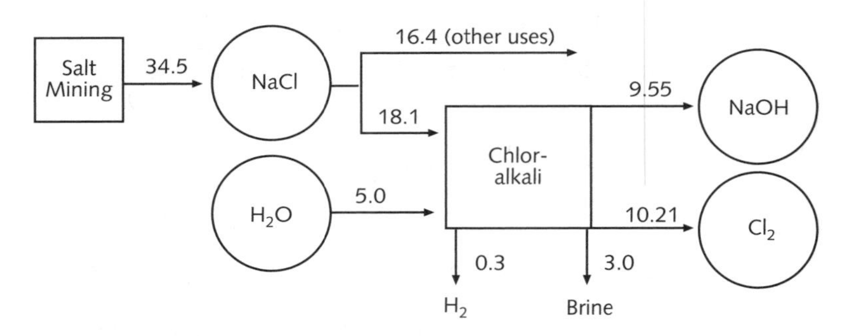


**Figure 2.11 Primary material flows (MMT) through chlor-alkali process.** (*Source of data*: Ayers and Ayers, 1999.)

the system. Allocating inputs and outputs among various co-products is most often done on a mass fraction basis (Boguski et al., 1996). For example, in the chlor-alkali process $18.1/34.5 = 52.5\%$ of the mass of NaCl mined is used in the chlor-alkali process. Thus 52.5% of the energy used, air emissions, and so forth associated with the salt mining operation should be included within the chlor-alkali life cycle. Other physical parameters for co-product allocation can be used if they better represent the partitioning of energy or material among the co-products (see Example 2.6). The use of economic value is discouraged since that is not a physically measurable parameter.

ISO 14041 (1998) recommends that the system's boundaries be delineated to exclude co-products. One way to do this is to subtract the co-product from the system by including in the boundaries an alternative method for producing the co-product. This approach is only applicable if an alternative process can be defined and all of its process flow streams can be quantified.

## Example 2.6 Methods for Co-Product Allocation

A manufacturing firm uses one paint booth to lacquer an equal number of two different products as shown in the accompanying figure. A significant amount of VOCs is released from the paint booth during this operation.

Product (a), a cube, 1.59 cm on each side, is then used as a component in the gadget for which you are completing an LCA. An equal number of product (b), 1 cm × 0.4 cm × 10 cm, is lacquered at the same time but is not used in your gadget. What fraction of the VOCs emitted from this paint booth should you include in the LCA for your gadget?

### *Solution*

Using a volume basis (assume they have the same densities so that volume and mass bases are equivalent):

$$\text{Volume (a)} = 1.59^3 = 4.0 \text{ cm}^3$$
$$\text{Volume (b)} = 1 \times 0.4 \times 10 = 4.0 \text{ cm}^3$$

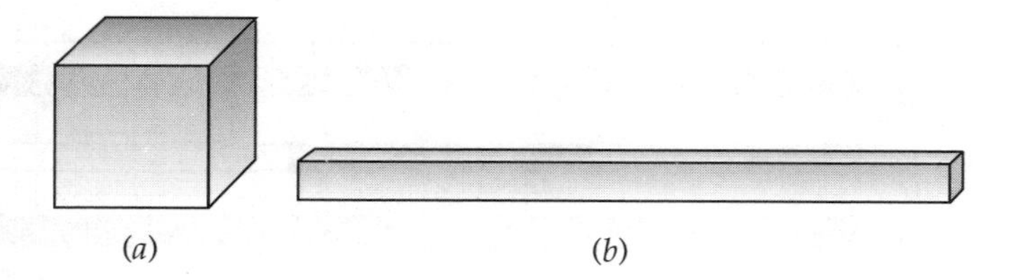


Based on equivalent volumes, 50% of the VOCs should be attributed to product (a). But, does this make sense? Product (b) appears to have a larger surface area and would need more paint than product (a). Therefore, we should consider allocating the VOCs based on surface area, since that is more physically realistic:

$$\text{Area (a)} = 6 \times 1.59^2 = 16.5\ \text{cm}^2$$
$$\text{Area (b)} = 2 \times (1 \times 10 + 1 \times 0.4 + 10 \times 0.4) = 30.8\ \text{cm}^2$$

Based on area,

$$\frac{16.5}{16.5 + 30.8} \times 100\% = 35\%$$

should be attributed to product (a). This could be a substantially lower estimate of VOC emissions relative to the estimate with mass allocation.

*Data Aggregation*

Results quantifying mass and energy flows for all processes within a life-cycle system represent a very large data set that is often too unwieldy for interpretation. Thus, the data must be aggregated normalized to the previously defined functional unit. Results of the paper LCI aggregate data in two ways (Figure 2.7). First, emission and energy data are lumped into the inventory categories listed on the $y$ axis of the figure. VOCs, for example, might lump together mass flows of several different chemicals. The inventory categories that are chosen for aggregation should reduce the level of complexity by grouping similar flow streams. "Similar" should be determined based on the effects that various flow streams have on the environment and human health. As discussed previously, it is not reasonable to aggregate all atmospheric emissions together because different pollutants have significantly different effects.

Life-cycle inventory data are also aggregated across unit operations within each life-cycle stage, as well as across all stages. The LCI for the paper life-cycle system (Figures 2.4 through 2.6) does not indicate which of the life-cycle stages contributes to each of the inventory categories. This approach is fine for the purposes of this example—to compare the overall mass and energy flows of two different paper systems. If the goal were to identify which steps in the life cycle could benefit the most from improvements to reduce these flows, then the aggregation of data across all life-cycle stages would be inappropriate. In this case, it would

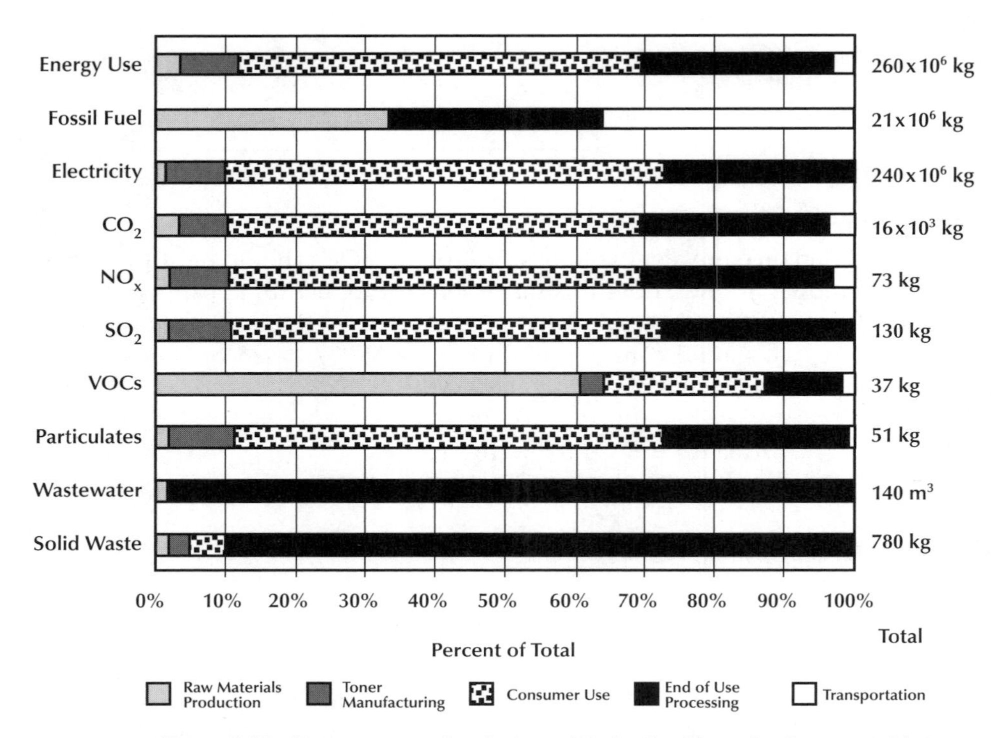


**Figure 2.12 Energy use and emissions data for the life cycle of xerographic toner (per Mton of toner).** (*Source*: Ahmadi et al., 2003.)

be better to present the data by life-cycle stage also and, possibly for the most significant stages, by unit operation as well.

Figure 2.12 presents aggregated LCI data for xerographic toner. The increased level of detail in the presentation of this data among life-cycle stages was important for identifying the assumptions made for each process. For example, raw materials could be produced in multiple ways; so it is important to include the detail and show how each material was produced. Also, for the end-of-use processing step, the steps involved and the environmental effects would not be clear without the level of detail shown.

The results of this study show that the consumer-use stage in the life cycle of toner particles has the highest energy use, generating significant air emissions. This suggests that efforts to reduce energy consumption in the photocopier should be the focus of efforts to improve xerography. The presentation of more detailed LCI data that are not aggregated across life-cycle stages helps substantially, especially with LCAs designed for engineering improvements.

## Life-Cycle Impact Assessment

Although many individuals use LCI results for interpretation and improvement, these data do not really define environmental "impact" and are often hard to compare when identifying priorities for improvement. The virgin paper life cycle (Figure 2.7) displayed similar mass releases of waterborne adsorbable organic halogens (AOX) as atmospheric emissions of hazardous air pollutants. Are these really the same? Are the resulting human exposures and health effects equivalent? Are ecosystem effects any different? An LCI does not answer such questions. Instead, a **life-cycle impact analysis** (LCIA) is needed to answer these questions. The impacts depend on the specific risks associated with each of the inventory categories. The inventory data are often considered to be "stressors." The LCIA attempts to establish a linkage between these stressors and their potential environmental effects. These effects often depend on the geographic region as well as temporal scale of the system of interest. For example, the release of $SO_x$ from the burning of fossil fuels creates acid rain, which can result in the acidification of lakes and the alternation of this habitat for acid-intolerant species. $SO_x$, which is often included as an inventory category, is certainly not an environmental effect; acid rain, for that matter, is not either.

As with the LCI, ISO standards (ISO, 2000) provide a template for conducting an LCIA. The LCIA must include:

1. Selection and definition of impact categories (e.g., global warming, eutrophication, resource depletion)
2. Classification to match LCI categories to the impact categories (e.g., $CO_2$ and $CH_4$ contribute to global warming)
3. Characterization—modeling the effects within each category

The following additional optional steps could also be included in an LCIA:

**Normalization** Effects are normalized for local or regional geographical, political, or environmental differences. Acidification because of acid rain is a good example of an environmental effect that varies substantially by region. Acid rain is not an important issue in the much of the southeastern United States because of the limestone subsurface that has high acid-neutralizing capacity. In parts of the Northeast, however, the granite bedrock has virtually no buffering capacity, and the result is significant acidification of lakes in the Adirondack and White Mountains. Thus, assigning a high priority to reducing $SO_x$ and $NO_x$ emissions to mitigate acidification depends on the region of interest for the LCA.

**Valuation** Weighting factors are applied to the various impact categories to compare their importance and to define priorities directly. This step is often considered the most subjective step in an LCA and is often not implemented because of the difficulty in justifying a particular approach on a scientific basis.

Impact categories define the *consequences* of the material and energy use and emissions on human health, ecosystem health, or the future availability of natural resources. Examples of environmentally relevant impact categories are:

- Global warming
- Eutrophication
- Stratospheric ozone depletion
- Terrestrial toxicity
- Human health

The selection of impact categories should be completed as part of the LCA scoping phase so that appropriate data are collected in the LCI. Table 2.3 presents some of the commonly used impact categories and characterization factors. Many of these equivalency factors are used to normalize the effect of a chemical as compared with the effect of a reference substance. For example, the global warming potential expresses the impact as kilograms of $CO_2$ (the reference substance) to kilograms of the chemical of interest. In selecting among these categories, the choices must be consistent with the overall goals and scope of the study. The choices must be justified, and all models used for characterization must be clearly cited.

Unlike USEPA, the ISO standards catalog no particular impact categories (ISO, 2000). They suggest several guidelines, including:

- Impact categories, category indicators, and characterization models should be internationally accepted.
- Categories should represent aggregated emissions that result in a specified environmental effect.
- Categories should be environmentally relevant.

Clearly, all of the categories included in Table 2.3 meet these criteria.

Knowledge of each of these environmental effects and the stressors that contribute to the problem are necessary for grouping the stressors together to associate them with their specific consequence. Some stressors contribute to multiple consequences. Chlorofluorocarbons, for example, contribute to both global warming and stratospheric ozone depletion. The assignment of stressors to a particular impact category is defined as the classification stage of the LCIA.

*Classification and Characterization*

Although many different stressors might contribute to a particular impact category, the magnitude of their effects often varies widely. The categorization stage requires the use of scientific understanding and modeling of the relationships between stressors and effects to add the contributions of many stressors directly to an individual impact category. Characterization factors, which are available for many environmental impacts, allow direct addition of the various contributors to a single impact category:

$$I_e = \sum_x C_{x,e} M_x \tag{2.8}$$

where $I_e$ = impact indicator for environmental concern $e$

**Table 2.3** Impact Categories and Classification Factors

| Impact Category | Scale | Common Characterization Factor | Abbreviation | Reference Point | Examples |
|---|---|---|---|---|---|
| Resource depletion | Global | Abiotic depletion potential (ADP) | ADP ($kg^{-1}$) | Worldwide availability | Mercury: ADP=$1.8\times 10^{-7}$<br>Lead: ADP=$1.3\times10^{-11}$ |
| Energy use | Global | Energy depletion potential (EDP) | EDP (MJ $kg^{-1}$) or (MJ $m^{-3}$) | Worldwide availability in energy content units | Crude oil: EDP=42.3 MJ $kg^{-1}$<br>Natural gas: EDP=35.7 MJ $m^{-3}$ |
| Global warming | Global | Global warming potential (GWP) | GWP (kg as $CO_2$ $kg^{-1}$) | 1 kg $CO_2$ | Chloroform: GWP=25 kg $kg^{-1}$ |
| Photochemical smog | Local | Photochemical oxidant creation potential (POCP) | POCP (kg $kg^{-1}$) | 1 kg ethylene | Benzene: POCP=0.189 kg $kg^{-1}$ |
| Acidification | Regional<br>Local | Acidification potential (AP) | AP (kg $kg^{-1}$) | 1 kg $SO_2$ | Ammonia: AP=1.88 kg $kg^{-1}$ |
| Stratospheric ozone depletion | Global | Ozone-depleting potential (ODP) | ODP (kg $kg^{-1}$) | 1 kg CFC-11 | Halon 1202: ODP=1.3 kg $kg^{-1}$ |
| Eutrophication | Local | Nutrification potential (NP) | NP (kg $kg^{-1}$) | 1 kg $PO_4$ | Nitrate: NP=0.1 kg kg-1 |
| Terrestrial toxicity | Local | $LC_{50}$ or ecotoxicity, terrestrial (ECT) | ECT (kg $kg^{-1}$) | 1 $m^3$ polluted soil | Cadmium: ECT=$1.3\times10^7$ kg $kg^{-1}$<br>Copper: ECT=$7.7\times10^5$ kg $kg^{-1}$ |
| Aquatic toxicity | Local | $LC_{50}$ or ecotoxicity, aquatic (ECA) | ECA (kg $kg^{-1}$) | 1 $m^3$ polluted water | Cadmium: ECA=$2.0\times 10^8$ kg $kg^{-1}$<br>Copper: ECA=$2.0\times10^6$ kg $kg^{-1}$ |
| Human toxicity | Global<br>Regional<br>Local | $LC_{50}$ or human toxicity factor (HT) | HT (kg $kg^{-1}$) | 1 kg human body | Cadmium: HT=580 kg $kg^{-1}$<br>Benzene: HT=3.9 kg $kg^{-1}$ |

*Sources*: Adapted from USEPA LCAccess website and van den Berg et al. (1996); additional values and the scientific explanation available from Hauschild and Wenzel (1998).

$C_{x,e}$ = characterization factor for emittant $x$ effect on environmental concern $e$

$M_x$ = normalized mass of emittant $x$ (per some suitable quantity)

The third column in Table 2.3 identifies some commonly used characterization factors that permit these comparisons. The reference book by Hauschild and Wenzel (1998) gives a thorough background on the scientific basis for these characterization factors and tabulates values for many environmental stressors. Example 2.7 illustrates how global warming potentials are used to characterize the net contributions from various greenhouse gases.

### Example 2.7 LCIA Classification and Characterization

Ethanol, generated from renewable biomass instead of fossil fuel, can be added to gasoline to reduce environmental effects. A variety of sources can be used as the feedstock for the ethanol production. In temperate climates, the sugar extracted from sugarcane is an excellent starting material. The plant matter remaining, termed *bagasse*, can also be converted into ethanol. (The alternative for disposed is open field burning.) Kadam (2002) conducted an LCA to determine the environmental effects of using bagasse for ethanol production with its use as an automotive fuel at 10% in gasoline (E10) versus burning and use of petroleum gasoline. Air emissions for these two alternatives are included in the following table. Note: The negative values represent credits because of differences in electricity use.

Determine the relative effect on global warming between these two cases.

Air Emissions for Ehtanol Production versus Burning of Sugarcane Bagasse

| **Air Emissions** | **Units**[a] | **Scenario 1: Burning + Gasoline Use** | **Scenario 2: Ethanol Production + E10 Use** |
|---|---|---|---|
| Carbon dioxide (from biomass) | kg | 1706 | 1625 |
| Carbon dioxide (from fossil fuels) | kg | 521 | −77 |
| Carbon monoxide | g | 69 | 23 |
| Hydrocarbons (excluding methane) | g | 8.7 | 10.2 |
| Lead | g | 30.9 | −0.14 |
| Methane | g | 8465 | −149 |
| Nitrogen oxides (as $NO_2$) | kg | 8.5 | 4.5 |
| Nitrous oxides | g | 20 | 21 |
| Particulates | g | 4195 | 148 |
| Sulfur oxides (as $SO_2$) | g | 2622 | 1774 |

[a]Per dry ton of bagasse.

#### *Solution*

General procedure: For an LCIA, we must first group together the air emissions that are considered greenhouse gases. Global warming potentials and equation (2.8) must then be applied to determine the equivalent greenhouse gas emissions for each alternative.

Research on greenhouse gases is required to show that the following air emissions in the LCI presented above are greenhouse gases: $CO_2$, $CH_4$, and $N_2O$. The other air emissions are not considered for this environmental impact category. The United Nations Intergovernmental Panel on Climate Change has published global warming potentials (GWP) for greenhouse gases (see USEPA, 2002):

| | | Impact Score for Global Warming ($I_{gw}$) | |
|---|---|---|---|
| **Greenhouse Gas** | **GWP (100-yr Time Horizon)** | **Scenario 1** | **Scenario 2** |
| Carbon dioxide | 1 | 2227 | 1548 |
| Methane | 21 | 178 | −3 |
| Nitrous oxide | 310 | 6.2 | 6.5 |
| | Total (kg $CO_2$ equivalents) | 2411 | 1551 |

Impact indicator scores are calculated by multiplying the GWP times the mass of each of the gases released (in kilograms). The results of this calculation are presented above. It is clear that:

- Carbon dioxide contributes most to the net greenhouse gas emissions.
- Carbon dioxide credits associated with scenario 2 substantially reduce the global warming effects.

Since global warming affects climate on a global scale, the results presented in Example 2.7 do not need to be normalized for regional effects. LCIA results for global warming can be considered complete at this stage, or advanced environmental models can be applied to determine better the net effect of the release of greenhouse gases. These mathematical models are decidedly controversial. Thus, many practitioners chose to consider an LCIA complete following the classification of inventory data in impact categories and the characterization of these data with factors to sum various contributors of each impact category. While this helps to simplify inventory data and provide some means of comparison, it still does not allow the significance of the environmental effects between impact categories, for example, global warming versus ozone depletion. (See Chapter 4 for advanced methodologies for these comparisons.)

*Normalization, Localization, and Valuation*

Normalization, localization, and valuation permit the expression of impact indicator scores in a way that can be compared among impact categories. Dividing by a selected reference value is generally used to normalize the results of the impact indicator to provide a basis for determining the significance of the added emissions or material use.

Some examples of reference values for normalization include (ISO, 2000):

- Total emissions or resource use for a given area, which may be global, regional, or local
- Total emissions or resource use for a given area on a per capita basis
- Baseline scenario

Depending on the choice of reference value and the geographic scale of the effects, normalization alone cannot provide a basis for the comparison of the significance among impact categories. For example, the effects of acidification cannot be directly compared with those of aquatic toxicity because the characterization factors were calculated using different scientific methods (USEPA LCAccess web site).

The Netherlands NSAEL method (Kortman et al., 1998) for LCIAs illustrates well the difficulty and extent of data required to complete the normalization, localization, and valuation steps. The name of this method stands for "no significant adverse effect level." It implies that the environment can safely handle a certain load of pollutants or other stressor (Figure 2.13). Application of this method for a specific geographic region requires knowledge of dose–response characteristics for each stressor and impact indicator scores for *all* sources of each stressor in this region. In this method, reference values for normalizing include:

- Sum of all impact indicator scores for all sources of the impact category in the region
- Assimilative factor defined by the current emissions and the assimilative capacity

$$I_{e(\text{localized})} = \frac{I_e}{\sum_{\text{region}} I_e} \cdot \frac{\left(\sum_{\text{region}} E\right) - E_{\text{NSAEL}}}{E_{\text{NSAEL}}} \tag{2.9}$$

where $I_e$ is the impact indicator score [equation (2.8)], $E$ is the environmental emissions of a particular stressor, and $E_{\text{NSAEL}}$ is the highest dose that does not significantly induce a negative response in a system (Figure 2.13). The sums of $I_e$ and $E$ and the $E_{\text{NSAEL}}$ are all specific to a geographic region. It takes a substantial amount of scientific understanding to define the $E_{\text{NSAEL}}$ values and bookkeeping to determine the regional emissions!

Because the $I_{e(\text{localized})}$ values include the assimilative capacity and, therefore, a measure of the significance of the effect to a specific region, this method allows these localized values to be summed over all

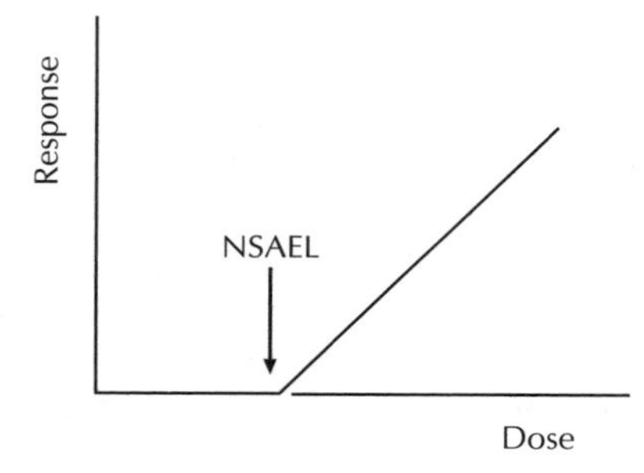


**Figure 2.13 Dose–response curve illustrating an NSAEL.**

impact categories to determine an overall indicator score. This is quite useful for comparing alternatives and avoids biases and the subjective nature associated with other weighting factors that are sometimes used to place a value on impact categories. This method was developed at the University of Amsterdam and has been applied to a few case studies in the Netherlands. Defining a suitable geographic boundary is critical for the application of this method. The boundary should be such that the required data are available yet should acknowledge that environmental stresses and responses are often not aligned with political boundaries. Example 2.8 illustrates an application of the normalization and localization steps for acidification in the Netherlands (Graedel, 1998).

### Example 2.8 Impact Indicator Scores—Normalizing and Localizing

Acidification has caused a great deal of damage to the forests in northern Europe. An assessment of the soil buffing capacity and forest tolerance to acid rain has resulted in an estimate that the $E_{\mathrm{NSAEL}}$ for acid deposition is 1400 $[H^+]$ $ha^{-1}$ $yr^{-1}$ in the Netherlands. Currently, 8200 $[H^+]$ $ha^{-1}$ $yr^{-1}$ are emitted in the Netherlands, indicating that the emissions already exceed the capacity of the forests to assimilate them. (*Note:* This assumes that none of the acid rain crosses the border of the Netherlands.)

#### *Solution*

A manufacturing process emits $NO_x$. Based on the inventory data and the LCIA characterization, a value of $I_{\mathrm{acid}} = 1.8$ was determined. For the region, $\Sigma I_{\mathrm{acid}} = 5.9 \times 108$.

Calculate a normalized impact indicator score for acidification:

$$I_{\mathrm{acid(localized)}} = \frac{1.8}{5.9 \times 10^8} \cdot \frac{8200 - 1400}{1400} = 1.5 \times 10^{-8}$$

This value can now be compared to the local acidification effects of other manufacturing processes and, because it includes a relative measure of the significance of this effect on the region, it can be added to the localized $I_e$ scores for other impact categories.

## Life-Cycle Interpretation and Improvement

The overall goal of any LCA should be to provide enough information about a process, product, or set of alternatives to enable better decisions about specific issues. The LCI and LCIA both provide the improved understanding about the life-cycle system. This next step in the LCA, interpretation and improvement, is to *use* that information. The results of the LCI and the LCIA can be used together, or just the results of an LCI can be used for interpretation. This stage must not, however, be considered a final step. The interpretation of data, data quality, and the nature of the interpretation that is possible should occur throughout the entire study in an iterative fashion so that the goal and the scope of the study can be adjusted as needed. In any case, efforts to interpret the data should follow the goals and scope of the study outlined at the beginning of the LCA process.

The ISO standards include two objectives of life-cycle interpretation (ISO, 1997):

- Analyze results, reach conclusions, explain limitations, and provide recommendations based on the findings of the preceding phases of the LCA and report the results of the life-cycle interpretation in a transparent manner.
- Provide a readily understandable, complete, and consistent presentation of the results of an LCA study in accordance with the goal and scope of the study.

As with the interpretation of any scientific or engineering study, the interpretation of the results should incorporate the limitations of the methods and data used. It is common to do a sensitivity analysis to help interpret results of the LCI and LCIA. Probability distributions for LCI data and/or ranges of values can be used to determine uncertainty in the results and conclusions (ISO, 1997). This analysis should determine the range of conclusions that could be drawn based on the uncertainty in the data and methods. Example 2.7 and the text related to Figure 2.12 have already illustrated the interpretation that can be drawn from life-cycle analyses.

## Alternative Approaches for LCAs

### Streamlined LCAs

The scope of an LCA can vary depending on the issue to be resolved (Figure 2.14). The standard ISO and USEPA approaches for conducting LCAs are quantitative and require significant amounts of data and time for data collection and analysis. These can be considered the "extensive"

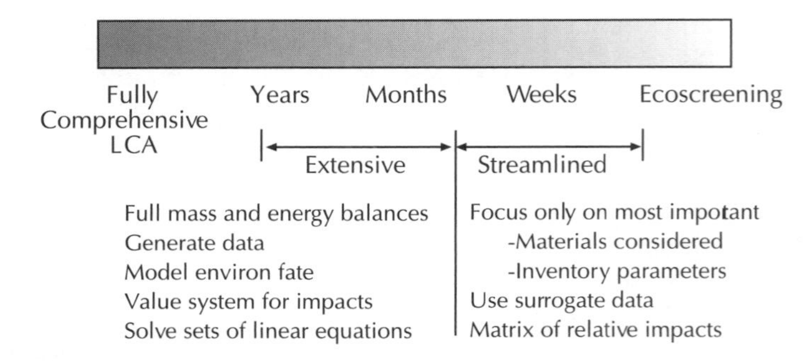


**Figure 2.14 Continuum of LCA studies.**

assessments shown on Figure 2.14. Streamlined approaches are available that address the value of managing and understanding a product or process system on a qualitative life-cycle basis. These approaches required much less data and analysis. The streamlined approaches can work well when comparing two alternatives or for a product in the development stage for which no data exist.

Graedel (1998) has written extensively about streamlined LCAs. He suggests that a streamlined LCA (SLCA) should include the following:

- All life-cycle stages
- All relevant environmental stressors in some manner
- The four basic steps (scoping, inventory, impact, interpretation), although not necessarily in a quantitative manner

Several methods have been developed for SLCAs (Graedel, 1998), generally based on a matrix approach with life-cycle stages on one dimension and environmental effects on the second dimension. Symbols or scores are placed in each matrix element to indicate the severity of the effect for each life-cycle stage.

Corporations that want a streamlined approach have developed methods that work best for their unique needs and goals. Motorola's SLCA approach, for example, includes social and business consequences as well as environmental consequences because they are important issues for decision making at Motorola. Others also include effects "not considered" (e.g., Table 2.3). These include safety, public perception, and noise.

The SLCA could be prepared as a matrix: life-cycle stages on one axis and environmental stressors on the other. Five life-cycle stages (*premanufacture, product manufacture, packaging and transport, product use,* and *recycle/disposal*) defining rows ($i = 1$ through 5) and five environmental stressors (*materials choice; energy use; solid residues, liquid residues* and *gaseous residues*) defining columns ($j = 1$ through 5) would yield a $5 \times 5$ matrix with 25 matrix elements $(i, j)$. The individual elements would be defined by their row and column indexes. For example, the energy used ($j = 2$) during product use ($i = 4$) would be defined by matrix element (4,2).

Each cell of the matrix is assigned a score from 0 to 4, where 0 indicates a significant effect with little effort to control the situation, and 4 indicates little or no effect. Summing the scores both across stressors for each life-cycle stage

$$\sum_{j=1}^{5} \text{stressor score} \quad \text{for each } i \tag{2.10}$$

and down the columns for each stressor

$$\sum_{i=1}^{5} \text{stage score} \quad \text{for each } j \tag{2.11}$$

provides an opportunity to identify the stages and stressors that are most significant. These semiquantitative results are useful for defining areas for improvement. The overall product score, which has a maximum value of 100, is defined as

$$\sum_{i=1}^{5}\sum_{j=1}^{5} \text{matrix score} \tag{2.12}$$

Graedel et al. (1995) applied this approach to compare the environmental effects of automobiles manufactured in 1950 to those manufactured in 1990. Data required for this assessment included the mass of materials included in each car, fuel efficiency, and whether or not the automobiles had air pollution control techniques and air conditioning. Example 2.9 illustrates the completion of one of the matrix elements.

### Example 2.9 Application of an SLCA for Comparison of Automobiles

The average gas mileage of 1990s' automobiles has been significantly improved over the 1950s' automobile. In 1950, cars received an average of 15 mpg. In 1990, this had increased to 27 mpg. Determine the matrix scores for energy use during the consumer-use stage of the automobile lifecycle (matrix element 4,2).

#### *Solution*

The guidelines for completing the matrix element are found in Graedel (1998):

Product matrix 4,2: Energy use during product use

If the following applies, the matrix element score is 0.

- Product use and/or maintenance is relatively energy intensive, and less energy intensive methods are available to accomplish the same purpose.

If the following applies, the matrix element score is 4:

- Product use and maintenance requires little or no energy.

If neither of the preceding ratings is assigned, complete the checklist below. Assign a rating of 1, 2, or 3 depending on the degree to which the product meets design for environment preferences:

- Has the product been designed to minimize energy use while in service?
- Has energy use during maintenance/repair been minimized?
- Have energy conserving features (such as auto shut off or enhanced insulation) been incorporated?
- Can the product monitor and display its energy use and/or operating energy efficiency while in service?

The 1950s automobile consumed a significant amount of gasoline. Less energy intensive methods probably were available but not incorporated because of cost.

Assign a score of 0.

The 1990s' car showed significant improvements in mileage. The score should be higher than zero, but certainly not 4. Use the checklist to help define a score:

**Design to Minimize Energy Use—Yes—** Designed to *reduce* energy use but not really to *minimize* it. Carburetors were redesigned, and weight of the car was substantially reduced to improve mileage.

**Maintenance and Repair—Not Really Important** Most energy is consumed during use.

**Energy Conservation Features** Not more than what was mentioned above.

**Monitor and Display Its Energy Usage** Not in 1990. Newer deluxe versions of late 1990s' models can do this. It helps to educate drivers about their driving skills as related to gasoline consumption.

The choice of a score is now subjective. Since more could have been done in the design and energy conservation aspects, many observers would choose a score at the low end of this middle range.

Assign a score of 1. Note: the published score is 2.

The difference between the score for energy use during the product-use stage selected by the authors of this text and that presented by Graedel et al. (1995) reflects uncertainty in this method. The uncertainty can arise from biases on the part of the person completing the matrix or different prior knowledge and experiences about the subject. Graedel et al. (1995) did not use a significant amount of data to complete the comparison between automobiles, a feature that makes this method attractive. The practitioner, however, must have extensive information about the manner in which various materials are extracted and used, the fate and consequence of many different types of air and water emissions, the nature of different manufacturing processes, and the materials and energy used, among other topics. Although much of this information is used in a qualitative versus quantitative manner, it is still critical knowledge. With sufficient general information about these issues and an effort to remain unbiased, the SLCA approach can provide a valuable tool to highlight the most significant aspects of a product that should be improved to reduce environmental effects.

The completed matrix for the 1990 automobile is shown in Table 2.4. The total scores for each life-cycle stage and environmental stressor are included. Each of these sums has a maximum score of 20, with the maximum score of 100 across the entire matrix. From the totals for the life-cycle stages, it appears that the product-use stage (10/20 points)

**Table 2.4** Environmentally Responsible Product Assessment Matrix Applied to 1990s Automobile

| | Environmental Stressors ($j$) | | | | | |
|---|---|---|---|---|---|---|
| **Life-Cycle Stage ($i$)** | **Materials Choice** | **Energy Use** | **Solid Residues** | **Liquid Residues** | **Gaseous Residues** | **Totals** |
| Premanufacture | 3 | 3 | 3 | 3 | 3 | 15 |
| Product manufacture | 3 | 2 | 3 | 3 | 3 | 14 |
| Product delivery | 3 | 3 | 3 | 4 | 3 | 16 |
| Product use | 1 | 2 | 2 | 3 | 2 | 10 |
| Refurbishment, recycle, disposal | 3 | 2 | 3 | 3 | 2 | 13 |
| Totals | 13 | 12 | 14 | 16 | 13 | 68 |

*Source*: Graedel et al. (1995); reprinted with permission of the authors—copyright holders.

has the most significant environmental effect. The very low score for materials use relates to dependence on a nonrenewable fossil fuel to power a car. Efforts to improve automobiles should focus on improving the gasoline mileage, perhaps changing the energy source for cars (e.g., electricity from natural gas, ethanol from biomass), and reducing air pollutants from the automobile. These changes would collectively raise the score for the product-use stage and, therefore, reduce the overall life-cycle environmental effect associated with automobiles.

## Economic Input/Output LCA

A primary limitation of the LCA approaches of ISO and USEPA is the need to acquire a substantial volume of information, the difficulties of which can result in the system boundaries incorporating only a few of the major suppliers of raw materials. An alternative approach, developed by researchers at Carnegie Mellon University, is based on the total flow of materials through the U.S. economy (Lave et al., 1995; Hendrickson et al., 1998). This economic input–output life-cycle assessment (EIOLCA, introduced in Chapter 1) thoroughly identifies economic flows and related environmental effects for the entire supply side of the life cycle of the system under consideration, including indirect as well as direct suppliers. It does not include the use and final disposal stages.

Figure 2.15 illustrates the broad range of industrial sectors that are included in the manufacture of paperboard. Compared with the discussion of boundaries associated with paper production (Figures 2.4 through 2.6), the boundaries included in the EIOLCA are far more extensive and illustrate the complexity of the supply chain for this commodity. If a limited standard LCA had only captured the primary processes and materials at each of the three life-cycle stages (corresponding to the three rows), only 15 to 67% of the total flow through the economy would have been considered primary inputs to these processes (Lave et al.,

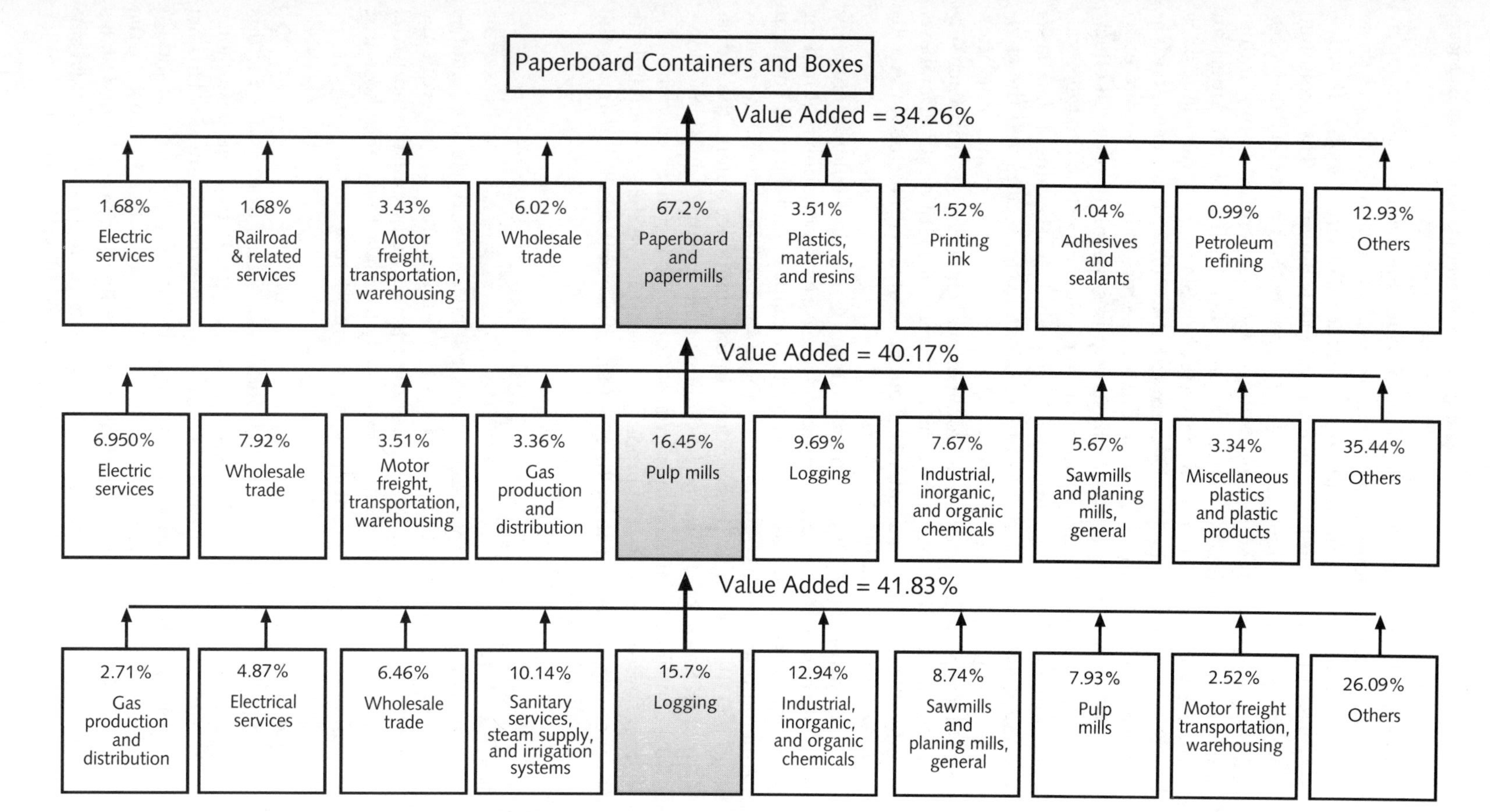


**Figure 2.15 Direct suppliers of paper products included in an EIOLCA.** Lave et al. (1995) report that a more standard LCA conducted in the mid-1990s would only include the three primary boxes (shaded). (From Lave et al., 1995. Reprinted with permission of the authors—copyright holders.)

1995). Substantial contributions required for the manufacture of paperboard could be outside the boundaries of a standard LCA, yet these other secondary inputs could have significant environmental effects.

The EIOLCA combines two analytical techniques. The economic input–output technique captures the material and monetary flows between economic sectors—including interdependencies among industrial sectors. Industrial Classifications, such as the North American Industry Classification System (NAICS) or Standard Industrial Codes (SICs), defined by the U.S. Department of Commerce are used to identify and quantify the different industrial sectors in the United States.

The second stage of this analysis requires the environmental effects for each sector to be determined based on the proportion (in dollar values) that each sector contributes to the whole. Government databases quantifying the movement of materials between sectors and the environmental effect for each industrial sector were used as the foundation of this approach. For example, conventional air emissions from each sector in the supply chain were estimated from the USEPA AIRS Database. Effects included the summary output generated with the EIOLCA software, shown in Table 1.2.

Software for EIOLCA analyses from Carnegie Mellon University (www.eiolca.net) includes both the database of information required for economic and environmental impact calculations and the mathematical methods to solve a complex set of linear algebraic equations. The matrix of equations permits the economic interdependence among the industrial sectors to be determined. Solution of these matrix equations allows the contribution of each industrial sector to the increased economic activity associated with the product of interest to be determined and the effects from each of these to be summed. Example 2.10 illustrates the use of this method.

### Example 2.10 Use of EIOLCA

The Energy Policy Act of 2002 supports the increased use of ethanol as a gasoline additive. Using a basis of $1 million in increased economic activity associated with the use of the ethanol, determine the industrial sectors that will contribute economically to this activity and those that will result in a significant discharge of greenhouse gases.

#### *Solution*

We must first determine which industrial sector to consider for this problem. A search of the possible SICs suggests that the best choice would be 2869, "industrial chemicals not elsewhere defined." This list includes "ethanol, industrial" among a list of 165 other organic and inorganic chemicals. An alternative approach would be to consider the feedstock for the ethanol. Corn is traditionally used in the United States for ethanol production, but other crops, such as sugarcane are even better (although not plentiful). A search on "corn" yields SIC 0115, which is not included in the EIOLCA database. "Sugar crops" are included, however (20502).

The SIC codes are entered into EIOLCA online, and the effects of concern are selected (economic and GWP). A table of results provides the information required. The following table summarizes some of the pertinent results from the EIOLCA software for both SIC 2869 and 20502.

| **$1 Million Activity—2869** | **Economic** | **Global Warming** |
|---|---|---|
| Total for all sectors | 2.313 $mill | 2487 MT $CO_2$ eq |
| Contributions among different sectors (%) | | |
| Industrial inorganic and organic chemicals | 55.8 % | 56.3 % |
| Crude petroleum and natural gas | 6.5 % | 0.9 % |
| Wholesale trade | 4.0 % | 0.7 % |
| Electric services (utilities) | 2.1 % | 26.7 % |
| Other sectors | 31.6 % | 15.3 % |
| **$1 Million Activity—20502** | **Economic** | **Global Warming** |
| Total for all sectors | 1.611 $mill | 742 MT $CO_2$ eq |
| Contributions among different sectors (%) | | |
| Sugar crops | 63.7 % | 38.7 % |
| Real estate agents | 4.6 % | 0.4 % |
| Pesticides and agricultural chemicals | 3.5 % | 3.3 % |
| Wholesale trade | 3.2 % | 1.3 % |
| Agriculture, forestry, and fishery services | 2.6 % | 0.3 % |
| Electric services | 0.9 % | 25.7 % |
| Other sectors | 21.5 % | 30.3 % |

Analysis shows that the results, at least in terms of total economic effect or global warming, depend greatly on the choice of the industrial sector for this particular problem. This is to be expected; $1 million of activity in either entire sector does not really relate to $1 million of ethanol production. This method is valuable for identifying the key sectors that influence the total impact. If our LCA goals include understanding GWP associated with ethanol from sugar crops, we want to be sure that the boundaries include at least pesticide and electric services sectors. These additions to the sugarcane sector, however, still account for less than 70% of global warming associated with the supply cane for sugar crops. Expanding the scope of the investigation to include sectors that contribute very little to the economic activity, but significantly to global warming, show that nitrogen and phosphate fertilizers contribute 10% of the total GWP (due mostly to $N_2O$, which has a high GWP), and trucking another 3%. Thus, it would be important to include these activities in an LCA also.

This example illustrates both the strengths and limitations of the EIOLCA method. The software interface is easy to use and provides rapid access to a large database of information about general industrial activities. The results include the entire supply chain; so, it is easy to determine which industrial sectors are the most significant contributors to the different environmental categories included in the software.

The software can be used effectively in two ways. LCAs that are intended to look very broadly at an industry can use this method directly and achieve results. For example, the EIOLCA has been used successfully for steel-reinforced concrete, asphalt paving, and paperboard—all wide-ranging industrial materials for which specific and suitable industrial sectors are defined (Lave et al., 1995; Hendrickson et al., 1998). As shown in the example, this technique is less suitable for providing direct answers for a narrowly focused study for which the SIC codes are less suitable. It is still valuable for developing boundaries. Results of the EIOLCA identify the most significant activities associated with each environmental category. Thus, it can be crucial for helping to establish boundaries that encompass the central activities on the supply chain side of a product's life cycle.

## Tools for Conducting an LCA

To facilitate LCAs, many diverse consulting firms have been established and databases and software have been developed. Firms, researchers, and governmental agencies in Europe led this development; more recent efforts in Asia and the Americas are catching up. The rapidly changing availability of these resources has led to the documentation of LCA resources on several reputable websites (e.g., USEPA's LCAccess and others listed at the end of this chapter). Some software that has been used successfully on different types of projects is described below.

PRé Consultants in the Netherlands developed LCA software in 1990 called *SigmaPro 5*. It provides a tool to collect, analyze, and monitor complex life-cycle environmental information for products and services. This software follows ISO 14040 recommendations and has been used by major industries, consultancies, and universities. Suggested uses include:

- Screening products for environmental improvements from the first stages of development to the realization phase to discover environmental hotspots and compare options for improvements.
- Calculating eco-indicator scores for commonly used components, materials and processes. Designers can use these scores to screen their designs.
- Incorporating LCA with an industry's environmental management system.

The French firm Ecobilan (*Ecobalance* in English), which is now associated with Pricewaterhouse Coopers, developed *TEAM*. This software allows the user to build and use a large database to model a life-cycle system and to calculate the associated life-cycle inventories and potential environmental effects. *TEAM* also complies with ISO 14040. The

Windows-based C++ calculation algorithms are linked to a comprehensive process and material database that was developed and is maintained by Ecobilan. The software also allows an industry's own inventory database to be incorporated into the assessment.

Few free resources are available for LCA. The *EIOLCA* software is an exception. Another exception is the Greenhouse Gases, Regulated Emissions, and Energy Use in Transportation (*GREET*) model that was developed at the Argonne National Laboratory. *GREET* evaluates energy and emission effects of transportation vehicles. This includes the fuel cycle from "wells to wheels" and the vehicle cycle through material recovery and vehicle disposal. It is available as a spreadsheet model in Microsoft Excel.

## Valuing the LCA in Business Decisions

Corporations use LCA for many different purposes such as identifying key processing steps that cause significant adverse environmental impacts, recognizing product innovation, strategic planning, policy-making, marketing, and to give a product a competitive edge. Performing an LCA can bring the environmental dimension to the business decision-making process. Once an industry decides to conduct an LCA, the real value to that industry comes from the use of the results. This section discusses what motivates an industry to use an LCA as a tool and what added value the LCAs can bring to industrial decision making.

### Motivation for Using LCA

Frankl and Rubik (2000) conducted a survey of the application of LCA in four countries to try to understand *why* companies would consider LCA as a business tool. Approximately 400 different companies from each of the countries were asked to fill out a questionnaire to compare and contrast different LCA application patterns. The results showed that many different factors motivated LCA use, and the factors varied among the countries (Figure 2.16).

Cost savings were important reason for LCA use in all the countries, but it is interesting to note that the role of cost savings as a driver was perceived differently in Sweden compared to the other three countries. The cost savings were mentioned explicitly as a driver in Germany, Italy, and Switzerland, but in Sweden, cost savings were indirectly related to cost avoidance because of future liabilities. Another important driver was the reduction of product-related environmental problems. The international management of parent companies largely influenced Italian companies; initiatives of research and development played a significant role for Swedish companies. Except for Sweden, environmental legislation was a significant prompt for the use of LCAs.

The implementation of a corporate environmental management system was an important factor for many companies to carry out LCA

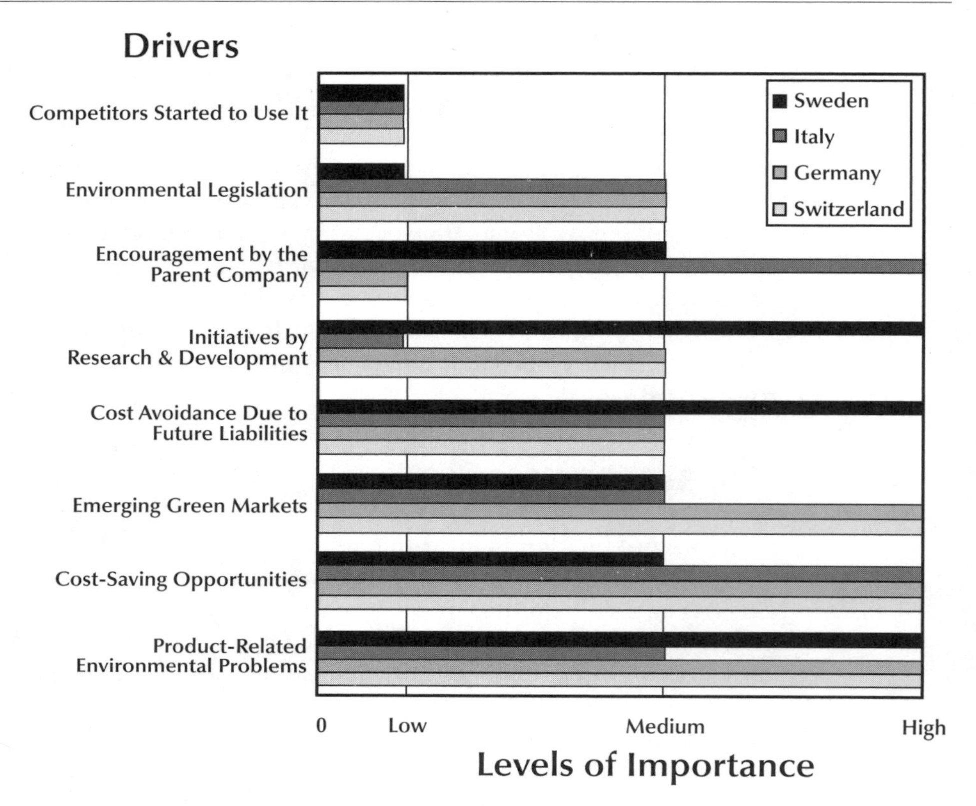


**Figure 2.16 Motivation for LCAs by country.**

activities. Many different factors and groups—the stakeholders—influenced companies, too. The survey found that companies incorporating LCA were more influenced by stakeholders than companies that did not. The consumer and public environmental policies were, of course, the critical stakeholders, but companies applying LCA said that the influence of environmental groups and consumer organizations were more important. This leads to the conclusion that external orientation and sensitivity may be an important factor for companies that choose to apply LCA (Frankl and Rubik, 2000).

## Applying the Results of an LCA

Corporations can use the results of an LCA in many ways to reduce the environmental effects of their products, processes, or services. Examples described below show how LCAs have been used to:

- Identify key steps in products and processes where significant environmental consequences were present.
- Improve marketing by comparing products to determine which was the most environment friendly.
- Modify or optimize products or processes, from an environmental standpoint.

*Identifying Key Steps*

Microelectronics and Computer Technology completed an LCA to identify the components and life-cycle stages of a computer workstation that would place the greatest stress on the environment from raw material usage, wastes, emissions, and energy consumption. In the LCA, inventory data were compiled for diverse components such as semiconductors, semiconductor packaging, printed wiring boards, computer assemblies, and display monitors. The company found that the operation of the monitor used the majority of energy over the life cycle. It was also found that the manufacture of semiconductors dominated hazardous waste generation and was a significant source of raw material use. This was an interesting fact since the weight of the semiconductors is a minuscule portion of a workstation (Allen, 1996).

An LCI focused on toner manufacturing performed in conjunction with a key toner manufacturing firm also yielded interesting results. The most energy-intensive part of the system was the customer's use. The energy required to transfer one metric ton of toner to paper (corresponding to 22 million copies at 135 pages per minute) consumed 57% of the total energy used during the life cycle of toner. End-of-use processing also consumed a significant fraction of the total energy use. Deinking by means of chemical and mechanical methods required considerable electrical energy. Customer use also generated significant air emissions, and deinking accounted for 98% of the total wastewater for the system.

The toner corporation was originally focused on the environmental effects of the manufacturing process. These effects were found to be minimal. Manufacturing contributed just a small proportion of solid waste, air emissions, and energy use over the toner's life cycle. The study illustrated that to improve the overall system, the manufacturer needed to focus on reducing the energy consumption in the photocopier to provide the greatest reduction in environmental impacts (Ahmadi et al., 2003).

*Comparison of Products*

In many cases, an LCA is used to gather information to compare competing products performing the same function or to evaluate product modifications to achieve the greatest potential waste prevention or reduction (Bishop, 2000). An example is fast-food chain McDonald's decision in 1990 to convert from polystyrene to paper-based quilt wrapping for its food products. McDonald's cited a 90% reduction in volume of packaging and reductions in energy consumption, water pollution, and air emissions as key factors in its decision. The motivator for this study was an outcry from consumers about polystyrene boxes, which followed an outcry over the use of paperboard boxes. In these cases, the public perceived that the paperboard was inferior because of the number of trees needed for the boxes, and the Styrofoam was considered harmful because it was made from toxic chemicals and packed landfill space. The final decision by McDonald's was based on the results of an LCI, however, not just public demands (Svaboda et al., 1993).

Ernst Schweizer AG conducted an LCA on mailboxes that it manufactures. The incentive was that many of the environment-oriented customers in public institutions wanted to reduce the consumption of plastics in building materials. Schweizer's mailboxes contained a high proportion of plastic, and the corporation knew that future marketing could be difficult because of their consumers' concerns. Schweizer used an LCA to determine whether products containing plastic were more of an environmental burden than others. It decided to compare its mailbox with one containing steel and the other containing aluminum. By performing the LCA, the company found that the original product was the best selection from an environmental perspective. The other two products caused more of an environmental concern because of high energy consumption during production. Based on these results, Schweizer did not change the product design (Frankl and Rubik, 2000) and had solid results to share with its consumers who were concerned with the plastic mailboxes.

*Modifying Products or Processes*

Bosch and Siemens Hausgerate (BSH) is an international corporation that produces electrical household appliances such as freezers, cookers, and washing machines. BSH performed an LCA to compare the steel soap containers in its washing machines versus soap containers made from polypropylene. Steel soap containers had always been used in all BSH washing machines, but when developing the new top-loader washing machine, the development department wanted to replace the steel containers with a polypropylene variant. In 1994, an LCA confirmed that the soap container made from polypropylene was superior to the steel variant for economic and technical reasons. BSH then replaced the steel soap containers, and in 1996 introduced the first washing machines equipped with polypropylene soap containers to the market (Frankl and Rubik, 2000).

FIAT Auto specializes in producing transportation vehicles and operates in 60 countries with 874 companies. FIAT Auto's environmental policy consists of a long-term strategic program that includes LCA as a core tool. The results of one LCA were surprising and unexpected and altered engineering design decisions that had been made based on preconceived notions about what is good and bad for the environment (Frankl and Rubik, 2000). FIAT's strategy had been to reduce the weight of its cars as much as possible to reduce fuel consumption. One of its main tactics was to substitute less dense aluminum for the cast-iron engine block. FIAT performed an LCA to compare the two different material engine blocks to learn if reducing weight provided a significant environmental benefit. The results from the study surprised the management, especially those who did not usually consider the entire life cycle of materials. Before the study, the corporation had thought any weight reduction in the cars was environment friendly because it would reduce gasoline consumption. The study demonstrated, however, that the use of aluminum was preferable only under certain conditions, and design decisions required careful thought on a large scale. This study had a major effect on the strategy of the corporation. Shortly afterward, it decided to slow the substitution of iron engine blocks with aluminum blocks (Frankl and Rubik, 2000).

### *Modifying How a Product Is Used*

If an LCA shows that the most significant environmental effects occur during the use stage, an industry can proceed in two ways:

- It can change the product to reduce that effect (e.g., redesigning computer monitors to consume less energy).
- It can encourage customers to alter how they use the product to achieve the same reduction (e.g., encouraging customers to turn off their monitors when not in use).

The Paper Task Force LCA, discussed above, recognized the need for consumers to use paper. Recommendations revolved around the consumer-use stage of the life cycle; that is, consumers should recycle their paper and buy paper with recycled content to minimize the environmental effects that occur over the life cycle of paper.

An LCA on polyester blouses showed that it was the use stage that causes the greatest consumption in energy. In the garment's life cycle, 82% of the energy was consumed in the cleaning process, mostly with machine drying. The assumption was that a woman wears a blouse 40 times, with 2 days of wearing the blouse between washings. Thus, a blouse manufacturing industry that is concerned about the environment might chose to suggest line drying on the care label of the blouse to reduce energy use and the associated environmental emissions (Allen, 1996).

Patagonia, an outdoor clothing manufacturer and distributor, has used LCAs to ensure that their products and processes are as environmentally benign as possible. One study focused on the energy consumption of the transportation and distribution of one of Patagonia's garments. In comparing standard shipping with overnight shipping, the company found that the energy used to ship the garment increased from 1% for truck and rail shipment to 28% for overnight shipping. Patagonia used these results to inform its customers of this difference and encouraged them to be patient with the slower delivery option (Allen, 1996).

### *Modifying Raw Materials*

An alternative approach to reducing environmental effects associated with a raw material is to change the supplier of materials—rather than switching to a different material. This change affects the purchasing department, not the design and manufacture of the product itself, as illustrated in the following example.

Scott Paper Company is one of the largest tissue producers worldwide. In the past, Scott Paper, like many other large producers, addressed environmental concerns at a local or site level. In the late 1980s,

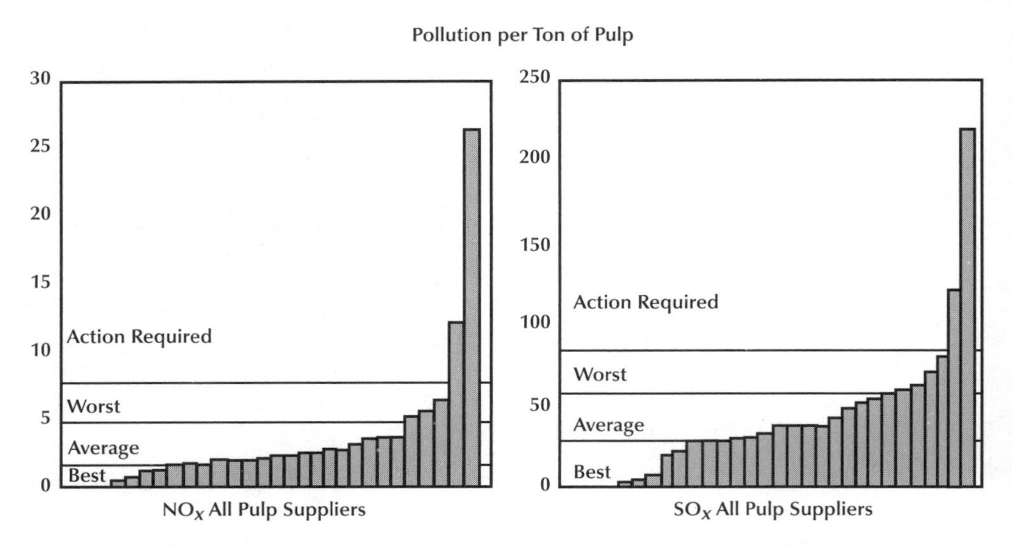


**Figure 2.17 Distribution of $SO_x$ and $NO_x$ among pulp suppliers.** (Adapted from Fava and Consoli, 1996.)

based on the evolving needs of the market, the company developed corporate positions on environmental matters and committed the company to pursuing an LCA approach. This position was taken after several internal studies demonstrated vastly different conclusions than had less comprehensive, single-issue perspectives. In 1991, Scott formalized its LCA-based policy and committed itself to understanding the life cycle of its products with regard to conducting supplier assessment, minimizing the effect of its operations, and providing training and communication for its employees (Fava and Consoli, 1996).

Scott Europe committed itself to developing and implementing a supplier assessment program that would include environmental criteria into its pulp purchasing decisions. Scott required all European pulp suppliers to complete a detailed qualitative and quantitative questionnaire. Some of the results from the assessment (for $SO_x$ and $NO_x$ emissions) are shown in Figure 2.17. It is interesting to note the considerable variations among suppliers, reflecting variations in environmental standards from country to country. Based on the assessment, Scott urged the poorest rated suppliers to improve performance. Some of the suppliers chose not to change, and Scott Europe dropped them as suppliers. Approximately 10% of Scott Europe's pulp supply was changed to a new supplier through this program (Fava and Consoli, 1996).

The results of this study were communicated to Scott's customers, and Scott went from being viewed as "one of the pack" in environmental issues, constantly playing catch-up, to being "preferred" from an environmental perspective among pressure groups and consumers. Based on this success, Scott continued to use LCA as a business and decision-making tool (Fava and Consoli, 1996).

## Suggested Reading/URLs

*Textbooks*

Allen, D. T. and D. R. Shonnard, 2002. *Green Engineering: Environmentally-Conscious Design of Chemical Processes*. Prentice Hall: Upper Saddle River, NJ.

Curran, M. A. 1996. *Environmental Lifecycle Assessment*. McGraw-Hill: New York.

Graedel T. E. 1998. *Streamlined Lifecycle Assessment*. Prentice Hall: Upper Saddle River, NJ.

Mihelcic, J. R., M. T. Auer, D. W. Hand, R. E. Honrath, Jr., J. A. Perlinger, N. R. Urban, and M. R. Penn. 1999. *Fundamentals of Environmental Engineering*. Wiley: New York.

Rubin, E. S. and C. I. Davidson. 2001. *Introduction to Engineering and the Environment*. McGraw-Hill: Boston.

Schenck R. C. 2000. *LCA for Mere Mortals*, Institute for Environmental Research and Education: Vashon, WA.

*Resources and Links to Additional Materials*

American Center for Lifecycle Assessment: http://www.lcacenter.org/.

Ecosite. The Worldwide Resource for LCA. http://www.ecosite.co.uk/.

SETAC: http://www.setac.org/lca.html.

Set of links for LCA, maintained by Dr. Thomas Gloria, ICF Consulting: http://www.lifecycle.org/.

United Nations Environmental Program Lifecycle Initiative: http://www.uneptie.org/pc/sustain/lca/background.htm.

University of Michigan, Center for Sustainable Systems: http://css.snre.umich.edu/index.html.

USEPA resource for LCA: http://www.epa.gov/ordntrnt/ORD/NRMRL/lcaccess/index.htm.

*Additional Software and Databases*

Carnegie Mellon Economic Input-Output LCA software: http://www.eiolca.net/.

EcoSite. A world wide resource for LCA. Provides online access to inventory data for several primary raw materials: http://www.ecosite.co.uk/.

GREET (U.S. DOE, Argonne National Labs): http://www.ipd.anl.gov:80/ttrdc/greet/index.html.

National Renewable Energy Laboratory of the U.S. Department of Energy is in the process of organizing an LCI database for energy-related studies: http://www.nrel.gov/lci.

Sigma Pro software (commercial, by Pre Consultants): http://www.pre.nl/simapro/default.htm.

Summary of databases (by G. A. Norris and P. Notten). http://www.sylvatica.com/Database%20summary%20May%202002.pdf.

TEAM (commercial, Pricewaterhouse Coopers/Ecobalance software): http://www.ecobalance.com/software/team/gb_teamidx.html.

USEPA lists a wide variety of LCA resources, including databases: http://www.epa.gov/ordntrnt/ORD/NRMRL/lcaccess/resources.htm.

U.S. LCI database (under development): http://www.nrel.gov/lci/.

## References

Ahmadi, A., B. Williamson, T. L. Theis and S. E. Powers. 2003. Lifecycle inventory of toner produced for xerographic processes. *Journal of Cleaner Production*, **11**(5):573–582.

Allen, D. 1996. Applications of lifecycle assessment. In M. A. Curran, ed. *Environmental Lifecycle Assessment*. McGraw-Hill: New York, Chapter 5.

Argonne National Laboratory's GREET model, http://www.ipd.anl.gov/ttrdc/greet/index.html. 2003.

Ayres, R. U. and L. W. Ayres, 1999. Use of material balances to estimate aggregate waste generation in the United States. In *Measures of Environmental*

*Performance*, P. C. Schulze, ed. National Academy Press: Washington, DC, pp. 96–156.

Bishop, P. L. 2000. *Pollution Prevention: Fundamentals and Practice*. McGraw-Hill: New York.

Boguski, T. K., R. G. Hunt, J. M. Cholakis, and W. E. Franklin. 1996. LCA methodology. In *Environmental Lifecycle Assessment*, M. A. Curran, ed. McGraw-Hill, New York, Chapter 2.

Ciambrone, D. 1997. *Environmental Lifecycle Analysis*. Lewis Publishers: Boca Raton.

Curran, M. A. 1996. The history of LCA. In *Environmental Lifecycle Assessment*. M. A. Curran, ed. McGraw-Hill: New York, Chapter 1.

Dorf, R. C. 2001. *Technology, Humans and Society: Toward a Sustainable World*. Academic: New York.

EDF (Environmental Defense Fund). 1995. The Paper Task Force: Recommendations for Purchasing and Using Environmentally Preferable Paper, Final Report. EDF: New York.

Fava, J. A. and F. J. Consoli. 1996. Applications of lifecycle assessment to business performance. In *Environmental Lifecycle Assessment*, M. A. Curran, ed. McGraw-Hill: New York, Chapter 11.

Felder, R. M. and R. W. Rousseau. 1986. *Elementary Principles of Chemical Processes*, 2nd ed., Wiley: New York.

Frankl, P. and F. Rubik. 2000. *Lifecycle Assessment in Industry and Business*. Springer: Berlin.

Franklin, P. M., C. P. Koshland, D. Lucas, and R. F. Sawyer. 2000. Clearing the air: Using scientific information to regulate reformulated fuels. *Environmental Science and Technology*, **34**:3857–3863.

Graedel T. E. 1998. *Streamlined Lifecycle Assessment*. Prentice Hall: New York.

Graedel, T. E., B. R. Allenby, and P. R. Comrie. 1995. Matrix approaches to lifecycle assessments. *Environmental Science and Technology*, **29**(3):134A–139A.

Greig-Gran, M., S. Bass, J. Bishop, S. Roberts, N. Robbins, R. Sandbrook, M. Bazett, V. Gadhvi, and S. Subak. 1997. Toward a sustainable paper cycle. *Journal of Industrial Ecology*, **1**(3):47–68.

Gullick, R. W. and M. W. LeChevallier. 2000. Occurrence of MTBE in drinking water sources. *J. AWWA*, **92**:100–113.

Hauschild, M. and H. Wenzel. 1998. *Environmental Assessment of Products, Vol. 2: Scientific Background*. Chapman & Hall: London.

Hendrickson, C., A. Horvath, S. Joshi, and L. Lave. 1998. Economic input-output models for environmental lifecycle assessment. *Environmental Science and Technology*, **32**(7):184A–191A.

ISO. 2000. Environmental management—Lifecycle assessment—Lifecycle effect assessment. ISO 14042:2000(E). ISO: Geneva.

ISO. 1998. Environmantal Management—Lifecycle assessment—Goal and scope definition and inventory analysis. ISO 14041:1998(E). ISO: Geneva.

ISO. 1997. Environmental management—Lifecycle assessment—Principles and framework. ISO14040:1997(E). ISO: Geneva.

ISO. 1996. Environmental management. ISO 14000. ISO: Geneva. http://www.iso.org/iso/en/iso9000-14000/iso14000/iso14000index.html.

Kadam, K. L. 2003. Environmental benefits on a lifecycle basis of using bagasse-derived ethanol as a gasoline oxygenate in India. *Energy Policy*, **30**:371–384.

Kadam, K., V. J. Camobreco, B. E. Glazebrook, L. H. Forest, W. A. Jacobson, D. C. Simeroth, W. J. Blackburn, and K. C. Nehoda. 1999. Environmental Lifecycle

Implications of Fuel Oxygenate Prodution from Califormia Biomass. Technical Report. NREL/TP-580-25688. Naional Renewable Energy Laboratory, Golden, CO.

Kortman, J. G. M., E. W. Lindeijer, H. Sas, and M. Sprengers. 1994. *Toward a Single Indicator for Emissions—An Exercise in Aggregating Environmental Effects.* Interfaculty Dept. of Environmental Science, University of Amsterdam, Netherlands, March [as cited by Graedel (1998)].

Kroschwitz, J. and M. Howe-Grant. 1991. *Kirk-Othmer Encyclopedia of Chemical Technology,* 4th ed. Wiley Interscience: New York.

LaGrega, M. D., P. L. Buckingham, and J. C. Evans. 1994. *Hazardous Waste Management.* McGraw-Hill: New York.

Lave, L. B., E. Cobas-Flores, C. T. Hendrickson, and F. C. McMichael. 1995. Using input-output analysis to estimate economy-wide discharges. *Environmental Science and Technology,* **29**(9):420A–426A.

OSHA database. North American Industry Classification System Search. http://www.osha.gov/oshstats/sicser.html.

Perry, R. H. and C. H. Chilton. 1973. *Chemical Engineering Handbook,* 5th ed. McGraw-Hill: New York.

Pulp and Paper Technical Association of Canada, Montreal, Quebec, Canada. http://www.paptac.ca/english/layout/index.htm.

Schenck R. C. 2000. *LCA for Mere Mortals.* Institute for Environmental Research and Education: Vashon, WA.

SETAC. 1993. *Guidelines for Lifecycle Assessment: A Code of Practice.* SETAC Workshop, Sesimbra, Portugal, 3 March.

SigmaPro 5 software. http://www.pre.nl/simapro/default.htm.

Svaboda, S. and S. Hart. 1993. *McDonalds Environmental Strategy.* National Pollution Prevention Center for Higher Education: University of Michigan, Doc 9-3, Ann Arbor.

TEAM software from Ecobalance. http://www.ecobalance.com/uk_team.php.

Technical Association of the Paper and Pulp Industry. http://www.tappi.org.

USDOE. Energy Information Administration. http://www.eia.doe.gov/emeu/cabs/usa.html.

USEPA. LCAccess website. http://www.epa.gov/ORD/NRMRL/lcaccess/lca101.htm#ch4.

USEPA. 2002. *U.S. Greenhouse Gas Inventory.* EPA 430-F-02–008, April. http://yosemite.epa.gov/oar/globalwarming.nsf/UniqueKeyLookup/RAMR5CZKVE/$File/ghgbrochure.pdf.

USEPA. LCA 101 web page. http://www.epa.gov/ordntrnt/ORD/NRMRL/lcaccess/lca101.htm#ch2.

van den Berg, N. W., G. Huppes, and C. E. Dutilh. 1996. Beginning LCA: A Dutch guide to lifecycle assessment. In *Environmental Lifecycle Assessment.* M. A. Curran, ed. McGraw-Hill: New York.

Vigon, B. W., D. A. Tolle, B. W. Cornary, H. C. Latham, C. L. Harrison, T. L. Bouguski, R. G. Hunt, and J. D. Sellers. 1993. *Lifecycle Assessment: Inventory Guidelines and Principles* (EPA/600/R-92/245). EPA Risk Reduction Engineering Laboratory: Cincinnati.

Winebrake, J., D. He, and M. Wang. 2000. Fuel-Cycle Emissions for Conventional and Alternative Fuel Vehicals: an Assessment of Air Toxics. ANL/ESD-44. Center for Transportation Research, Argonne National Laboratory: Argonne IL.

# CHAPTER 3

# Extending Product Responsibility in the United States: Life-Cycle Opportunities

HILARY G. GRIMES

**Extended *producer* or *product* responsibility** is a term applied to a recent policy movement to capture a variety of approaches with the common goal of avoiding the impacts of waste on the environment. Various forms of this approach are now evolving in the United States and in the global markets. These forms differ in the composition of industrial and nonindustrial actors, incentives, and design and manufacturing activities that they employ, but the objective of all is to manage the environmental consequences of products from their design to their end of life—that is, throughout the product life cycle. In the United States, extended (or shared) product responsibility (EPR) requires that actors along the product chain, driven by market conditions, share the physical and financial responsibility for reducing their product's total life-cycle environmental impacts (Eastern Research Group, 1996).

Extended product responsibility evolved from mandatory producer-focused product take-back and recycling quotas initiated among European and Asian governments that wished to reduce solid waste. U.S. consumers, corporations, and government agencies have responded with interest to the producer responsibility example, learning valuable lessons of the costs of implementation, the issues of state and federal sovereignty, and the role of consumer participation and culture. Subsequently, an initiative in the United States led by private corporations has developed incentives for participation, relied on state and local governments to impose product and waste-related regulations, and expanded the meaning of *responsibility* to include material characteristics, hazards of use, and life-cycle-wide partnerships in managing products and their end of life.

**Objectives of This Chapter**

Characterize the Status of EPR in the U.S.

Describe corporate activities such as Design for Environment (DfE) tools, green materials selection, product take-back, remanufacturing, and supply chain management.

Discuss the motivating forces of regulatory pressure, consumer demands, potential cost savings, and market differentiation.

Present case studies of companies attempting to shape life-cycle environmental management to suggest the capabilities, limitations, and systems of measure associated with existing U.S. extended responsibility programs.

Draw lessons and conclusions of the feasibility and future of EPR in the United States and beyond.

Note: The effectiveness of this approach to a national-scale initiative in the United States cannot yet be assessed because the lack of specific or accepted bounds, metrics, and goals for extended responsibility have made it an activity for corporate competitive benchmarking rather than organized public policy.

The extended responsibility efforts being enacted in the United States and abroad continue to evolve in form and scope as decision makers attempt to determine how best to improve the environmental performance of corporate production while maintaining the best economic performance possible. The issue of who should pay (take responsibility) for the degradation caused by production, consumption, and disposal of materials and goods will only increase in importance as demands on the planet and its resources grow with population and development. Today's students, learning of the history of EPR, could be tomorrow's decision makers in the manufacturing plant, in the government office, or in the marketplace considering the costs and benefits of extended responsibility for a more sustainable society.

## Extended Product Responsibility in the United States

U.S. corporations have been responding to state and federal environmental regulations on industrial wastes for several decades. For much of this time, corporate environmental management was characterized by efforts to minimize the extent of government intervention and the costs of remediation in business operations. At the same time, their customers, shareholders, and the American public increasingly scrutinized industrial activities for adverse environmental and economic effects to the air, water, and land.

In comparison, extended responsibility marks a new approach: the **voluntary** initiatives of manufacturers to shoulder the environmental consequences of products outside the bounds of the manufacturing stage. Some U.S. manufacturers and consortiums have led efforts of their own accord to improve design, process technology, and business operations such that responsibility for potential life-cycle environmental effects and the social costs become internal to the company's decisions.

This extension of environmental responsibility to address effects of material extraction, production, use, and end of life can be attributed in part to rising government attention to the quantity, toxicity, and environmental persistence of materials used in consumer products and the disposal of these products. Manufacturers selling to the global market are obligated to meet imposed or impending regulations in European and Asian countries on recycle potential and disposal of packaging, electronics and electric appliances, and automobiles. These obligations collectively are known as *extended producer responsibility* (EPR).

**Aspects of EPR**

Design for environment

Greener materials and cleaner processes

Product take-back for recycle and remanufacture

Supply chain management

End-of-life take-back

Ecolabel certifications

Recyclable and reusable parts and products

Greener designs (Eastern Research Group, 1996; OECD Business and Advisory Committee, 1997)

**Aspects of End-of-Life Take-back**

Reusable products

Remanufacturing returned components

Recycling postconsumer materials

Sound disposal and energy recovery practices

Individual states in the United States and Canadian provinces have seen legislation proposed requiring that manufacturers take back electronics, packaging, and household hazardous wastes, and banning hazardous materials. These proposals have designated specific roles to government, industry, and entrepreneurs in meeting outlined environmental goals (Davis et al., 1997; Cutter Information Corp., 1998; Driedger, 2001). The growth of voluntary corporate extended responsibility in North America may be attributed in part to manufacturers' desire to avoid promulgation of similar programs through federal legislation, fearing that these requirements would be less flexible and more costly than the programs that they have established of their own volition.

Demonstrating environmental responsibility over the life cycle of a consumer product is also becoming important for corporations to maintain competitive performance in certain markets. Environmental and consumer advocacy groups as well as some government agencies have developed special labels to distinguish products that are designed to meet environmental and human safety criteria. These *ecolabels* restrict hazardous materials use, dictate energy conservation mechanisms, and may set minimal recycling and recycled content for some consumer product categories. Increased attention from media and advocacy groups to product life-cycle consequences and corporate reputations is motivating corporations to differentiate their products and services by demonstrating improved environmental performance.

**Example Ecolabels**

Energy Star [U.S. Environmental Protection Agency (USEPA)]—energy efficiency standards for 38 product categories (www.energystar.gov)

Green Seal certification (not-for-profit)—life-cycle-based performance standards for over 40 product categories (www.greenseal.org)

Blue Angel (Germany)—ecological performance, health, safety, and quality standards for over 80 product categories (www.blauer-engel.de)

Nordic Swan (Nordic countries)—life-cycle environmental, quality, and performance standards for over 50 product categories (www.svanen.nu)

Eco Flower (European Union)—ecological and quality performance standards for 19 product categories (www.eco-label.com)

Hermann et al., 2001

## Global Mandates

Extended producer responsibility policies were enacted in Europe well before such practices were seriously discussed in the United States. Europe's environmental and political history supports product waste directives issued by government and carried out by industry under the "producer pays" model. In 1991, Germany issued its Ordinance on the Avoidance of Packaging Waste, often credited for setting legal precedence to "internalize the externalities" of the environmental effects of consumer product wastes. Rather than leaving the responsibility for waste disposal to the public, Germany requires the distributors and manufacturers of retail products to provide recovery and recycling programs for packaging materials (later setting a similar standard for electronics) that would otherwise go to a public landfill (Lifset, 1993). The Swedish government also initiated early programs to manage the end of life for packaging, consumer electronics, and vehicles. It had formally recognized the need for manufacturer responsibility for chemical process and product wastes more than a decade earlier (Lindquist, 1992; Fishbein, 1998). Italy and Switzerland were among the first to enforce industry take-back for electronics and appliances (Cutter Information Corp., 1998). Some programs made use of voluntary covenants or negotiated agreements rather than mandates, such as in the Netherlands' packaging waste programs (Hanisch, 2000) and France's 1998 proposed appliances recycling. A few supported shared responsibility among a broader set of industrial players (Cutter Information Corp., 1998).

The producer pays model was employed to the greatest extent, however, with the producer defined as the entity engaged in product development and/or sale. The expertise for dealing with the environmental effects of manufacturing processes and the resulting consumer products during and after use seemed most correctly to belong to the organization that developed the product—the manufacturer. Figure 3.1 illustrates the producer pays model (Figure 3.1*a*) and compares it with a suggested model for "shared responsibility" (Figure 3.1*b*).

The need to control products at end of life by changing the manufacturer's role and the early acceptance of such controls was influenced by environmental disasters and their social costs that affected many European countries. These included atomic reactor accidents in the 1950s, record oil spills along the French and British coasts in the 1960s, contaminated waterways and fish consumption bans in the 1960s and 1970s, and a 1986 chemical plant release to the Rhine River that "shocked many [German Federal Environmental Agency] projects forward" (www.umweltbundesamt.de). These calamities were highly visible among the heavily industrialized and densely populated European countries—separated by little other than political borders. Europe's population density also made limited landfill space and growing consumer waste volumes a management priority much earlier than such issues became a public concern in the United States.

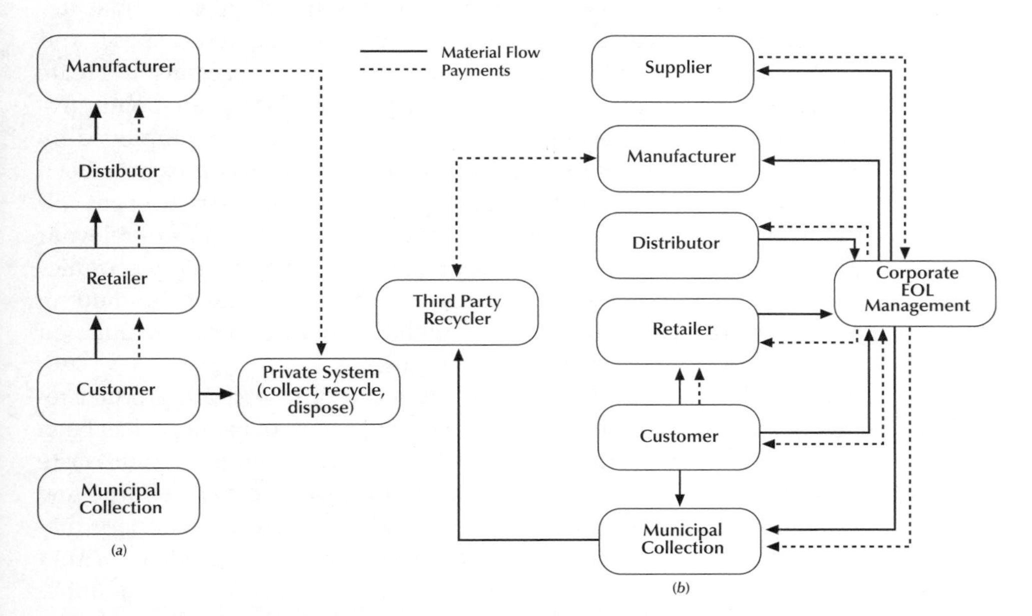


**Figure 3.1 Take-back and recycle models.** (*a*) Producer pays take-back and recycle model. (From Ertel, 1994. © 1994 IEEE; reprinted by permission.) (*b*) Shared responsibility take-back and recycle model (some economic flows are two-way due to value of returning materials to each party). (Adapted from Ertel, 1994.)

Additional factors that fostered Europe's pioneering EPR were its greater representation of socialist principles in politics in the 1990s and its stronger adherence to the "precautionary principle" in setting environmental policy to prevent harm even at a higher cost. This principle can be compared to the practice of informing policy with cost/benefit analysis, used to choose a management approach with the greatest benefit for least cost (Harremoes et al., 2001).

It has been suggested in Europe that EPR could be used as a national policy to generate funds for environmental and waste management programs (Lifset, 1993). Fines would be levied against noncompliant manufacturers, and all manufacturers would be required to pay for government-run take-back systems. As EPR policy evolves for all member countries of the European Union (EU), the goal is to create common elements for waste minimization and recycling of specific products sold among the EU's member countries. The government of the EU, the Commission of European Communities, signed a 1994 Packaging Directive for recycling quotas, relying on member countries to establish their own methods of compliance (Davis et al., 1997). An EU-level electronics recovery, recycling, and hazardous materials reduction directive was passed in 2002 (Applebaum, 2002). The draft acknowledged that inconsistent national take-back and recycling policies created unfair competitive advantage for electronics and electrical appliance manufacturers in countries with weak or nonexistent legislation and imposed large costs on manufacturers in countries requiring more stringent recovery and recycling as well as reporting. An EU-wide management plan would better address the internationally dispersive nature of air and water pollution associated with electronics manufacturing and disposal and might provide the appropriate economies of scale necessary to create a viable recycling program (Commission of the European Communities, 2000). The approaches, goals, and reporting metrics employed by each European country for producer responsibility programs have been blamed by manufacturers for redundant costs, excessive time spent collecting and reporting different data, and unclear definitions of relevant products (Bell, 1998). The EU waste packaging and waste electronics directives offered member countries a mechanism to meet common recovery and recycling quotas, although individual countries maintained the freedom to monitor their own progress.

Participation in EPR programs, often through separate product recovery systems redundant with existing waste management, has been costly to producers (Hanisch, 2000). Upon startup of the German Waste Packaging Ordinance and other early European recovery efforts, few markets had been established for the sudden flood of secondary materials. Europe experienced a supply unmet by demand either within its borders or as an export, resulting in prices much lower than anticipated to pay for the collection and processing (Goddard, 1993). The successes and setbacks of European producer responsibility activities had an effect on the U.S. manufacturers, consumers, and governments meeting them, observing them, and considering their adaptation for the U.S. market.

## Responding to Environmental Awareness

The reluctance of U.S. producers and the federal government to embrace the producer pays model for managing consumer waste caused conflict on the world market and at home among environmental, regulatory, and industrial factions. Nonetheless, the United States was not without its share of existing regulations forcing greater environmental liability on industry. Federal legislation passed from the mid-1970s to the early 1980s governed many aspects of waste management (see box).

### U.S. Legislative Efforts to Control Waste

Hazardous and nonhazardous waste management through permits and reporting—1976 Resource Conservation and Recovery Act (RCRA) (Pohanish and Greene, 2000). RCRA and later amendments also facilitated industrial recycling initiatives and set standards for federal facilities' preferred purchase of green goods and services, superseding policies in the Office of Federal Procurement set in the 1970s for energy conserving goods and recycled goods (Burman, 1992).

Establishment of a Superfund to finance the remediation of abandoned hazardous waste sites—Comprehensive Environmental Response, Compensation and Liability Act of 1980 (CERCLA) (Pohanish and Greene, 2000).

Amendment to CERCLA requiring industrial reporting of chemical usage, releases, and associated hazards—Emergency Planning and Right to Know Act (EPCRA, or SARA Title III) (Pohanish and Greene, 2000).

The 1990 Pollution Prevention Act required reporting of chemical use reduction and recycle by larger industrial polluters publishing their releases to USEPA's Toxic Release Inventory (McCarthy and Tiemann, no date).

During the late 1980s and early 1990s interest in a healthy economy and interest in a healthy environment, often viewed as opposing forces, seemed to cohere. Consumers and manufacturers both were coming to terms with rising costs. Environmental problems, moreover, correlated with the new crises of ozone depletion and the reported phenomenon of global warming. In 1990, USEPA reported that half of all U.S. landfills would close within 10 years (Donnelly, 1990), stimulating widespread municipal recycling and interest in environment-friendly low-waste product choices to avoid rising disposal costs. The twentieth anniversary of Earth Day in 1990 showcased corporate environmental programs that were in large part an effort to repair public relations damage caused by events such as the massive 1989 *Exxon Valdez* oil spill in Alaska and the antienvironment reputation of the Reagan administration (Kline, 2000; Kleiner, 1990b).

The Brundtland Commission report "Our Common Future" (see Introduction to this book) explored the simultaneous improvement of worldwide living standards and sustainable development, and its release was credited with inciting corporate environmentalism. Though the Brundtland report suggested measures for government-led changes in energy, safety, and resource conservation, U.S. companies understood it to say that responsible business activity could preserve the planet for future generations (Kleiner, 1990a).

## Extended Product Responsibilty as U.S. Government Policy

Increased grassroots interest in recycling and in legislation abroad may have led to the proposed 1992 RCRA amendments requiring producers to reduce waste effects of some packaging, glass, and paper products through redesign, reuse, or recycle. The proposal attempted to circumvent the high operating costs and low-value secondary materials resulting from the separate collection and recycling programs inherent in the German and EU packaging regulations. It encouraged producer innovation in developing new markets for recovered products (Lifset, 1994). The bill failed but set the stage for further stakeholder discussions on a U.S. policy for product responsibility. These stakeholders—academic, governmental, environmental, and industrial parties—joined forces at a workshop in 1996 sponsored by the President's Council on Sustainable Development (PCSD) (Eastern Research Group, 1996). PCSD was created to investigate the response of U.S. manufacturers and consumers to producer responsibility programs and to advise on the future direction for U.S. policy. It suggested an alternative to legislating producer take-back and recycling by proposing that the costs of product life-cycle management be shared among parties in the industrial product chain according to their relative contributions to environmental effects. Thus, it coined the term *extended product responsibility*. With this version of EPR, the potential for business benefits, rather than regulatory requirements, would motivate corporations to participate.

Brand-name manufacturers such as Xerox and SC Johnson demonstrated their environmental and economic success in cooperating with suppliers, recyclers, and consumers to manage the effects of product manufacture, use, and disposal. They, and others, explored potential material and manufacturing cost savings, competitive advantage, and other business benefits gained by a shared responsibility that was driven by market demands for environment-friendly products. Participants cautioned that U.S. antitrust and liability laws, existing facility-based environmental regulations, and traditionally isolated industrial supply chain members were just a few factors inhibiting a successful corporate-led shared responsibility initiative, Yet, American manufacturers and USEPA initially embraced EPR as a largely voluntary alternative to producer responsibility (Eastern Research Group, 1996). The manufacturers and USEPA favored the "voluntary, shared responsibility within the product chain" to meet goals of waste reduction and recycling, reduced burdens on local waste management systems, and encourage producers to reflect these goals in design and manufacture (Dillon, 1999 p.199). USEPA's Office of Solid Waste hoped "voluntary action will be spurred by enlightened self-interest" and that industry would be inclined to make investments to "forestall [the] spread of producer responsibility mandates" in the United States. The USEPA encouraged state government action to drive EPR instead of national specifications (Cotsworth, 1999). Federal environmental regulations are largely enforced by state governments, but states often may enact additional measures as needed.

**Incentives for EPR at State Level**

Design for materials reduction and recycle: California and Oregon recycled content container law

Disposal bans for specific materials: Massachusetts CRT (cathode ray tube) disposal bans to avoid lead contamination

Information disclosure policies: New Hampshire mercury reporting requirements for manufacturers

Taxes on virgin material use

Subsidies for secondary material use

**Consumer Incentives to Buy Environmental Products**

Variable disposal fees

Extra taxes on some products

Deposit/refund systems

Incentives and education for municipal waste handlers

Improved communication among manufacturers, consumers, recyclers, and government

Cotsworth, 1999

By the late 1990s, some corporations were motivated to participate in U.S. shared responsibility endeavors because customers increasingly demanded environment-friendly products; simultaneously, they eschewed any extra costs. There is no bigger customer account in the United States than the federal government itself. In 1998 the Executive Order for Greening the Federal Government (USEPA, 1998) directed federal agencies to develop strategic purchasing plans and reporting systems when shopping for office supplies, equipment, or even services to consider environmental attributes such as recycle potential, recycled content, reduced waste and virgin materials, increased bio-based products, and energy-conserving mechanisms.

The desire to satisfy domestic consumer demands for environmental products and to meet federal procurement standards spurred several consumer products manufacturers and manufacturing consortiums in the United States to take a greater role in managing product life cycles and in developing useful tools and models for assessing the resulting business effects. At the turn of the century, EPR and shared responsibility policies continued to evolve and undergo heavy debate throughout the global market over their costs, fair enactment, and ultimate environmental benefits.

## Extended Product Responsibility in the Product Life Cycle

### Participants and Activities

Corporate strategies to capture anticipated business benefits of EPR include product differentiation along environmental or safety lines; influencing industry standards or government regulation; saving production, regulatory, or liability costs through alternative environmental materials and designs; managing environmental risk; and redefining business models (Reinhardt, 1999).

**Activities That Characterize the Range of Manufacturers' EPR Efforts in the United States**

Design for Environment (DfE, product redesign to reduce resource consumption)

"Servicizing" (providing the desired function while maintaining manufacturer ownership of the physical good)

Institutionalizing procurement policies

Product take-back for recycle, reuse, or remanufacture

Engaging in multistakeholder partnerships

Managing environmental and product performance for other members of a supply chain

Some corporate groups such as Monsanto, DuPont, and the Chemical Manufacturers Association (CMA, an advocate for the industry) have successfully raised industry standards and reduced the risks associated with chemical production by requiring stronger environmental management systems and promoting improved environmental health and safety aspects of chemical and engineered products (White et. al., 1999). IBM, Xerox, and other electronics manufacturers have enacted take-back programs for products, components, and packaging. These efforts are supported by life-cycle design or DfE initiatives that allow new opportunities for reductions in material input and waste, improved recovery of by-products and end-of-life goods, greater postlife values through material choices, and improvements on non-waste-related effects such as energy consumption (White et al., 1999).

Other businesses have tried other approaches such as converting to leasing or service-based systems from the model of one-time product sale. Their goal has been improving environmental performance of end-of-life products while reducing repetitive costs of maintenance and replacing or repairing the product for both the customer and the service provider. One such company is Interface Flooring Systems; it provides carpet renewal and recycling services rather than carpet sales (Eastern Research Group, 1996; Anderson, 2000). Manufacturers such as Ford, GM, DaimlerChrysler (White et al., 1999), and SC Johnson Wax (Eastern Research Group, 1996) have engaged in cooperative partnerships among themselves and with other organizations providing supportive services to enable greater recycling and waste management throughout their product chains.

A breakout session in the 1996 PCSD/USEPA workshop on EPR outlined a business case for voluntary proactive life-cycle management activities by identifying feasible benefits that included increased collaboration, risk reduction, less resource use, and marketing advantage. Critical management factors needing further analysis included customer loyalty, the importance of price and performance, the delicacies of forming information and cost-sharing partnerships throughout the product chain, risk reduction, and resource efficiency. The PCSD hoped that EPR "as a product related principle, ensures that the connection [between ecosystem, commerce, and our social system] operates within the principle of sustainability" (Eastern Research Group, 1996).

*Extending EPR*

The emphasis placed on incorporating upstream and downstream life-cycle contributors when managing a manufactured product's life cycle in the United States suggests that industrial sectors other than the producer often play a critical role in generating a product's environmental effects as well as realizing best environmental performance. See Figure 3.2. Incorporating their roles in an EPR model is, therefore, necessary for successful product life-cycle management.

Extracting, generating, and refining materials from which a consumer product will be devised, such as metals, petroleum products, and engineered composites, can be an energy-intensive process with high waste rates and air and water pollution. A great deal of expertise in dealing with these effects from a life-cycle perspective, therefore, resides with the raw materials suppliers. Suppliers of secondary inputs such as electrical subcomponents and formed parts in a traditional industrial transaction sell their goods to producers but retain much of the available information on those goods' environmental, structural, and technical characteristics. As an example, a motor bought from an outside equipment manufacturer to be incorporated in a large appliance will be constructed of several different metals, wires, plastics, and assembly mechanisms that may not be obvious to the appliance maker.

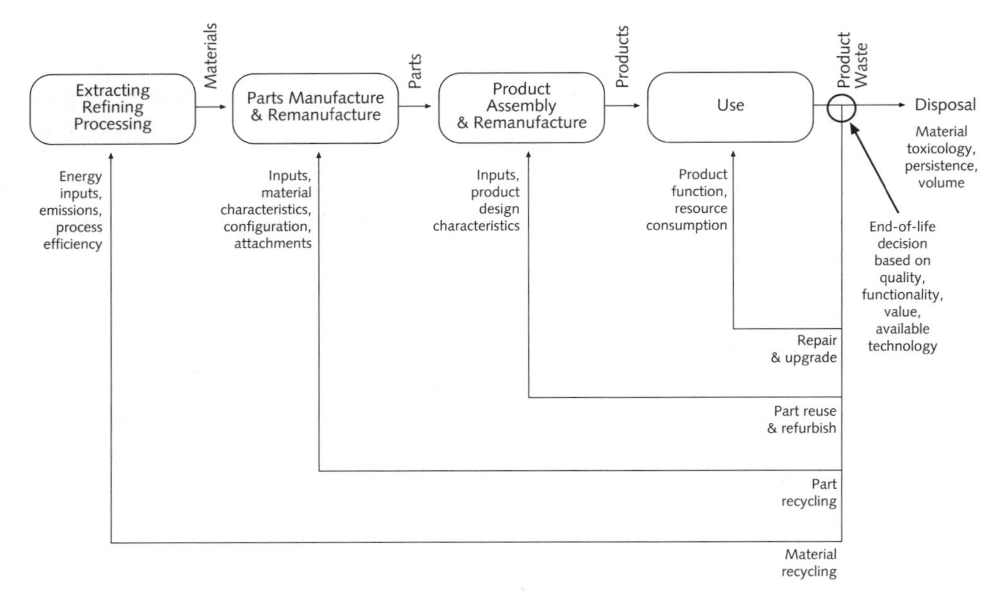


**Figure 3.2 Product life-cycle considerations and end-of-life options.** (Adapted from Sundlin et al., 2000, © 2000 IEEE; reprinted by permission.)

This lack of knowledge limits recovery opportunities if an EPR strategy neglects suppliers in its incentives to participate.

After the sale of the finished product, consumer decisions during and after use can determine the product's downstream environmental effects in terms of resource consumption over the lifetime and end-of-life wastes. The ability of producers to influence consumer decisions depends in part on the information provided in marketing, distributing, and maintaining the product, but also in part on the incentives provided to users to engage in preferred behavior during their ownership. Customers may handle and maintain products differently if provided with a rebate or discount on future purchases in exchange for entering leasing contracts and/or returning used goods. In the case of deposit/refund programs, customers have the incentive to return products after their use to get back their money deposit. However, if product return, recycle, or other environment-friendly behavior creates financial or environmental advantages that the customer considers important or worthwhile, that information can also provide an incentive to act according to the producer's interests.

Finally, end-of-life options can succeed only if the following knowledge is available throughout the industrial product chain and is maintained by appropriate signals to engage in cooperative, mutually beneficial relationships:

- Material and part characteristics
- Product design and functionality
- Aspects of use and maintenance
- Feasible recycling and disposal technology

The U.S. version of extended product or shared responsibility suggests that these relationships and linkages can be developed through industrial and social policy instruments. They can also be developed through the use of market metrics such as operating and product costs, environmental risk, market share, and cost/benefit analyses. Examples of industrial policies that can promote EPR in the United States include adopting supply chain management programs, commitment to provision of service as a new framework for business, and membership in industry-wide organizations and multi-industry consortiums with proactive environmental standards for entry. Nonindustrial interests promote life-cycle environmental responsibility activities through award programs and media campaigns that target both producers and consumers. Metrics can be developed to gauge the effectiveness of these policies in supporting EPR in the United States once there is wider consumer and producer acceptance of cost/benefit analyses that acknowledge externalities in sale prices and operating costs. These analyses will necessarily consider the benefits of reduced environmental risk and increased market share. Examples later in this chapter present corporate policies and improved metrics intended to facilitate broader environmental management activities as well as to encourage participation throughout the product supply chain.

## International Corporate Standards

For the most part, extended responsibility in the United States has arisen as a way to gain a competitive edge and differentiate products and technologies from other market players. Before a sustainable model of EPR can be developed for industry, however, standard product and operational aspects that can be communicated through environmental performance metrics may be required. Variations in the data considered relevant for assessing EPR activities inhibit the creation and retention of shared responsibility roles. Common technology and corporate practices among and across industrial sectors could create the infrastructure to foster sharing of environmental objectives.

The corporate environmental management system (EMS) is one example of an internal and external "yardstick" to assess progress and shortcomings in the environmental arena. The EMS describes accepted company procedures for evaluating environmental effects and safety risks of processes, products, and any corporate activities; monitoring and reporting on environmental compliance and performance; and the roles of company employees in doing so. (An EMS is described in detail in Chapter 5.) The EMS has come to the fore in life-cycle environmental performance programs and is growing in popularity throughout industry. EMSs have been faulted for being too specific to internal production processes and associated emissions; too rigid to encompass new corporate values for intellectual innovations, conversion to service models, and technological improvements by which environmental issues are being internalized; and too often based on regulations rather than operations and functions (Allenby, 1999). Implementation of a comprehensive EMS, however, can provide a party in an industrial product chain with retrievable information allowing its own and other parties' assessment of environmental programs and effects. An EMS can also foster the impetus to improve a company's performance voluntarily while simultaneously realizing market gains or cost savings (Klassen and McLaughlin, 1996). A set of international management system standards from the International Standards Organization (ISO) now serves many industrial facilities worldwide as a benchmark EMS and certification of environmental performance (see box).

**ISO 9000 series: Certifications for quality assurance programs** means high levels of product and process consistency and functionality with mechanisms for measuring and encouraging improvements. A company so certified has greater potential to improve material and operational productivity and reduce wastes through lower defect rates than an organization without a plan for quality management. The ISO 9000 certification also indicates a competitive advantage to many customers.

**ISO 14000 series establishes environmental management systems** with procedures for environmental audits, meeting environmental labels, and performing life-cycle analyses. ISO 14000 standards are anticipated to raise the bar on environmental activities, even to replace other regulatory reporting requirements, because participants must commit to compliance with environmental laws, establish an internal system to monitor environmental effects, and seek to improve beyond meeting their legal liability.

The diffusion of ISO 14000 standards in global corporations sets a framework for broadening environmental management to include life-cycle effects not typically considered in traditional U.S. compliance-based management models. See Figure 3.3. As the EMS grows in scope to account for environmental performance beyond regulations and incorporates decisions and concerns throughout the corporation rather than focusing on individual business divisions, a greater understanding of life-cycle environmental concerns and industry roles can develop. The emphasis on setting environmental objectives and collecting the data necessary to gauge performance should motivate the development of appropriate metrics that reflect information relevant to internal corporate practice and external industrial performance. ISO 14000 could also provide a structure from which common language on environmental issues, practices, and measures of improvement could be established and incorporated within extended responsibility policies. Environmental and operating data collection, reporting, and progress measurement

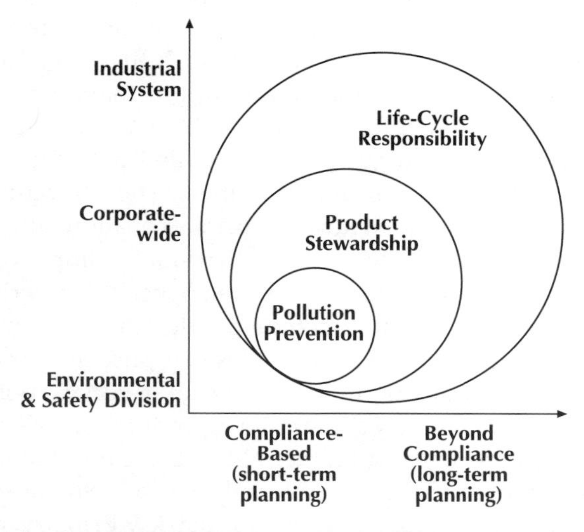


**Figure 3.3 Extent of environmental management strategies.** (Adapted from Darnall, 2000, © Nicole Darnall, reprinted with permission.)

lack coherence among members of the product life cycle; this inconsistency creates a barrier to the sharing of information and distribution of roles under U.S. extended product responsibility. ISO certifications on quality and environmental management could help to overcome these barriers to the benefit of extended product responsibility.

## Rationale for Employing EPR

Any corporate organization undertaking new projects and policies requires a strong business rationale for doing so. Though shared life-cycle responsibility can be promoted by corporate managers as simply "the right thing to do" regardless of cost, the long-term success of such policies depends on the comparative weight of associated social, economic, and environmental costs versus the potential benefits. Gains in competitive advantage and corporate reputation may be sound reasons for companies to pursue extended product responsibility programs, but they are difficult to quantify. Making a case based on cost-savings opportunities, customer survey data, and market regulations allows use of measurable data that has specific meaning to corporate decision makers.

### Reduced Material Intensity

By reducing the total quantity of physical resources and energy required to develop a finished product or service, suppliers, manufacturers, distributors, and even customers stand to reduce their product costs. One way to achieve reduced materials intensity is by decreasing size and weight of parts, products, and packaging during the design process; another is to improve the efficiency of packing and transport. For example, aluminum packaging has undergone redesign to reduce weight without sacrificing volume or reliability such that over 50% more cans are produced from one pound of aluminum than were produced in the 1970s (Can Manufacturers Institute, 2002). Designing smaller products can also reduce material intensity while improving convenience and costs to both producers and consumers. The computer initially filled a room with equipment. It now performs even more functions but fits on a desk or even in the hand. The rule that smaller improves environmental performance has exceptions however. Specialty materials engineered to replace the strength and functionality that thicker, larger, simpler materials provided in products can cause complex environmental problems in their production and in their disposal or recycle potential. Reducing the quantity of a given material in a product may also reduce inherent incentives to recover that material for its postuse value.

Reuse, remanufacture, and recycle of products, components, and materials can also reduce materials intensity. Primary resource and energy consumption and waste associated with virgin production can be avoided by reintroducing high-quality postuse products, scavenged and refurbished parts, and recovered processed materials (plastics, metals, glass, and paper) to the industrial life cycle. The improvement is not guaranteed unless recovered products, parts, and materials retain market value, are amenable to the available remanufacture or recycle technology, and require fewer inputs to return them to a state equivalent to their virgin counterparts.

For each recovery option, other conditions must hold to achieve true environmental and economic gains. Reuse and remanufacture, for example, generally are preferred recovery options because they bypass several stages of production inputs (and avoid production costs) while delivering the same product. The separate collection and storage of products for remanufacture or reuse, however, require an additional energetic burden that will be wasted if there is no demand for the remanufactured products; the remanufacture process can be labor and material intensive if the product has a high rate of malfunction after use or was not originally intended to be remanufactured. Recovery for material recycle in the same product type or in different products can save costs or provide some monetary returns for a participating organization as well, but the benefits depend on the quality of the secondary materials and their marketability. Recovered aluminum, polyethylene terephthalate (PET), and high-density polyethylene (HDPE) plastics have a high market value when the quality of their sorted supply is high, and sorted collected glass is in high demand by manufacturers because its use generates sizable energy savings when remelted to form a product. On the other hand, other types of plastics and paper have a low or negative value in the secondary materials market because of problems maintaining supply integrity, demand, and lack of cost-effective reprocessing technology (Breen, 1990).

The various recycle options may also differently affect desired environmental improvements. Recycle of recovered plastic containers into carpeting and fleece clothing may reduce virgin inputs to those production operations and divert plastic container waste from the landfill. Such options do not, however, reduce the consumption of resources and energy in the plastic container industry. Because an equal number of plastic containers must be newly manufactured to fill the market need, this form of plastic recycling does not create a "closed loop" for the industry sector of concern.

## End-of-life Management

In certain situations, a manufacturer can profitably accept the external environmental costs of disposal—costs usually paid by the public through taxes and fees. Staying ahead of product take-back regulations can mean managing costs of new technology, administration, and monitoring rather than reacting to them and can mean added benefits of unveiling new-product responsibility features to the market ahead of competitors. Product end-of-life management is, therefore, often enacted to anticipate legislation or satisfy customers.

Several states have considered take-back and recycle legislation; California, Massachusetts, and Connecticut have passed items such as landfill bans on certain electronics parts and electronics recycling programs (Hileman, 2002; Raymond Communications, 2002). Under RCRA, users and discarders of large amounts of electronic products containing CRTs must handle them as hazardous waste—although a USEPA rule proposed in early 2002 may relax this regulation to allow flexibility in developing collection and recycling programs. The National Electronics Products Stewardship Initiative (NEPSI) is a consortium of state and federal agencies, nongovernmental groups, and manufacturers working on an electronics reuse and recycle effort for the United States. It may draft federal legislation that would set up and finance the collection program (Lindsay, 2002).

Requirements for avoiding consumer solid waste and handling end-of-life products create different types of costs for the producers and for the government agencies managing the programs—in administering the rules, collecting reports, and monitoring the activities. Greater emphasis on recycle will create a need for better separation and recycling technology, or more training for the labor force undertaking the process of separation for reuse or remanufacture. Some corporations have already analyzed these added costs in changing their operations to meet European and Asian laws and to provide similar practices in the United States, for example, 3M, DuPont, Ford, and IBM (Shapiro and Stoughton, 1999). Their experience may provide only limited insight into the economic effects of potential U.S. EPR requirements, however, because these analyses are often conducted internally and may be based on proprietary information that cannot be shared in an independent study.

Customer satisfaction is an often arbitrary measure based on surveys and market data, but it is an important metric for production and sales of manufactured products. By meeting the needs of customers with product features and corporate policy, a corporation will be able to keep customers and maintain or improve market position; a dissatisfied customer will likely turn to a competitor. Unfortunately, available decision-making tools for product and process life-cycle management often neglect to consider the consumer's role. Over a lifetime of product use, the consumer incurs energy consumption, waste production, and even hazardous and air quality emissions effects. The direct and indirect costs associated with use of the product may be significant. Ironically, many consumers indicate willingness to change their product use behavior and help to manage these issues: Product surveys show that customers demand and seek energy efficiency, reductions in hazardous materials and packaging, and ecolabels on products (Brouwers and Stevels, 1995; Azar et al., 1995).

Materials refining operations and subassembly manufacturers may see their business success in the hands of manufacturing customers, but assembly and sales organizations that transfer the product to consumers measure their business success in consumer satisfaction (Darling, 1995; Barich and Kotler, 1991). Collectors, recyclers, and service representatives also attempt to meet and exceed their own clients' expectations. Although life-cycle environmental attributes are a high priority for many products and services today, functionality, quality and reliability, price, and enhanced features are the long-term standards that dictate the strength of an industrial player's market share (Stocum, 2001).

## Avoid Exclusion

Exclusion from markets is a significant concern for businesses evaluating the feasibility of participating in product life-cycle management. To some degree, a decision not to pursue EPR would eliminate some market sectors from selling in countries with strict producer responsibility laws and might even affect their sales in the United States to government customers. To a larger degree, differing product-based regulations among different regions may cause trade to shift heavily toward markets that are lax about product end-of-life management. The competitive advantage gained by the corporations selling their products at lower costs to markets without product take-back regulations, as opposed to corporations taking on the extra cost of take-back and recycle, is a problem that has come up among industry discussions in both Europe and the United States. It should be noted that the U.S. EPR approach suggested by the PCSD would impose different costs and consequences on industrial stakeholders than do existing EPR programs. Without coordination of market requirements for EPR efforts and clarity on the required roles and targets for redistribution of product environmental costs, issues of competition and unfair advantage will remain a concern for global trade.

## Implementing EPR in the United States

The U.S. version of life-cycle management for consumer products has been characterized as corporations acting on economic and market incentives. State and municipal legislation designed to invoke greater product responsibility among identified industrial players suggests that, in the future, voluntary corporate activities will be supplemented by public policy tools throughout the product system. Incentive-based instruments include special information provision, economic reward in the form of subsidies or credits, or penalty in the form of taxes and fees to guide environmental performance at least cost and greatest flexibility (Harrison, 1999; Harlan, 2000). Both industry and government are interested to know whether these efforts will be sufficient to achieve EPR goals for the United States or if they will more efficiently achieve waste reduction and ecoefficiency than a program based on mandates and stricter regulation.

### Policy Tools

Policy makers in the government and advocacy worlds have a range of tools that they can use to drive social change while allowing varying degrees of cooperation and flexibility among themselves and target organizations. Cooperative policy instruments include government inaction (least coercive or most voluntary), exhortation, and regulation (least voluntary). With inaction, government intervention may be a threat; nongovernmental groups can engage in agreements with industrial organizations to establish and meet EPR objectives. The Rainforest Alliance and the Forest Stewardship Council are two groups that have applied public pressure on forest products industries to certify products and to adapt harvesting/manufacturing practices to what these organizations see as improved environmental and ecological standards.

Government or policy group challenges, contracts, and education programs constitute cooperative exhortation instruments. The desired action can be evoked with research assistance, awards, and good publicity; unsuitable behavior can be discouraged with threats, public warnings, or admonitions. USEPA's WasteWise program enlists business organizations of all sizes and types in a 3-year waste management and reduction program. It provides technical assistance and promotional tools in return for waste audits and improvements. Beyond admonitions, government regulations derived through stakeholder consultation, and/or in conjunction with technical advice to avoid violations, are also considered a cooperative approach to achieve compliance (Harrison, 1999).

Buyers of consumer products have prompted changes in product life-cycle environmental performance both by responding to and in some cases creating information-based policy instruments. Consumer advocacy groups have developed ecolabel criteria and product certifications that are useful in dictating minimum expectations for producers and occasionally materials suppliers in creating environment-friendly designs. Broader procurement policy can also focus attention on expected environmental attributes and activities not only for product manufacturers but also in some cases for material suppliers and providers of services such as maintenance, waste disposal, or recycling. Consumer information policy instruments such as ecolabels and procurement standards could be modified to include specifications on other goods and services during stages of a given product life cycle that are rarely considered during a product's purchase. These types of product standards are gaining the attention of the scientific community. The scientists' assessment of the effectiveness of these standards in meeting product environmental goals will color consumers' and manufacturers' reliance on ecolabels for production and purchasing decisions.

Market-based instruments are those that drive preferred behavior with economic incentives. They are transferred by either a central authority (often through mandated participation or through fees or fines) or a "self-run" marketplace of voluntary participants. Governments have used taxes and direct charges with some success to modulate air pollutant emissions in identified regions. Subsidies could be employed to encourage emissions reductions or perhaps indicate preferred materials. Deposit/refund schemes have worked to manage postconsumer products in the United States as well as Europe by developing collection markets (Field, 1997).

Charges, taxes, and subsidies all require careful monitoring of the emissions or improvements at specific sites to see if their value is equal to the marginal cost of abating the targeted pollutant (Tietenberg, 1990). Such imposed costs may be less effective, or at least difficult to apply, given conditions of varying environmental effects for the same pollutant under different conditions. For example, effects associated with solid waste volumes being considered for a disposal charge will be different in a heavily urbanized area than in a rural area. Solid waste containing plastics and heavy metals will create different environmental effects than food waste and paper, and one charge will likely not provide sufficient incentive to abate of both types.

On the other hand, the markets for tradable discharge permits (or reduction credits) are still being developed and tested for a limited number of air and water pollutants. Regulators have created these instruments to promote least-cost pollution reductions with the expectation that industrial participants will base their emission management strategies on the lowest cost alternative. Tradable discharge permits are useful in situations where it is difficult to identify any single polluter and its share of adverse environmental effects in a widespread pollution problem. In the case of tradable permits, the governing authority sets a cap on the total emissions allowed for a defined region. The variable abatement costs from polluter to polluter stimulate trade among them. The price of the individual permits is set by the supply of permits available from organizations that found low-cost ways to reduce their emissions and the demand created by organizations whose emission reductions costs are high. Opportunities for using tradable market mechanisms to cap hazardous materials use, promote recycling technology through credits for recycled products or volumes, or cap disposal of certain materials may be on the horizon. The British government proposed in 1997 to allow trading of recycling "surplus" between companies with greater recycling capability and those that could not meet recycling targets set by UK waste packaging regulations (*Financial Times*, 1997).

Partnerships and coalitions among members of a supply chain, within industry sectors, between government agencies and corporate management, or with corporations and public-interest representatives are examples of cooperative relationships that can lead to coordinated decisions for production, sales, and management activities. Such arrangements have also been employed to set some forms of environmental policy at the governmental and corporate level. By pursuing collaborative relationships to address collective problems or behavior or to influence regulatory requirements, participants create a mechanism by which change can be communicated, monitored, and enforced through the power of peer pressure.

Industrial-interest groups have developed coalitions to protect member firms from disadvantageous policies and to present a more favorable united image to the public and to regulators. The Electronics Industries Alliance, a consortium performing research and advocacy roles for its industry members, has developed a material declaration guide and common reporting requirements to track quantities of hazardous chemicals being used in the electronics industries (O'Connell and Brady, 2002). The Chemical Manufacturers Association (CMA) developed internal regulations for members' pollution prevention and environmental health and safety programs in the late 1980s under its Responsible Care Initiative (Reinhardt, 1999). These efforts at self-regulation can enhance environmental performance among the membership as well as separately in the respective industries largely by validating that they deserve improved reputations among participants and to the public.

Self-regulation can also be subject to opportunistic participation if specific penalties for noncompliance are not enforced. Without a method of ascertaining compliance, participants are not prohibited from joining in name only to hide undesirable activities, or "free-riding" (Palmer and Walls, 2002). Organizations participating in such associations may have no additional incentive to improve their environmental performance over their nonparticipating competitors (King and Lenox, 2000).

Material characteristics and quantities are also the subject of landfill bans and manufacturers' reporting requirements. Such mechanisms can have the effect of reducing toxic materials in the environment. While disposal bans and material phase-out and label warnings have been implemented in some states through legislation, their details can include voluntary aspects. In New Hampshire, for example, requirements that manufacturers submit regular reports of the quantities and uses of mercury do not directly impose limits or eliminate mercury from many existing products (State of New Hampshire General Court, 2000). Knowledge that the reported information may be communicated to the public, however, often motivates manufacturers to replace toxic materials in their products where alternatives are feasible. Landfill bans can also catalyze initiatives involving municipal, collection, consumer, and manufacturing organizations. Such a coalition might develop a coordinated approach to the problem of banned materials and their end-of-life management and ultimate elimination from a product system in ways not specified by the regulations imposing the ban.

## Case Studies in Industry

To prepare for EPR mandates, U.S. manufacturers have used a variety of cooperative approaches. We discuss three examples of corporate voluntary efforts to internalize the social costs of products' life cycles. They describe the adaptation of producer-led activities to include broader representation of supply chain members in their decisions and management role.

### Hewlett-Packard Company

Hewlett-Packard (HP), a worldwide electronics manufacturer known for computer equipment, issued a corporate environmental policy embracing life-cycle responsibility of products, services, and its own business operations in 1992, according to a 1997 case study by Patricia Dillon of the Gordon Institute. Motivated by emerging market and regulatory forces that would put environmental issues, especially management after useful life of consumer products, in the forefront of global attention, HP circulated its new policy on life-cycle responsibility through its many levels of business:

- At the product level with DfE principles
- At the management level in the form of internal material consumption, energy, design, and manufacturing waste metrics and globally located product stewardship teams
- At the market level by meeting ecolabel standards and creating the infrastructure for active return and reuse, resale, or recycle of used workstations and a few smaller products and supplies
- At the supply chain level by establishing standards (measurable through a scoring system) for suppliers' performance in quality, cost, and local and global environmental practices (Dillon, 1997; Maxie, 1994).

Some of the associated activities provided a direct financial benefit to the company; others captured the benefits of improved environmental reputation. HP's life-cycle commitments were focused on efforts that met greatest demand at the least cost.

*Product-Level Activities*

The DfE concepts implemented for a benchmark personal computer (PC) product line helped to cut the product weight nearly in half and greatly reduced both the number of different materials and the complexity of assembly and disassembly. Disassembly time was reduced 90% over other models. Heavy metals and other materials of concern in the packaging, manual production, the product, and batteries were reduced or eliminated. Recycled content became an important attribute of packaging and the product documentation, and recycling codes for most sizable plastic parts were implemented according to ISO standards. Packaging required to ship this and other new PC models was also reduced by one-third because recyclable polypropylene foam was built into the body of the computer to provide additional protection in place of metal internal construction.

Hewlett-Packard received the strict German Blue Angel certification for recycle potential and for commitment to end-of-life recovery, heavy metals, and energy consumption with this product. Additionally, over 100 HP models have been developed to meet USEPA Energy Star requirements.

Although these environmental design accomplishments were entirely an HP initiative for a specified product, the process by which HP could ensure that its computer parts were recyclable and that materials of concern were avoided required coordination of efforts with their suppliers. This coordination undoubtedly was more freely gained from the voluntary nature of HP's commitment to life-cycle responsibility, and the mutual benefits that could be derived from that commitment, than would have occurred if HP had been under pressure to comply with product regulations aimed at the producer alone.

*Management-Level Activities*

The product design changes that facilitated HP's extended responsibility program could not have happened without strong messages to and from managers at every business level. Every product line gained an environmental steward who had access to a global support network for translating the corporate commitment to responsibility activities within the bounds of market forces and forthcoming legislation. A global product stewardship council consulted on environmental issues and needs affecting the entire HP corporation. A self-audit was developed to verify that individual operations and products met these issues and needs.

Hewlett-Packard's high-volume computer products business provided the testing area for new environment responsibility tools that would be incorporated throughout the company. The effort included a set of internal performance metrics chosen to indicate progress in improving material conservation, waste recovery, energy efficiency, and disassembly and end-of-life materials management among products, consumable supplies, and packaging. Manufacturing processes were gauged by wastes produced, their reuse or recycle, and emissions.

By providing management with concrete tools and an organized support structure, HP clearly communicated the feasible business benefits of meeting environmental requirements on a voluntary basis. This demonstrated success and internal confidence facilitated the extension of life-cycle responsibility ideas to organizations outside the company.

*Market-Level Activities*

Acquiring ecolabel certifications and satisfying customers' requests for environmental features and end-of-life management required HP to commit to product recovery programs that would be cost-efficient and not compromise its new products' competitive positions. The quest for product recovery and reuse or recycle, therefore, emphasized the following:

- Recovery for reuse
- Remanufacture of high-end workstations for sale to secondary markets and internal or partnering organizations
- Material recycle to recoup value or processing costs and build demand in the recycled materials market

Recovery and refurbishment of parts for service applications began as early as 1987. With these activities, HP sought to improve costs and timely availability of service parts that would otherwise take months to procure as newly built items. The recovery organization also provided an opportunity to maintain supplies of older products' parts long after production had stopped. Customer sites and service representatives collectively supplied about 40% of the 9000 tons/year of HP equipment recovered this way. The remainder returned from internal business divisions and excess inventory. Of the total, 30% was either resold as components or reused in service; less than 1% was sent to landfill according to Dillon's 1997 study.

Material recycling was, therefore, the management option for nearly 70% of returned HP equipment and the most significant aspect of HP's responsibility policy. The company engaged suppliers, its internal recycling organization, its printer business, and resin and molded parts suppliers to build a high-quality product with reliable recycled content consisting of both postconsumer purchased material and internal process wastes. The supply of recycled plastic resin remained a quality and cost concern for HP throughout its efforts to develop recycled plastic parts with consistent color and strength. A measure of its success was its DeskJet 850 printer that achieved up to 25% recycled content. This product alone provided an opportunity to recycle up to 6 million pounds/year of plastic given full integration across this one product platform alone. HP also initiated a toner cartridge-recycling program at no cost to the customer, allowing HP to reuse some parts and materials in new cartridge manufacture, recycling of materials, and minimal disposal to landfill. It also prevented capture of HP cartridges for refill and resale on the "gray market" (Dillon, 1997).

*Supply Chain Level—and Beyond*

Hewlett-Packard is careful not to dictate environmental compliance procedures to its many suppliers, but its use of "TQRDC-E" scoring communicates the company's expectations on environmental policy, performance metrics (especially for materials of concern such as ozone-depleting substances), and improvement plans to suppliers and vendors at other stages in HP's supply chain. Besides assessing a potential supplier's technological leadership (*T*), consistent quality (*Q*), flexibility or responsiveness (*R*), timely turnaround for product delivery (*D*), and costs (*C*), HP rates the three environmental criteria with a numeric ranking system. HP managers must be careful to meet certain criteria when deriving environmental expectations of suppliers: An environmental issue must have worldwide relevance and criticality, should not be associated with existing legislation and specifications, must be economically measurable, and must be reviewed with suppliers before going into effect (Maxie, 1994).

**Considerations for Widespread Policy Initiatives**

Complexity of communicating compliance and technology needs to and from multinational organizations and their broad supply chain networks

Need for clear and measurable compliance requirements that apply within various stakeholders' scope of work

Avoidance of creating unfair advantages and trade restrictions to motivate participation

Involvement of industry stakeholders in shaping flexible and practical policies to foster innovation and benefits beyond compliance rather than submitting to prescriptive specifications with heavy administrative burdens

Bast et al., 1995

Experience with engaging its supply chain in a life-cycle responsibility program has prompted HP to develop suggestions for guiding new-product legislation and environmental policy on the global market in a way that is designed to benefit the environment, the public, and even the companies responding. The myriad environmental issues and their associated effects must be ranked using tools and scientific knowledge that incorporate the economics of avoiding those risks to find the most effective application of resources to solve the most critical problems. This approach should maximize the benefits of investing in environmental improvements for the greatest progress in the shortest time. Voluntary life-cycle management initiatives must engage and reward the private sector and comprehensively consider technology and resource changes that could generate conflicting effects across multiple media. Such initiatives should facilitate the most efficient and effective implementation of product responsibility legislation for environmental improvements (Bast et al., 1995).

*Follow-up*

Nearly 10 years after implementing its new-product life-cycle policy, HP's social and environmental sustainability report corroborates the company's continued commitment to product life-cycle management. HP will continue to emphasize recovery and recycle, and its policy puts a new focus on dematerialization—reducing material consumption while improving functionality of products. The company reports environmental performance metrics for manufacturing that no longer include recycling or reuse but retain emissions, hazardous and nonhazardous waste disposal (including "landfill diversion rate"), and compliance as measured in penalties for violations. Environmental features continue to be offered for HP products and services. Energy efficiency gains have been brought about through improvements in printer functionality and by developing energy-saving business models. HP produced the world's first inkjet printer to qualify for the German Blue Angel ecolabel, redesigned printer components, and created a new computer line, saving thousands of tons in materials used each year as it reduced costs for the printer line. Changes to inkjet cartridge packaging resulted in reduced plastic and cardboard use for retail sale, annually saving over 100,000 gal of hazardous materials, 100,000 Btu, and 2 million pounds of cardboard in addition to more than $2 million (Hewlett-Packard Company, 2002).

## Xerox Corporation

Xerox Corporation is another world-renowned electronics company committed to EPR. The company established its policy on "waste free factories and waste free products" in the early 1990s. Having experience with cost-saving recovery and refurbishment of used copy machines through a long-standing lease-based business model, Xerox was well positioned to meet the demands of European take-back legislation and a growing number of U.S. customers requesting postuse recovery. However, it required a new corporate vision to integrate life-cycle environmental thinking throughout its many equipment and supplies manufacturing organizations, its sales and marketing forces, service organizations, and beyond to its customers, vendors, and even indirectly its competitors around the world.

Xerox implemented this new vision with three focus areas:

- New asset recovery management program

- DfE standards and new tools to support DfE
- Engaging suppliers in the product life-cycle management process

These efforts were coordinated through Xerox's benchmark environmental business machine, the DC 265 line, in a test of the economics and business opportunities that were anticipated from a life-cycle responsibility product strategy.

*Asset Recovery*

Xerox's *asset recovery management* program was redesigned in accord with the company's new policy and incorporated in the early stages of product development. According to Xerox managers in 1993, the company's objectives were to implement an integrated recycle process in which end-of-life products were managed through a hierarchy of reuse, remanufacture, and recycle options and to disseminate the design-for-recovery principles and tools among product engineers. The new asset management strategy allowed recovery for reuses and recycle to begin upon product launch, instead of waiting to accumulate used equipment from the market. The new program employed concurrent development of new-build and remanufacturing lines during the concept and design phases. This approach minimized storage costs by avoiding large inventories of used machines and provided immediate access to the still-functional parts and components being returned to support the continued production of new machines. Thus, a new approach to asset recovery set the stage for greater life-cycle management within early product design (Berko-Boateng et al., 1993).

**Components of Xerox's Internal "Policy Deployment Process" for Asset Recovery Management**

An index to compare recommended materials' environmental characteristics

Codes that indicated remanufacturing procedures and recycling or disposition instructions for all parts used in Xerox machines

Design model to inform engineers of tradeoffs associated with component and material changes

Cost effects of recycling, remanufacturing, or reuse decisions

Assessment of the recovery viability for returned machines

Guidelines on existing regulations and customer requirements, preferred materials, and design features such as attachments, color, and accessibility.

Berko-Boateng et al., 1993

The *integrated recycle process* communicated Xerox's priorities for returning copier products, supplies such as toner cartridges, components from serviced machines and customer-replaceable units, and the materials these items were composed of:

- Direct reuse or conversion to other product models of machines and parts was preferred because such activity required the least in additional resources and processing yet retained the most value in the returned machine.
- Sales to secondary markets would also require little additional expenditure of time and resources but would bring lower revenues to the company.
- Parts repair and/or reuse would remove items with the greatest value but result in lost opportunities for other portions of the stripped machine.
- Equipment and component remanufacture would require disassembly, cleaning, repairing, and qualifying to a "like new" standard but could take the place of its newly built equivalent on the market while reducing the consumption of new resources and the associated processing costs and effects.
- Finally, used products could be disassembled and their materials recycled for use in Xerox products or other markets, in which case product functionality and value would be lost, but this option would be preferable to landfilling.

The integrated recycle process allowed the greatest value to be recouped from returning machines, reduced the need to purchase extra service parts, and provided ready-made, quality components that could be installed in new-build machines to reduce their production costs.

*Design for Environment*

According to Xerox in 1995, the DC 265 product line was a "clean sheet" design of a multifunction machine, its peripherals, and supplies to satisfy environmental goals encompassing its manufacture, use, and end of life. These goals were checked at each phase of the DC 265's development. They included maximum use of recovered parts and recycled materials; reliable use of recycled paper in the machine; minimum noise, volatile emissions, and energy consumption; clean and efficient toner containers; and recycled and reduced packaging. Each aspect supported Xerox's "zero waste to landfill" vision.

The goals reflected the "voice of the customer," product take-back regulations, safety laws, and major ecolabel requirements. Prototypes were tested during design to ensure that asset recovery and environmental strategies would be enabled; compliance with hazardous material and emissions specifications, the recovery codes, and plans to improve postconsumer materials use was verified within Xerox and among its suppliers. Xerox developed a diagnostic tool to determine the life and performance potential of some returned electrical components, giving decision makers additional information with which to predict the environmental and economic benefits of their recovery options (Azar et al., 1995).

Added to the assessment tools and processes associated with the asset recovery management strategy, the DfE approach provided product and supplies engineers with the mechanisms by which to incorporate life-cycle responsibility in the DC 265 product and future products with both economic and environmental success. Some of the success stories included reusable recyclable toner containers for many product lines and modular design that greatly reduced the costs of product assembly and allowed for functional fast-wearing parts to be incorporated together into "customer replaceable units," reducing the likelihood of scrapping whole machines and reducing service time.

*Supplier Engagement*

The Xerox vision of "zero to landfill" applied not only to its products but became integral to its offices, manufacturing floors, distribution lines, and beyond. Performance on solid waste and recycling and on energy and water consumption became as important in corporate cafeterias as it was on the production lines. Internal and external suppliers' environmental performance, especially on compliance, was checked with a self-audit.

The DC 265 development process provided the infrastructure for suppliers' involvement in environmental design aspects and postuse recovery operations. Project managers "sold" the potential benefits in reduced raw materials requirements (or at least a long-term account with Xerox) to the supplier organizations. According to members of the DC 265 project team in 1999, slow acceptance and implementation of the supplier engagement effort delayed the start of the remanufacturing operation. Even customers were included in the message that their sites could aspire to be waste-free, with Xerox's products to help.

One important life-cycle entity to the product responsibility approach, usually external to a manufacturing corporation, is the recycling organization and its capabilities. Xerox studied its product's recycle potential and current recycling rates in 1994. It then engaged a consultant to improve its recycling program for scrap metal, waste toner, and PC/ABS (polycarbonate and acrylonitrite butadiene styrene) plastic. The resulting recycling program provided 24 different categories of materials for recycle, maintained recycled materials collection through vendors, and generated profits while drastically reducing the amount of waste being disposed. The more sophisticated separation program also benefited the recycling vendors by providing higher quality material stocks with less opportunity for contamination by incompatible materials.

**Xerox Improves Its Recycling**

Assessed technical feasibility of material separation given that components were also being recovered for reuse

Identified recyclers to buy the separated materials

Performed a cost–benefit analysis

Xerox saw value in communicating its policy on environmental responsibility to groups not directly affected by its products. It used Earth

Day celebrations for employees and members, partnerships with organizations and municipalities where sites were located, and media releases describing millions of dollars saved as well as the benefits of being a "good corporate citizen" while extending its responsibility, and the responsibility of its supply chain, for its products' environmental effects.

*Follow-up*

Xerox's 1999 *Environment, Health and Safety Progress Report* maintains the waste-free concept for products, supplies, and company operations in general. Nearly 150 million pounds of equipment and parts were reused or recycled (90% of Xerox equipment can now be remanufactured), with the majority being recycled. The process used is depicted in Figure 3.4. Fuji Xerox, a partner operating in Asia and Australia, achieved a 20% part reuse rate in the previous year and was aiming for 50% by 2000. Toner cartridge returns (which include toner and the photoreceptor belt) passed 60% in 1998 for the United States (or 1.2 million cartridges), and almost 3 million toner bottles were returned.

In all returns of supplies, Xerox diverted almost 9.5 million pounds of material from the landfill. Xerox increased the number of products with ecolabel certifications, and 10 products met German Blue Angel requirements. Total nonhazardous recycling rates for the company have increased by almost 20% since 1993. Yearly hazardous waste generation at one U.S. manufacturing site reduced from about 2500 to 1500 tons in the same time frame. Ongoing savings from reuse and reduction under

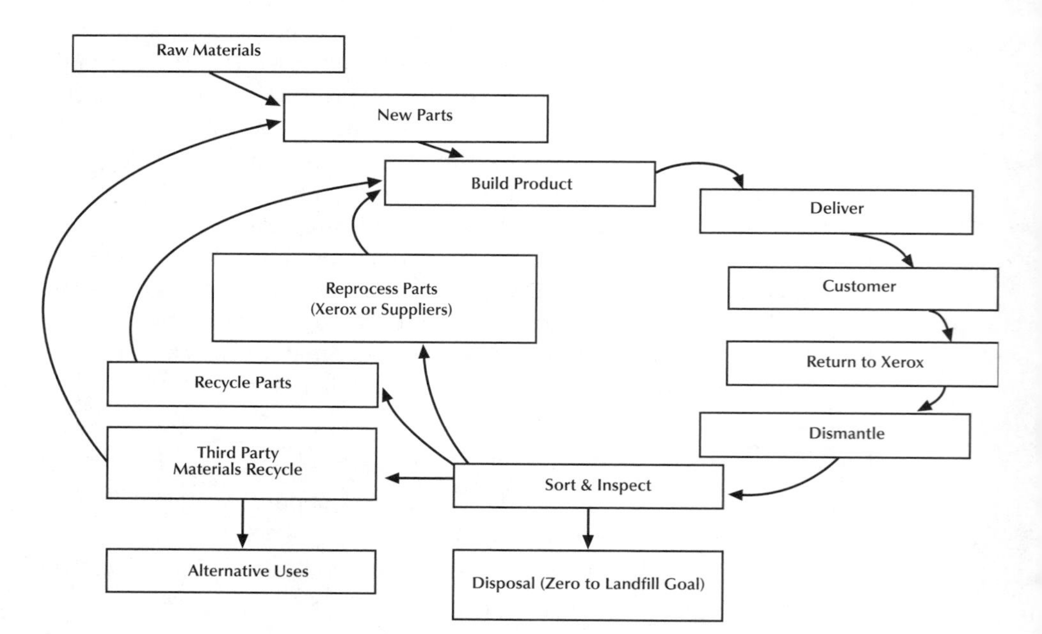


**Figure 3.4 Xerox remanufacture and parts reuse/recycle management process.** (Adapted from Xerox 2002 Environment, Health, and Safety Progress Report, © 2002, reprinted with permission.)

the waste-free factory concept were calculated at $45 million (Xerox Corporation, 1999).

In 2002, Xerox announced that a decade of EPR activities had resulted in over $2 billion savings, with the equivalent of 1.8 million machines reused or recycled (Environmental News Network, 2002). New-product designs for the U.S. market incorporate the profitable environmental features developed for the DC 265, but recovery programs for some lesser-valued parts and supplies are an expense now avoided unless specifically requested for customer accounts. Xerox encourages customers to recycle some toner containers in their local recycling program rather than returning them (Xerox Green World Alliance website, www.xerox.com/greenworld); customer rebates for returned supplies have been eliminated for many products. Promotional materials for the DC family of products on Xerox's U.S. corporate website in 2002 showcase the increased speed and multiple office product functions offered. The brochure notes that the DC products are designed to be energy and paper efficient and produce little waste, that customer-replaceable units are recovered by Xerox for recycling, and that models can contain both new and recycled components. Xerox's continued effort to enacting product life-cycle responsibility through voluntary initiatives and business incentives transferred throughout the product chain in the United States has been supported by its highly competitive position in the market and its documented cost savings and measurable business benefits.

## SC Johnson and Son, Inc.

SC Johnson and Son, makers of consumer chemical products worldwide, faced widespread public concern over the use and recycle of aerosol spray cans in the late 1980s and early 1990s. In response, it included recyclers and customers in its campaign to improve management of its packaging after use. The company had a reputation as a manufacturer of safe reliable products, but many of its customers were reluctant to buy, use, and dispose of aerosol cans because of beliefs that the sprays were propelled by ozone-depleting chlorofluorocarbons (CFCs) and that empty cans sent to recycling created an explosion or fire hazard due to residual contents under pressure. As early as 1977, USEPA had banned use of CFCs from nonessential applications, specifically in aerosols (Train, 1976), and SC Johnson had voluntarily eliminated CFCs in aerosols 2 years earlier (SC Johnson, 2002).

Nonetheless, in 1991, only one municipal recycling program in the United States allowed aerosol cans, according to Tom Benson of SC Johnson Wax in 1997. The company studied the safety of crushing empty steel aerosol cans. It also worked with the Steel Recycling Institute to investigate the risks of recycling aerosol cans as opposed to other containers. The results demonstrated that aerosol cans, designed to be completely emptied of both propellant and product, could be safely crushed and recycled. This information was communicated to community recyclers. SC Johnson and the Steel Recycling Institute also undertook their own consumer awareness campaigns to reassure the public of the safety of using and recycling spray cans, and they promoted the environmental benefits of doing so. SC Johnson also worked with WMX, at the time the world's largest waste management corporation, to acquire accurate information from the aerosol industry and to develop can recycling programs among WMX services (Eastern Research Group, 1996).

Today, SC Johnson calls itself "a champion of environmental stewardship." It has developed a classification program for the environment-preferred materials that it uses in its detergents, propellants, insecticides, and resins. It is employing USEPA's PBT (environmental persistence, bioconcentration potential, and aquatic toxicity) risk model to test current and future product ingredients.

SC Johnson has achieved a 90+% recycling rate on production waste at its largest U.S. manufacturing facility, amounting to over 17 million pounds of waste in 2001. To meet this level of recycling, it developed a partnership with Goodwill Industries' employment program. Goodwill acted as consultants to locate secondary markets and develop internal recycling innovation. Other environmental savings achieved over the last 10 years include over 460 million pounds of waste eliminated or diverted through efficient designs and recycling, avoiding annual costs of $125 million. The company has reduced its virgin packaging use by almost 30% since 1990 and has eliminated 99% of polyvinyl chloride (PVC) used in its bottles (SC Johnson, 2002).

## Removing Barriers

The case studies indicate where long-standing industry barriers needed to be addressed to facilitate corporate EPR activities in the United States:

- Increasing supplier involvement in the product design process
- Building manufacturers' trust in the recycled material market through direct influence and demand development
- Dealing with public and government reluctance to handle hazardous materials

Hewlett-Packard and Xerox influenced supplier cooperation in providing and accepting environmental specifications and "trade secrets" of designs and processes, in part through their business reputations and the size of their potential accounts. They also spent time, however, considering suppliers' roles and developing appropriate systems of measurement that would communicate expectations and progress in environmental performance. Anticipating the U.S. markets for environment-responsible product features and the business benefits of being early adopters of such features also allowed both companies to plan their strategies and how suppliers would fit, instead of simply reacting at short notice. Both companies had already been subjected to product regulations in Europe, experience that helped them to prepare corporate policies that would address perceived shortcomings of the producer responsibility mandates and provide greater business opportunities under the U.S. model.

SC Johnson used information sharing and demonstrated success with a technological problem to engage downstream members of the industrial system in participating in improving life-cycle management options for its waste products. The steel recycling market also provided incentives for municipalities and recyclers to collect the millions of aerosol cans being produced by U.S. industry, once fears about safety were allayed. Sharing information in unthreatening ways between manufacturers and with upstream and downstream businesses in conjunction with demonstrable cost savings or increased income increases the likelihood that shared responsibility programs will succeed in the United States.

When HP and Xerox initiated product recovery for recycle, supplies of recycled materials, especially plastics, were variable and often of questionable quality for strict manufacturing specifications. Both companies contributed to the demand for parts made with recycled materials and simultaneously strengthened the secondary materials market with their efforts to improve collection and ease of separation. SC Johnson also faced a recycling-related barrier—its used products were essentially banned from collection facilities. The company's response was to help to develop and test recycling methods and to improve the recycling community's perception of its products' safety and recycle potential. These experiences suggest that the willingness of manufacturing industries to contribute to both aspects of the recycled materials market—building demand and generating reliable supply—improves the viability of recycling as a life-cycle management option. State and federal regulations continue to limit options for the handling and disposal of hazardous materials, but all three manufacturers reduced their use of hazardous materials significantly.

## Considerations for Effective EPR

The success of cooperative policy tools and corporate activities that have been employed in product life-cycle responsibility programs in the United States relies largely on two forces outside the control of the industry participants:

- Consumer participation
- Regulatory foundations affecting product markets

The programs will fail without consumer acceptance of product environmental features, end-of-life recovery activities, and the way these activities affect consumers' own behavior. Some businesses have provided economic incentives such as reduced prices, rebates, or prepaid takeback to encourage consumers to buy, use, and return their environment-friendly products. For example, Xerox's marketing division was able to promote sales of the DC 265 product and supplies line because these products' prices were equal to or lower than other models. At the same time product rebates and prepaid return channels were provided. Xerox had to weigh the cost to the company of providing incentives for the millions of unit sales they made against the cost to the company if postuse assets critical to its program were not returned or if customers became less satisfied with the brand if no incentives were offered. It also had to consider the consumer's own costs of responding to incentives. The space to store packaging and return information while the product is being used and the time spent preparing a used product for return may not always be justified by the size of a rebate or free shipping. Similarly, buying a machine with energy conservation features may result in increased downtime at the customer site, or a machine using recycled paper may have a reputation for malfunctions—negatively influencing a customer's decision to buy the environment-friendly product.

Some corporations could use the costs of nonparticipation to their advantage when marketing an environment-responsible product. The costs of energy and supplies can be calculated for a given product's lifetime and compared with a competing product without these features. This information can be communicated to the consumer to demonstrate the benefits of the environment-responsible product. The cost of waste disposal is a growing concern for consumers and industry sectors alike and is well documented for given regions and waste types. An assessment of savings on customers' waste disposal costs achieved when returning products through take-back programs, especially for large customers using and discarding many products regularly, can provide a strong incentive for participation in EPR efforts.

Existing regulations on waste management may preclude some products from recovery and reuse or recycle. RCRA sets stringent rules on who can be authorized to handle hazardous waste and how such waste may be ultimately disposed of. However, some electronics, such as CRTs, are listed as hazardous waste because they contain heavy metals. Therefore electronics at end of life, characterized as solid waste from nonresidential large generators (such as schools and facilities that already report other hazardous wastes), may not always be collected and stored with the intent of recycling. This discourages entrepreneurs from developing innovative technologies and applications for the growing volumes of electronics waste. The confusion over permitted activities and the efforts of USEPA and industry representatives to amend the RCRA rules to address concerns over waste from electronics products, indicate the necessity of developing clear regulations with easily obtained compliance measures to address evolving environmental issues.

The Basel Convention and treaty to monitor transport of wastes across boundaries, ratified by over 150 countries in 1989, has also been a concern for manufacturers worldwide when developing private take-back and recycling programs. The Basel Convention was intended to stop dumping of industrial wastes in less developed countries with lower environmental standards, but it also bars end-of-life products in one country from being sent to recycling operations in another country (Basel Action Network, 2002). The United States was among several industrialized nations opposed to the ban and does not recognize the Basel Convention as binding. Multinational companies with U.S. operations may still be bound by the Basel ban from creating end-of-life management operations in some of their foreign business locations or from increasing returns by collecting end-of-life products from other countries to improve economies of scale for recovery facilities.

## Consequences of Implementing EPR in the United States

To the extent that corporations and government organizations have promoted EPR among stakeholders, businesses and society have benefited. The improvements in costs, profits, and environmental and safety performance for all stakeholders may not be attributable to the voluntary initiatives and product-based policies alone, but the effects of many different regulatory, market, and cultural forces put efforts to manage a greater extent of product life cycles in the forefront of corporate activities. Economies gained through more efficient designs, avoided disposal and liability costs, reduced materials inputs, and internal recycling loops have been demonstrated for committed proactive companies. Capture of new and broader markets with environmental products and programs can be expected as environmental awareness spreads through the population. Shared responsibility efforts seem to be supporting reductions in raw materials use for reasons of cost during the manufacturing stages, although, absent technological advances, increasingly scarce and dispersed resources around the globe require greater inputs of energy to obtain in large quantities and result in greater generation of wastes. Trends in the consumption of land, water, and minerals are also likely to continue increasing in the United States and around the world as developing regions improve their technological capabilities and purchasing power and the U.S. economy benefits, unless environmental concerns become critical and force drastic lifestyle changes or technology advances. Still, as marketed products provide more and more environmental features to their buyers, the "reduce, reuse and recycle" message may become more of a fixture in American consumerism.

Energy consumption and process emissions have been scrutinized as their costs to the manufacturer and multiple social effects have become more apparent. In many cases the design and operations changes that bring about an environment-responsible product have achieved significant energy conservation and pollution reductions for the participants. Responsibility policies imposed by corporate organizations have often simultaneously emphasized responsibility for safety and health of their workers and end users, even as design changes eliminating or reducing hazardous materials or materials of concern, material mass, and product emissions have improved the safety of product manufacture, assembly, transport, use, and disposal.

A question to be addressed, however, is the ultimate environmental consequences of collecting widely dispersed used products for recovery rather than promoting local recycling and responsible disposal. The collection process imposes new demands for fuel and transport facilities, especially when collection, cleaning and remanufacturing, or recycling facilities are not co-located. The result may be greater consumption of fossil fuels, increased vehicle emissions, and more traffic congestion. Decision makers considering product take-back programs must have analytical tools at their disposal that can assimilate the environmental effects imposed as well as effects averted by expected recovery rates and management plans for end-of-life products. They must be willing to base their analysis on identified priorities for their company and the environment. Life-cycle analysis tools are increasing in scope and ability to reconcile with economic measures, and they may help to organize a corporation's environmental goals to provide the most meaningful benefits.

In the United States, barriers created by corporate, governmental, economic, and social conditions will likely continue to stand if extended responsibility is viewed as a model for an elite few corporate organizations and as life-cycle responsibility activities are seen as an added expense unjustified by the current market or regulatory arena. Nonetheless, examples of practical application of EPR strategies and life-cycle programs that combine aspects of design, materials, product manufacture, sales, use, and end of life—under influences both internal and

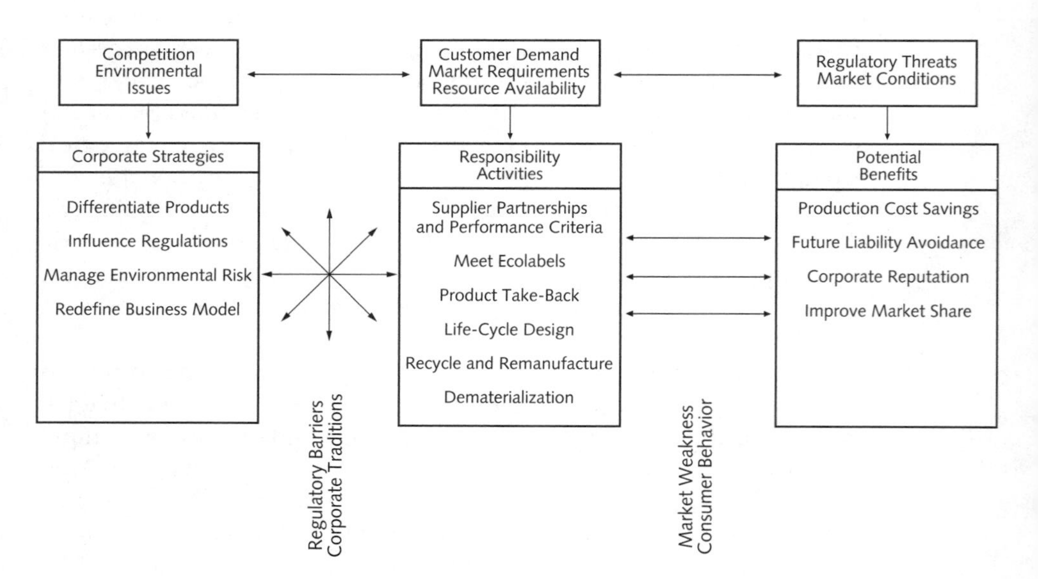


**Figure 3.5 Influences on U.S. corporate responsibility.**

external to the industrial system—have demonstrated that such conflicts are worth addressing with continued attention to extended responsibility projects (Figure 3.5). Positive examples of life-cycle responsibility initiatives serve as incentives to continue investigating collaborative environmental performance throughout U.S. industry.

## Lessons Learned, Experience Gained in the United States

### *Mandated Producer Responsibility*

Some conclusions about the accomplishments and shortcomings of the U.S. EPR point of view can be drawn from assessments of European producer responsibility. The producer mandates implemented in some countries focused attention and effort, by definition, on diverting post-consumer waste from the landfill, making upstream design changes and downstream collaboration necessary. The impetus for involving parts and materials suppliers, service organizations, and recyclers in the changes to enable end-of-life management was placed with the product manufacturers and distributors, without the help of regulatory incentives or threats aimed at the other life-cycle stages. An effective program sufficient to induce the necessary improvements in the product life-cycle performance probably must go beyond applying take-back mandates and recycling quotas only to producers. Some have argued, however, that producer responsibility creates incentives exactly where needed to leverage action (Lifset, 1999).

The costs of funding a separate collection program as required by some countries and the EU for packaging have been high, because of the monopolistic behavior of early collection services and the costs of reporting recycling rates according to multiple national requirements; the dual collection effort was seen as needless in areas where recycling programs were already operating. Another concern over separate product recovery programs conducted by third parties is that manufacturers pay for the recovery but do not benefit from the additional profits of recycle and thus have few incentives to change their design for better recovery and recycle potential (Palmer and Walls, 2002). Corporations have complained that basing performance on common quotas restricts their opportunities to gain competitive advantage: They can provide more environmental features than their competitors, they say, in addition to the product take-back, but the mandates do not recognize such features.

Consumer participation in the mandated take-back programs was higher than expected but resulted in a flood of secondary materials throughout Europe with little capacity for processing. Market prices dropped too low to entice businesses to develop new technology (Dragicevic, 1995). Some producer take-back programs, such as Germany's, were begun on a voluntary basis but became mandatory when little progress was made among manufacturers to incorporate end-of-life management. U.S. manufacturers are considering all of these lessons from their European markets as they develop their own approaches to life-cycle responsibility at home.

*Shared Responsibility for Product Life Cycles*

In aspiring to realize a set of truly shared responsibility activities that fairly distributes costs among life-cycle parties with the greatest contributions to a product's environmental effects, the U.S. approach has its own successes and problems. An effective product recovery and reuse or recycle initiative requires planning during the product development process when materials and design features can be influenced by environmental objectives. Recovery programs assembled after a product is on the market risk higher management costs associated with processing parts and materials not designed for separation, resulting in a less significant degree of recovery.

Although manufacturers often house the designers who devised the product specifications, they may out-source some parts to suppliers with greater expertise or lower costs of production than they could provide internally. With out-sourcing comes a loss of autonomy over details that could prove critical in managing the product at end of life. Similarly, manufacturers may not have much reason to research recycling markets and available technology as they engineer their product, but the production decisions made could greatly affect the viability of recycling programs later. Therefore, aspects of take-back and reuse or recycle of products, parts, and materials may best be handled, or at least consulted on, by the members of the supply chain who will manage them.

Finally, U.S. manufacturers have learned that despite desires to be good corporate citizens or socially responsible companies, sustainable business benefits are the ultimate justification for their life-cycle responsibility activities. The extra costs of collecting and processing some used products have proved too high to be maintained by any resulting revenues or cost savings, especially given the high cost of labor in the United States and the relatively low disposal fees for nonhazardous wastes. The "low hanging fruit"—projects easily identified and enabling small changes and investments in return for big environmental and economic paybacks—may have already been plucked bare by the manufacturing industry looking for ways to be environment friendly. More intense efforts to improve environmental performance must be balanced with other resource demands for the company, and they must meet more stringent business standards for approval, such as a faster investment turnaround, by cautious corporate leaders. It is hoped, therefore, that the intense environmental programs that will be necessary to reduce the strains of consumer demands on materials, energy, and ecosystem can be allocated task by task to the upstream and downstream members of the manufacturing system. Innovative ideas that will transform product concepts without harming the corporate bottom line can form with improved relationships among suppliers, manufacturers, distributors, customers, the public, recyclers, waste management, and government organizations.

## Conclusion

Extended product responsibility policies around the world are new and evolving. U.S. industry now struggles with the paradox that while it resists government intervention on manufacturing products and processes, some regulatory influence may even the playing field by providing justification for the expenses of life-cycle responsibility activities such as major technology changes or the infrastructure to collect and process returned products. In a market where environmental features succeed based on consumer acceptance, these features must be provided at no extra cost to the customer, and they must meet or exceed their competitor's features. A lack of clear incentives that can be communicated through all stages and contributors to the product system may prevent broad implementation of shared responsibility, but incentives directed largely at the producer have proven effective for the U.S. consumer products industry.

Therefore, the question of choosing among voluntary or mandatory approaches, producer-led or shared programs, is best answered by the constraints, conditions, and objectives for a given product, in a given environment, with a given set of life-cycle participants. This chapter has characterized the boundaries and general costs and benefits associated with two major extended responsibility strategies on the market: mandated producer responsibility and shared product responsibility. Further monitoring of responsibility activities worldwide as they develop will continue to foster understanding of the circumstances under which product policies may drive environment-responsible, perhaps even sustainable, manufacturing.

### Acknowledgments

Research was supported by the National Science Foundation through the Environmental Manufacturing Management program at Clarkson University, under grant DGE 9870646 of the Integrative Graduate Education Research and Training (IGERT) program, Wyn Jennings and Larry Goldberg, program directors.

### Suggested Reading

Davis, G., C. Wilt, P. Dillon, and B. Fishbein. 1997. Extended product responsibility: A new principle for product oriented pollution prevention. Center for Clean Products and Clean Technologies, University of Tennessee at Knoxville, and the US EPA Office of Solid Waste, June. Online at http://eerc.ra.utk.edu/ccpct/index.html.

Fishbein, B., J. Ehrenfeld, and J. Young. 2000. *EPR: A Materials Policy for the 21st Century*. INFORM, Inc. www.informinc.org/eprbook.htm.

Lifset, R. J. 1993. Take it back: Extended producer responsibility as a form of incentive based environmental policy. *Journal of Resource Management and Technology*, **21**(4):163–175.

Reinhardt, F. 2000. *Down to Earth: Applying Business Principles to Environmental Management*. Harvard Business School Press: Boston.

*Research and Technical Organizations*

International Institute for Industrial Environmental Economics at Lund University (Sweden) conducts applied research focused on "understanding and development of policies, strategies, and instruments to promote preventive approaches to environmental problems in society," http://www.lu.se/IIIEE/research/index.html.

Raymond Communications, Inc., covers state and international recycling issues and laws; some material requires a paid subscription, but many articles are accessible. The website also offers a free monthly recycling policy bulletin, http://www.raymond.com/index.html.

Resources for the Future, "think tank conducting independent research rooted in economics and social sciences on environmental and natural resource issues," http://www.rff.org/Default.htm.

Tellus Institute Business & Sustainability Group "provides strategic thinking and management tools to advance corporate sustainability," http://www.tellus.org/b&s/index.html.

*Applications of Life-Cycle Environmental Strategies in Industry*

Gungor, A. and S. M. Gupta. 1999. Issues in environmentally conscious manufacturing and product recovery: a survey. *Computers and Industrial Engineering*, **36**:811–853.

Ulhoi, J. 1997. Industry and the environment: A case study of cleaner technologies in selected European countries. *Journal of Engineering and Technology Management*, **14**:259–271.

## References

Allenby, B. 1999. EMS: A tool whose time has passed? *Green Business Letter*, May, p. 8.

Anderson, R. 2000. 2000 Ceres Report, Interface, Inc.; online at www.interfacesustainability.com/ceres.html.

Applebaum, A. 2002. Europe cracks down on e-waste. *IEEE Spectrum*, **39**(5):47–51.

Azar, J., V. Berko-Boateng, P. Calkins, E. DeJong, J. George, and H. Hilbert. 1995. Agent of change: Xerox design for environment program. Proceedings of the IEEE International Symposium on Electronics and Environment, May 1–3, Orlando, FL, pp. 51–61.

Barich, H. and P. Kotler. 1991. A framework for marketing image management. *Sloan Management Review*, **32**(2) Winter: 94–104.

Basel Action Network. 2002. The Basel ban—A triumph for global justice. Briefing Paper No. 1, March, Seattle, WA.

Bast, C., C. Johnson, T. Korpalski, and J. Vanderstraeten. 1995. Globalization of and guiding principles for environmental legislation and public policy development. Proceedings of the IEEE International Symposium on Electronics and Environment, May 1–3, Orlando, FL, pp. 220–224.

Bell, V. 1998. How manufacturers are responding to extended producer responsibility programs and how these programs can be made more effective. OECD Workshop on Extended and Shared Responsibility Products: Economic Efficiency/Environmental Effectiveness, December 1, Washington, DC.

Berko-Boateng, V., J. Azar, E. DeJong, and G. Yonder. 1993. Asset recycle management—A total approach to product design for the environment. Proceedings of the IEEE International Symposium on Electronics and Environment, May 10–12, Arlington, VA, pp. 19–31.

Breen, B. 1990. Selling it! The making of markets for recyclables. *Garbage*, **2**(6) (Nov/Dec):44–49.

Brouwers, W. C. J. and A. L. N. Stevels. Cost model for the end-of-life stage of electronics goods for consumers. Proceedings of the IEEE International Symposium on Electronics and the Environment, May 1–3 1995. Orlando, FL.

Burman, A. 1992. Policy letter 92-4 on procurement of environmentally sound and energy efficient products and services. Executive Office of the President, Office of Management and Budget. Online at www.arnet.gov/Library/OFPP/PolicyLetters/.

Can Manufacturers' Institute. 2002. Cans, a visual history. Online at www.cancentral.com/brochure/.

Commission of the European Communities. 2000. Proposal for a directive of the European Parliament and the Council on waste electrical and electronic equipment and on the restriction of the use of certain hazardous substances in electrical and electronic equipment. COM (2000)347 final, June 13, Brussels, Belgium.

Cotsworth, E. M. 1999. Take It Back! conference in Washington, DC. Extended product responsibility. USEPA Office of Solid Waste: Washington, DC.

Cutter Information Corp. 1998. Product Stewardship Advisor, February 1998. The progress of take-back legislation for electronic equipment around the world. Online at http://cutter.com/psa/psatabl.htm.

Darling, C. 1995. Do firms benefit from electing a voluntary approach to environmental compliance? Proceedings of the IEEE International Symposium on Electronics and the Environment, May 1–3, Orlando, FL, pp. 24–28.

Darnall, N., D. R. Gallagher, R. N. L. Andrews, and D. Amaral. 2000. Environmental Management Systems: Opportunities for Improved Environmental Business Strategy. *Environmental Quality Management*, **9**(3):1–9.

Davis, G., C. Wilt, P. Dillon, and B. Fishbein. 1997. Extended product responsibility: A new principle for product oriented pollution prevention. Center for Clean Products and Clean Technologies, University of Tennessee at Knoxville, and USEPA Office of Solid Waste, June.

Dillon, P. 1999. Recycling infrastructure for engineering thermoplastics: A supply chain analysis. Proceedings of the International Symposium on Electronics and the Environment, May 11–13, Danvers, MA.

Dillon, P. 1997. Product stewardship at Hewlett Packard Company. In *Extended Product Responsibility: A New Principle for Product Oriented Pollution Prevention*. Center for Clean Products and Clean Technologies, University of Tennessee at Knoxville and USEPA Office of Solid Waste, June, pp. 3.9–3.20.

Donnelly, J. 1990. Degradable plastics. *Garbage*, **2**(3)(May/June):42–47.

Dragicevic, M. 1995. German waste exports. *TED Case Studies*, **4**(2)(June): online at www.american.edu/TED/class/all.htm.

Driedger, R. 2001. From cradle to grave: Extended producer responsibility for household hazardous wastes in British Columbia. *Journal of Industrial Ecology*, **5**(2)(Spring):89–102.

Eastern Research Group, Inc. 1996. *President's Council on Sustainable Development/USEPA Proceedings of the Workshop on EPR*, October 21–22. Downloaded from www.whitehouse.gov/PCSD/Publications/ EPR.html#opening.

Environmental News Network. 2002. Xerox saved $2 billion through eco design and manufacturing. Wednesday, May 1. Online at http://enncom/news.

Ertel, J. 1994. Current technologies for the valorization of PCBs and electronic waste. Proceedings of the IEEE International Symposium on Electronics and Environment, May. Piscataway, NJ.

Field, B. 1997. *Environmental Economics, An Introduction*, 2nd ed. Irwin McGraw-Hill: New York.

*Financial Times*. 1997. European Commission to unveil proposals requiring car manufacturers to take back vehicles they sell in the European Union for recycling. *Financial Times*, London Edition, 970120, p. 21.

Fishbein, B. 1998. EPR: What does it mean? Where is it headed? *Pollution Prevention Review*, **8**(4)(October):43–55.

Goddard, H. 1993. The benefits and costs of alternative solid waste management policies: A critical analysis of current problems and proposed solutions with recommendations for public policy. *Balancing Economic Growth and Environmental Goals, A Policy Symposium*. Sponsored by the American Council for Capital Formation Center for Policy Research, September 29, Washington, DC.

Hanisch, C. 2000. Is extended producer responsibility effective? *Environmental Science & Technology*, **34**(7):170A–175A.

Harlan, J. 2000. Environmental Policies in the New Millennium: Incentive-Based Approaches to Environmental Management and Ecosystem Stewardship, A Conference Summary. World Resources Institute, April 3.

Harremoes, P., D. Gee, M. MacGarvin, A. Stirling, J. Keys, B. Wynne, and S. Guedes Vaz. 2001. Late lessons from early warnings: The precautionary principle 1896–2000. *Environmental Issue Report 22*. Office for Official Publications of the European Communities and the European Environmental Agency, Copenhagen. Online at http://reports.eea.eu.int/.

Harrison, K. 1999. Talking with the donkey: Cooperative approaches to environmental protection. *Journal of Industrial Ecology*, **2**(3):51–72.

Hermann, F., H. P. Urbach, and H. Wendschlag. 2001. Eco-labeling and the IT industry. Proceedings of the IEEE International Symposium on Electronics and Environment, May 7–9, Denver, pp. 1–3.

Hewlett-Packard Company. 2002. *Social and Environmental Sustainability Report*, July, Palo Alto, CA. Online at www.hp.com.

Hileman. 2002. Electronic waste. *Chemical & Engineering News*, **80**(26)(July 1):15–17.

History of Environmental Protection. 2001. Provided online by the German Federal Environmental Agency, accessed August 19, at www.umweltbundesamt.de.

King, A. A. and M. J. Lenox. 2000. Industry Self Regulation Without Sanctions: The Chemical Industry's Responsible Care Program. *Academy of Management Journal*, **43**(4):698–716.

Klassen, R. and C. McLaughlin. 1996. The effect of environmental management on firm performance. *Management Science*, **42**(8):1199–1214.

Kleiner, A. 1990a. Brundtland's legacy. *Garbage*, **2**(5):58–62.

Kleiner, A., 1990b. PR's changing face. *Garbage*, **2**(6):56–57.

Kline, B. 2000. *First Along the River: A Brief History of the U.S. Environmental Movement*, 2nd ed. Acada Books: San Francisco.

Lifset, R. 1999. Linking source reduction and extended producer responsibility. OECD Workshops on Extended Producer Responsibility and Waste Minimization, Paris, France, May 4–7.

Lifset, R. 1994. Extending producer responsibility in North America: Progress, pitfall, and prospects in the mid-1990s. Proceedings of the Symposium on Extended Producer Responsibility, November 14–15, Washington DC, pp. 37–55.

Lifset, R. J. 1993. Take It Back: Extended Producer Responsibility as a Form of Incentive Based Environmental Policy. *Journal of Resource Management and Technology*, **21**(4):163–175.

Lindquist, T. 1992. Mot ett förlängt producentansvar—analys av erfarenheter samt förslag (Toward an extended producer responsibility—analysis of experiences and proposals) for the Ministry of the Environment and Natural Resources, Lund, Sweden.

Lindsay, C. 2002. Product stewardship. Presented at the Electronics Products Recovery and Recycling Conference, National Safety Council, March 12–13, Washington, DC.

Maxie, E. 1994. Supplier performance and the environment. Proceedings of the IEEE International Symposium on Electronics and the Environment, May 2–4, San Francisco, pp. 323–327.

McCarthy, J. and M. Tiemann. No date. Solid Waste Disposal Act/Resource Conservation and Recovery Act. Congressional Research Service Report RL30022, distributed by the National Library for the Environment. Online at www.cnie.org.

O'Connell, S. and T. Brady. 2002. Product Material Declaration Program at Intel. Proceedings of the IEEE International Symposium on Electronics and Environment, May 6–9, San Francisco, pp. 113–116.

OECD Business and Advisory Committee. 1997. Shared product responsibility. *BIAC* Discussion Paper for the OECD Workshop on Extended Producer Responsibility, 2–4 December, Ottawa, Ontario, Canada.

Palmer, K. and M. Walls. 2002. The product stewardship movement; Understanding costs, effectiveness, and the role for policy. *RFF Report*, November. Resources for the Future.

Pohanish, R. and S. Greene. 2000. *Hazardous Chemical Safety Guide for the Plastics Industry*. McGraw-Hill: New York.

Raymond Communications. 2002. States, Congress Move on mercury. *Recycling Policy Newsbriefs Email Bulletin*, July 24.

Reinhardt, F. 1999. Bringing the environment down to earth. *Harvard Business Review*, July–August, pp. 149–157.

SC Johnson. 2002. Public Report 2002, Continuing the Legacy. Online at www.scjohnson.com.

Shapiro, K. and M. Stoughton. 1999. Making the business case for extended product responsibility. *Environmental Perspectives*, No. 14 (February), Tellus Institute.

State of New Hampshire General Court. 2000. *House Bill 1418-FN-Local on Mercury-containing Products*. The House Committee on Environment and Agriculture.

Stocum, A. 2001. Personal communication, March 16.

Sundin, E., M. Bjorkman, and N. Jacobsson. 2000. Analysis of service selling and design for remanufacturing. Proceedings of the IEEE International Symposium on Electronics and Environment, May 8–10, San Francisco, pp. 272–277.

Tietenberg, T. 1990. Economic instruments for environmental regulation. *Oxford Review of Economic Policy*, **6**(1):17–33. Reprinted in *Economics of the Environment*, 4th ed. R. N. Stavins, ed., Harvard University. W. W. Norton: New York.

Train, R. 1976. Statement at a public meeting on chlorofluorocarbons. USEPA, December 3. Online at www.epa.gov/history.

USEPA. 1998. Executive Order 13101—Greening the Government through Waste Prevention, Recycling and Federal Acquisition. *Federal Register*, **63**(179): 49641–49651. Online at www.epa.gov/fedreg/.

White, A., M. Stoughton, and L. Feng. 1999. *Servicizing: The Quiet Transition to Extended Product Responsibility*, Tellus Institute, May.

Xerox Corporation. 1999. *Environment, Health and Safety Progress Report: Toward Sustainable Growth*. Xerox Corp.: Webster, NY.

## CHAPTER 4

# Pollution Potential Approach to Environmental Optimization—The Case of CFC Replacements

THOMAS P. SEAGER AND THOMAS L. THEIS

## Summary

This chapter introduces a new environmental metric called *pollution potential*. It is a measure of the change in chemical composition of the environment. Unlike risk-based environmental metrics, pollution potential is not a biological measure of health or expected increased incidence of disease. It is expressed as the exergy per mole required to remove a pollutant from the environment in an ideal thermodynamic process; it is analogous to economic measures expressed in dollars per kilogram of treatment costs. The value of the pollution potential metric is that it evaluates all chemical pollution on a single scientific scale, allowing comparison of different environmental priorities.

Pollution potentials for chlorofluorocarbon (CFC) replacements are presented here to demonstrate how the pollution potential approach can be used to compare global warming and stratospheric ozone depletion effects, both of which are difficult to evaluate with conventional risk-based metrics. A pollution potential approach may help to balance different policy objectives, such as those expressed in the Montreal Protocol (which limits chlorinated fluorocarbon production) and the Kyoto Protocol (which proposes to limit release of greenhouse gases). Some of the chemicals, called *hydro*chlorofluorocarbons, prohibited by the Montreal Protocol have global warming characteristics that are more favorable than the unregulated alternatives, from the perspective of the Kyoto Protocol. Pollution potential provides a rationale for ranking the trade-offs.

One of the most important industrial applications involving these trade-offs is in refrigeration, both in the compressor working fluids and the foam insulation. The last section of this chapter shows how the pollution potential approach can be applied in life-cycle assessment and compares pollution potential with total equivalent warming index (TEWI) in the design of foam freezer insulation panels. It turns out that the optimal design depends on which metrics are considered.

## Introduction

The most highly developed scientific methods for quantifying sustainability are found in economics and thermodynamics. Economists and physicists can assess the potential of two systems or technologies either to generate profits or to perform useful work. They can express a preference for one over another based on a single number representing something tangible (and perhaps credible, albeit narrow) such as dollars or joules. Environmental engineers, scientists, or policy makers, on the other hand, must juggle multiple measures of environmental quality, ecosystem health, or social justice. **Pollution potential** is a broad-based environmental metric that raises consideration of environmental criteria to the same level of decision making at which economic or thermodynamic criteria reside. Unlike human and toxicological risk assessment, pollution potential is not tied to a particular ecological or human health end point. As a result, it may be used advantageously in instances where the specific or cumulative hazards associated with pollutants are unknown or have yet to be manifest at detectable scales. (See Chapter 1 for a more complete description of the multiple dimensions of sustainability.)

The usual task of the environmental engineer is to convert one type of pollution (e.g., sewage) into another (e.g., sludge and carbon dioxide) that is either handled more economically, may be sequestered from the environment more reliably, or may be regarded less harmful (even if not entirely benign). The life-cycle perspective shows how classic end-of-pipe treatment approaches can be interpreted as merely displacing the adverse consequences of pollution to a different medium, type of impact, location, or time, making it difficult to compare incommensurate environmental criteria. Consequently, engineers are called upon to make trade-offs involving the mitigation of one type of adverse effect in favor of the potential exacerbation of others. Figure 4.1 shows the multiple dimensions of sustainability and how several different possible metrics might be employed to judge the efficacy of any design decision; this is only a partial list.

To overcome the difficulty of managing so many different kinds of information, many manufacturing companies have developed empirical metrics to gauge overall improvements in environmental performance. Typically, these measures apply subjective value judgments regarding the importance of various characteristics to allow comparison of technologies, materials alternatives, or material and energetic trade-offs. These methods probably fail, however, to capture future risks that have yet to be manifested in the environment or realized on a detectable scale. They may be too narrowly focused (e.g., on acute human toxicity), rely on controversial monetization schemes (e.g., economic externalities), fail to capture the disparate qualitative characteristics of various substances (e.g., materials or energy intensity indices), or be applicable only to a narrow set of products, industries, or a single company.

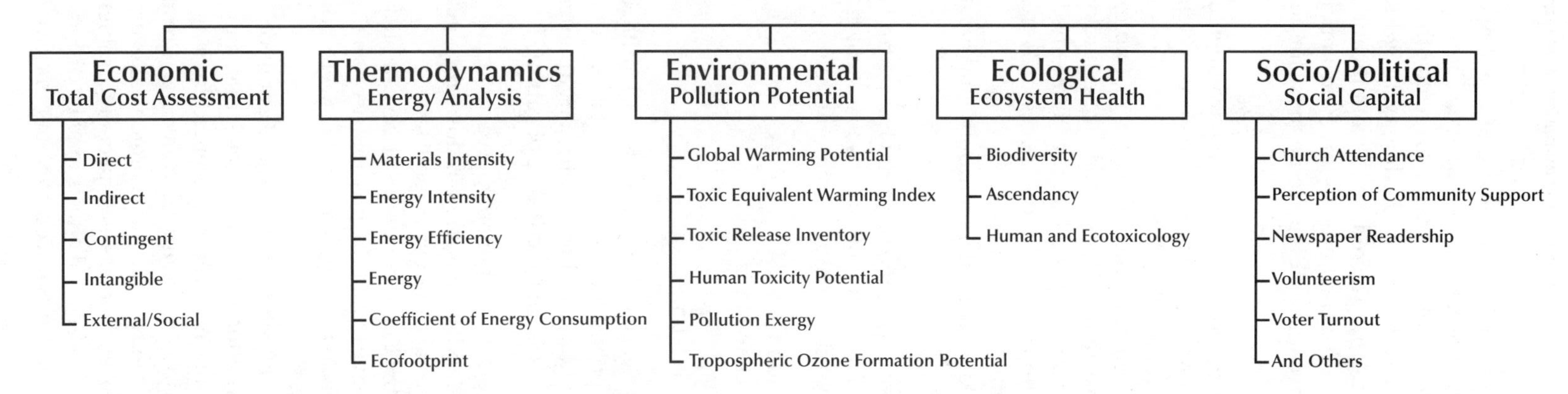


**Figure 4.1 Multiple dimensions and metrics of sustainability.** Pollution potential is a broad-based environmental metric that raises consideration of environmental criteria to the same level of decision making at which economic or thermodynamic criteria reside. Unlike human and toxicological risk assessment, pollution potential is not tied to any particular ecological or human health end point. As a result, it may be used advantageously in instances where the specific or cumulative hazards associated with pollutants are unknown or have yet to be manifest at detectable scales. (See Chapter 1 for a more complete description of the multiple dimensions of sustainability.)

Consequently, environmental management and policy objectives have been chiefly reactionary rather than proactive. The solutions promulgated to ameliorate one problem sometimes create others.

A single scientific assessment methodology by which all kinds of pollution problems may be ranked could significantly improve the environmental decision-making process or even anticipate pollution problems *before* they become evident. To be most helpful, a successful methodology should:

- Facilitate a broad-based approach to environmental management that lends itself to optimization of industrial systems for minimal total adverse environmental effects.
- Introduce new regulatory flexibility, for example, by governing new chemicals created or demanded by technological advancements or by allowing cross-pollutant or cross-medium emissions permit trading programs (without necessarily relaxing current health-based or other constraints).
- Be capable of ranking potential environmental hazards for further research.

One approach to comparing the potential human and ecological consequences of different management options is toxicological risk assessment (TRA). TRA is based on a four-step methodology:

1. Hazard identification
2. Pollutant source characterization
3. Exposure modeling
4. Dose–response modeling

Toxicological risk assessment provides the scientific basis for calculating human toxicity potential, and it represents a valuable and established tool for characterizing and estimating environmental risks. TRA also has several weaknesses in practical application:

- The methods of data gathering are time consuming and expensive, for example, in establishing sublethal dose–response relationships for multiple species at different trophic levels.
- The hazardous effect end points must be established in advance of study, which leaves open the possibility of environmental "surprises" or unintended consequences that were not considered in the original study.
- Indirect ecological effects (in which the effect upon one species ripples throughout the food web to species that may not have been directly exposed) are not well represented by single-species laboratory studies of dose–response relationships (Preston, 2002).
- Comparing TRA metrics for management alternatives with dramatically different end points is especially problematic. Adverse ecological and environmental effects may be dislocated temporally and spatially. Different species, ecosystems, or demographic groups

may be perceived as bearing increased risk inequitably. Establishing a preference for one alternative in comparison to another requires juxtaposing incommensurate decision criteria, complicating the communication of recommendations to inexpert audiences.

The *pollution potential* approach represents a more precautionary measure in that the end points are based upon only the presence and mobility of pollutant in the environment. Because pollution potential includes only source and fate characteristics, it may be determined with greater certainty than ecological or human health effects. This approach may be advantageous when end points are uncertain, unknown, or incommensurate, but it has the disadvantage of communicating no information about specific manifestations of risk (such as increased cancer incidence).

Pollution potential is based upon the concept of *exergy* (or entropy) and is a thermodynamic measure of changes in the chemical composition of the environment. It is analogous to the portion of Gibbs chemical potential that is due solely to mixing; that is, the free energy of mixing—albeit with environmental rather than standard reference states. To understand pollution potential, therefore, it is first necessary to explain some preliminary thermodynamic concepts.

## Exergy

Like the first law of thermodynamics, the second law (or entropy law) applies to all physical processes at any scale. However, entropy-based measures improve on mass and energy approaches because they can distinguish between high-quality and low-quality resources. While mass and energy are conserved, entropy is not. In effect, the economic system may be modeled as a thermodynamic process that captures high-quality resources from the environment, refines or upgrades them at the expense of fossil fuels, and then degrades (or consumes) them into low-quality wastes before redistributing them (for the vast majority of materials) back to the natural environment from which they were extracted. [See Ayres (1998) or Ayres et al. (2003) for a critical review of the role of thermodynamics in economic theory and further treatment of exergy as a factor of production.] Second-law approaches, therefore, are capable of creating a more complete resource and waste accounting and may be a useful scientific basis for broad-based approaches to environmental assessment such as those undertaken in industrial ecology. Although it has been argued that thermodynamic optimization of industrial processes can be justified on environmental or moral grounds (Wall, 1997), current applications are generally limited to industries where thermodynamic and economic criteria are closely aligned. Electric power generation, refrigeration, and distillation exemplify current applications (Bejan, 1996).

**Sustainability Implications of the Second Law**

Thermodynamics is the branch of physics concerned with practical applications and *useful work*. On a sensory level, the second law seems intuitive, even obvious. After all, water flows downhill and spilled paint will never find its way back into the can. Richard Feynman's test for irreversibility was simply that when you run the movie backwards, people laugh (von Baeyer, 1998). Yet the theoretical implications of the entropy law have almost impenetrable depth. Even experts sometimes misunderstand or misapply the basic principle that the entropy of a closed system can never decrease.

Partly this is because of the curious historical development of thermodynamic concepts, which at times have been the subject of intense controversy. [For interesting examples, see Mirowski (1992), Lindley (2001), or von Baeyer (1998).] Partly, this is because thermodynamics has been a difficult subject to teach. Many engineering students are required to take only one basic course (typically during their sophomore year) that barely scratches the surface of practical applications. Students in the natural or social sciences or liberal arts probably receive no instruction in thermodynamics. In either case, the necessity of many special thermodynamic variables of state (e.g., Gibbs free energy, Helmholtz energy, enthalpy, exergy, entropy) is rarely revealed, and none is directly observable (as is temperature or pressure). In fact, teaching thermodynamics has the effect of removing it from all experience, raising it to a level of such mathematical abstraction that it seems to have little to do with reality.

Among the common misconceptions regarding entropy is the notion that because the quality of energy (i.e., exergy) is always degraded in any real process, the resources lost in performance of useful work can never be replaced. On a cosmological scale, this may be true. However, Earth is *not* a closed system. We receive a constant stream of renewable exergy from the sun that far exceeds the needs of humankind. Most of this is reradiated to space in the form of infrared light (the energy wave that is absorbed by carbon dioxide and other gases and responsible for global warming). Capturing only a small portion of the solar insolation available at the surface of Earth and directing it toward human needs could hypothetically supplant the need for combustion of fossil fuels and leave a surfeit of exergy available to stoke further economic growth. The political, economic, and technical obstacles to realizing this vision may seem insurmountable, but it remains the ultimate application of the industrial ecology analog (see Chapter 1)—and perhaps cause for long-lasting optimism.

Many different metrics for second-law analysis incorporate entropy to describe the maximum work that may be performed by a thermodynamic system under ideal conditions: Gibbs free energy, Helmholtz energy, availability, exergy, essergy, and others. *Exergy* was first suggested partly because it translated readily into the many languages used in Europe (Rant, 1956). For practical purposes, the change in *chemical* exergy between two systems is identical to the change in Gibbs free energy:

$$\Delta B_{\text{chemical}} = \Delta G = \Delta H - T\,\Delta S \tag{4.1}$$

where $\Delta B_{\text{chemical}}$ represents change in chemical exergy, $\Delta G$ change in Gibbs free energy, $\Delta H$ change in enthalpy, $T$ absolute temperature, and $\Delta S$ change in entropy between two thermodynamic states.

Computation of Gibbs free energy of formation has been the subject of extensive research. Standard tables are available as supplements or appendixes to textbooks that describe the general principles in application to environmental problems (Stumm and Morgan, 1996). The standard reference state is most commonly taken to be a pure elemental zero-valence form (e.g., $O_{2\,(g)}$, $Cl_{2\,(g)}$, $Ag_{(metal)}$) and the free energies of formation of all other compounds may be computed by comparison to these. Unfortunately, the highly reactive state of some standard forms (such as chlorine and hydrogen) makes them an inappropriate reference state for exergetic analyses incorporating environmental considerations. Moreover, pure reference conditions do not represent the state in which chemical compounds commonly occur in the environment.

The principal advantage of exergy compared to Gibbs free energy is that exergy employs a system of environmental reference states that identifies the chemical characteristics of three different reference environments for computation of standard chemical exergies: the atmosphere, the ocean, and Earth's crust (Ahrendts, 1980). In many cases, the most oxidized form of an element serves as the appropriate reference state in each environment; however, consideration must also be given to the molar concentration of a compound in the specified environmental sink. This is of the utmost importance because it is well recognized that:

- Natural systems are not at thermodynamic equilibrium; rather, they must be approximated as ongoing quasi-steady-state reactions limited by both kinetic and energetic considerations.
- Pure substances released into the environment eventually dissipate to background concentration levels.

The environmental reference states typically employed in exergy calculations, therefore, imply a context that Gibbs free energy lacks.

Computation of chemical exergy at standard pressure and temperature is accomplished from standard Gibbs free energies in a simplified general formula:

$$B_P^0 = G^0 + \sum_i \left( \frac{n_1}{n_p} B_i^0 \right) \tag{4.2}$$

where $G^0$ is the free energy of formation (kJ/mol) of the compound from the elements, $n_i$ is the number of moles, and $B_i^0$ is the standard chemical exergy (kJ/mol) of the $i$th reactant required to form $n_p$ moles of the product compound; $n_i$ and $n_p$ are determined by the stochiometric balancing numbers of the appropriate chemical reaction. The formation of ammonia gas from nitrogen and hydrogen provides a simple example. The chemical reaction is represented by:

$$\tfrac{1}{2}N_{2\,(g)} + 1\tfrac{1}{2}H_{2\,(g)} \rightarrow NH_{3\,(g)} \tag{4.3}$$

$$-16.48 + \tfrac{1}{2}\,(0.72) + 1\tfrac{1}{2}\,(236.1) = 338\ \text{kJ/mol} \tag{4.4}$$

The Gibbs free energy of formation of ammonia gas is given as −16.48 kJ/mol. The standard chemical exergy, however, is found to be 338 kJ/mol by substituting thermodynamic data from Table 4.1 into equation (4.2).

Although the Gibbs free energy represents the ideal thermodynamic work required to synthesize pure ammonia from pure elements, the standard chemical exergy represents the maximum work that could be obtained under ideal conditions from pure ammonia gas, making exergy the thermodynamic variable of choice for studies related to sustainability.

## Configurational Exergy (or Exergy of Mixing)

Among the compelling hypotheses motivating exergy research is the intuitive notion that waste exergy emissions must somehow be correlated to quantitative measures of environmental or ecological impact. Additionally, a corollary hypothesis defines exergy consumption, efficiency, or source renewability as a measure of sustainability. However, these hypotheses have not yet been validated by analytical methods. Wherever they have been rigorously examined, some aspect or exception has demonstrated that the relationships are only approximate or inconsistent. In the former case, this is primarily because waste exergy may be embodied in many different forms; in the latter, it is because sustainability is a multidimensional concept that cannot be completely captured in a single measure.

Total waste exergy is a crude aggregated measure that is an improvement on mass- or energy-based measures of waste, but it may be significantly improved in its relation to environmental impacts by some refinement. Barring nuclear processes, all waste exergy may be generally characterized as being embodied in a material or energetic form, and within these broad categories, further specificity may be defined (see Figure 4.2). Waste exergy embodied in a material stream may be manifest in the chemical bond or in the unusual composition of industrial wastes; that is, the concentration gradient between concentrated pollutants and a dilute reference environment. Waste exergy embodied in various forms of energy may be characterized as thermal, physical

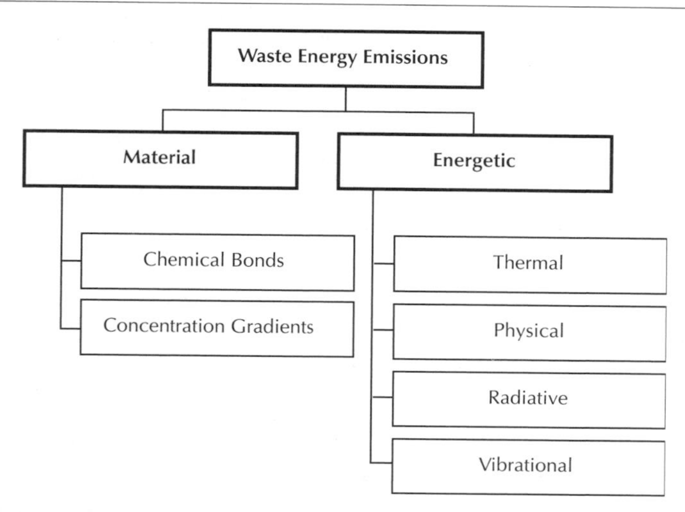


**Figure 4.2 Different forms of waste exergy.** (*Source:* Seager and Theis (2002); used with permission from Elsevier.)

(as in pressure–volume type work), radiative (including light), or vibrational (at the bulk molecular scale such as sound or shock waves—e.g., noise).

Different forms of exergy may be dissipated in the environment at vastly different rates. It may be postulated that different quantitative relationships exist between different forms of waste exergy and environmental impact. In particular, the popular notion of chemical pollution as an environmental impurity is best captured by that portion of chemical exergy that is solely due to concentration gradients and will be referred to as *exergy of mixing*, which is the principal focus of this section. This exergy may be very small compared to total chemical exergy, even negligible in comparison to the exergy related to the heat of combustion (e.g., embodied in covalent chemical bonds). They are released to the environment by different phenomena and create different effects, however.

Pollutant degradation (often by oxidation) takes place over widely varied time frames, depending on the pollutant under consideration. High total exergy alone is not a sufficient condition for chemical reaction in the environment (stability also depends on kinetics) and, therefore, does not suggest a *prima facie* case for disruptive environmental consequences. It can be said, however, that ecological systems have little evolutionary experience with substances that are only sparingly cycled or available in the natural environment, and that changes in the chemical environment may lead to disruption of biological functions or changes in evolutionary pressures. Particularly when the environmental abundance of rare (or entirely human-made) chemicals is increased, ecological systems become exposed to chemical boundary conditions that can change much faster than evolutionary responses. A loss of biological and ecological diversity (and concomitant functions) is the likely result. Chemical exergy of mass transfer (mixing), therefore, may be more

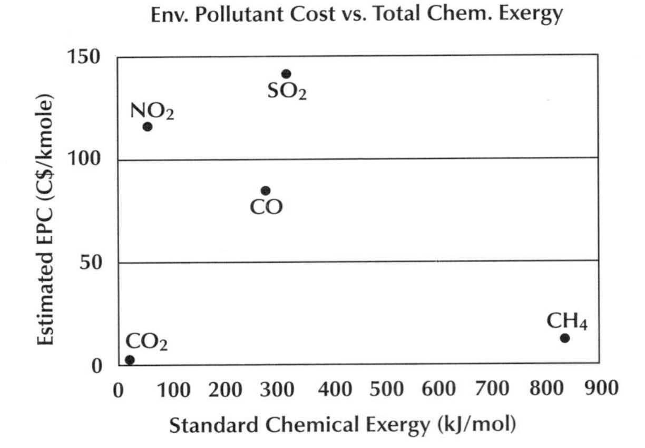


**Figure 4.3 Total chemical exergy vs. environmental pollutant cost (EPC).** EPC is an estimate of the total financial cost of pollution, including external social costs. No correlation to total chemical exergy is discernible. However, compare this graph with that depicted in Figure 4.4. (Adapted from Gunnewick and Rosen, 1998. Reproduced with permission of John Wiley & Sons, Limited.)

closely related to adverse environmental and ecological effects than the chemical exergy of heat transfer (e.g., in chemical reaction).

Figure 4.3 illustrates the importance of distinguishing between total chemical exergy and exergy of mixing. Environmental pollutant cost (EPC) is an estimate of the total financial cost of pollution, including external social costs. Greater EPC is presumed to equate to more severe environmental effects, but no relationship to total chemical waste exergy is discernible. When the graph is modified to show only the exergy of mixing (Figure 4.4), a correlative or predictive relationship is suggested.

The pollution potential hypothesis relies on this observation that entropy production in chemical systems is a manifestation of two separate phenomena: heat transfer and mass transfer. The former is mathematically described by Clausius's formulation $\Delta S = q/T$, where the change in entropy is defined as the ratio of heat exchanged to absolute temperature. In certain systems entropy may be produced without heat transfer (such as in the diffusive mixing of two ideal gases at constant temperature and pressure). To distinguish the entropy of mass transfer from that of heat transfer, the *entropy of mixing* may be mathematically modeled using Boltzmann's statistical interpretation:

$$S_f = nR\ln(y) \tag{4.5}$$

where $S_f$ is entropy of mixing (J/K); $n$ is the total number of moles; $y$ is the molar concentration of a particular chemical species in an ideal thermodynamic system; $R$ is the ideal gas constant: 8.314 J/K/mol. (Entropy caused by heat transfer accompanies the dispersion of non-ideal gases, for example, by disrupting the weak chemical associations between neighboring gas molecules. The framework proposed here

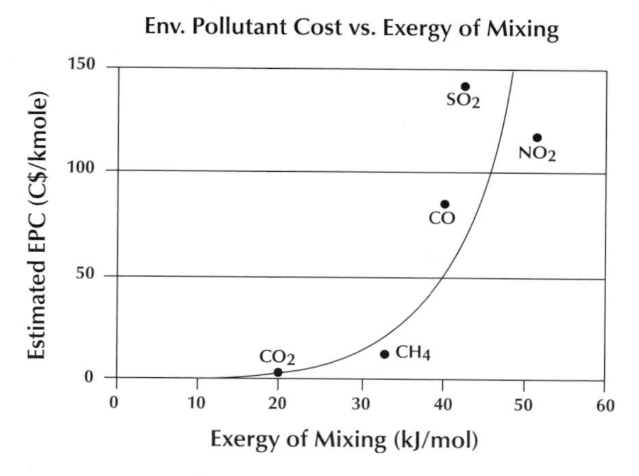


**Figure 4.4 Chemical exergy of mixing vs. EPC.** For most pollutants, the portion of total chemical exergy due solely to mass transfer (i.e., mixing or configuration exergy) is a small fraction of total pollutant exergy and is a better predictor of EPC. Exergy of mixing calculations are for $T = 298.15$ K, with reference atmospheric concentrations (mol/mol) of 331 ppm for $CO_2$, 1.7 ppm for $CH_4$, 90 ppt for CO (Seinfeld and Pandis, 1998) and 35 ppt for $SO_2$, and 1 ppt for $NO_2$ (estimated from Szargut et al., 1988). (Adapted from Seager and Theis (2003); reprinted with permission from Elsevier.)

requires a separate accounting of the entropy due to heat and mass transfer, these having different qualitative characteristic effects in the environment. To the extent that nonidealities are manifested in entropy of heat transfer, they may be neglected for determining pollution potential.)

A comparison of methane versus coal combustion helps to illustrate the hypothesis. Two equilibration steps characterize these chemical processes as they relate to the environment. The first of these is chemical reaction (e.g., combustion to form carbon dioxide and water). The second is dilution or dissipation of the residual reaction products throughout the environment (e.g., the atmosphere). The thermodynamic work available from heat transfer is usually far greater than that from mass transfer, making the former the focus of practical study. For methane combustion:

$$CH_4 + 2O_2 \rightarrow CO_2 + 2H_2O \tag{4.6}$$

$$\begin{aligned} \Delta G^0 &= -[(-50.79) - 2(0)] + [-394.37 + 2(-228.57)] \\ &= -800.72 \text{ kJ/mol} \end{aligned} \tag{4.7}$$

The standard change in Gibbs free energy for the reaction is given in equation (4.7) and is computed from the thermodynamic data in Table 4.1 by subtracting the stochiometric equivalent standard free energies of formation of the reactants from those of the products. The standard chemical exergy is computed with equation (4.2) as 831.7 kJ/mol from the formation reaction as in the ammonia example.

**Table 4.1** Thermodynamic Data ($R = 8.314$ J/mol/K, $T = 298.15$ K, $P = 101.325$ kPa)

| | $G^0$ (kJ/mol)[a] | $B^0$ (kJ/mol)[b] | $y_i^0$ (mol/mol)[b] | $-RT\ln(y_i^0)$ (kJ/mol) |
|---|---|---|---|---|
| $N_2$ | 0 | 0.72 | 0.7479 | 0.720 |
| $H_2$ | 0 | 236.1 | — | — |
| $NH_3$ | −16.48 | 337.9 | — | — |
| $CH_4$ | −50.79 | 831.65 | 1.7 ppm | 32.9 |
| $O_2$ | 0 | 3.97 | 0.201 | 3.97 |
| $CO_2$ | −394.37 | 19.87 | 0.000331 | 19.87 |
| CO | −137[b] | 275.10 | 90 ppb | 40.2 |
| $H_2O$ | −228.57 | 9.49 | 0.0217 | 9.49 |
| $NO_2$ | 51.3 | 55.6 | 1 ppb (est.) | 51.4 |
| $SO_2$ | −330.2 | 313.4 | 35 ppb (est.) | 42.6 |
| $C_{(graphite)}$ | 0 | 410.26 | — | — |

[a]Stumm and Morgan (1996).
[b]Szargut et al. (1988).

The difference between the two would seem to contradict equation (4.1), which states that the change in chemical exergy and Gibbs free energy should be equivalent. The *standard approach* to analyzing a reaction such as (4.6) is different in exergy and Gibbs methods, however. Exergy assumes the reactants and products to be at environmental concentrations; Gibbs assumes they are pure. The difference of 31.0 kJ/mol between the standard chemical exergy and the negative Gibbs free energy is due to the net exergy available from dissipation of the products (taken as pure carbon dioxide and pure water vapor in accordance with the Gibbs standard states) throughout the environment, minus the exergy required to concentrate pure oxygen at the site of the reaction. From this example, it is clear that the standard chemical exergies $B^0$ of the reference compounds $O_2$, $CO_2$, and $H_2O$ may be determined solely by their exergies of mixing computed from equation (4.5), assuming environmental reference activities $y_i^0$ shown in Table 4.1 and that the results are consistent with those given by equations (4.2) and (4.1).

From an economic standpoint, the net exergy of mixing is valueless; nevertheless, it embodies the thermodynamics of the final step in which waste products are introduced into and dispersed throughout the environment and may provide a specific link between exergetic life-cycle assessment (LCA) and a holistic measure of the potential for unintended or adverse environmental consequences. Because the work available in the chemical reaction step is generally much larger than that of the dilution step, engineers to date have nearly uniformly ignored dissipative considerations.

The case of coal, which is a mixture of hydrocarbons, oxygen, sulfur, nitrogen, moisture, ash, and trace elements, provides a basis for comparison. Mass fractions typical of moist coal are given in Table 4.2. The total chemical exergy is estimated as 23,583 kJ/kg; therefore, approximately 35.3 g of coal are required to provide the same 831.7 kJ exergy as 16 g (1 mol) of pure methane gas. A simplified chemical reaction (ignoring

**Table 4.2** Properties Typical of Bituminous Coal

| | Mass Fraction (kg/kg) | Molar Weight (g/mol) | Molal Fraction (mol/kg) | $n_i$ (mol/35.3 g) |
|---|---|---|---|---|
| C | | 12 | 48.1 | 1.70 |
| $H_2$ | 0.041 | 2 | 20.5 | 0.724 |
| $H_2O$(liquid) | 0.1 | 18 | 5.56 | 0.200 |
| $O_2$ | 0.112 | 32 | 3.50 | 0.124 |
| S | 0.013 | 32 | 0.406 | 0.015 |
| $N_2$ | 0.007 | 28 | 0.250 | 0.009 |
| ash | 0.15 | — | — | — |

*Source:* Szargut et al. (1988) used with permission from Elsevier.

trace elements and ash) is shown below. The stochiometric balancing numbers represent the number of moles of each substance involved in the reaction of 35.3 g of coal:

$$\left(1.70\,C + 0.724\,H_2 + 0.200\,H_2O_{(l)} + 0.124\,O_2 + 0.015\,S + 0.009\,N_2\right)_{coal} + 1.862\,O_{2(g)} \rightarrow 1.70\,CO_2 + 0.924\,H_2O_{(g)} + 0.009\,NO_{2(g)} + 0.015\,SO_{2(g)} \quad (4.8)$$

The exergy of mixing of the exhaust gases may be computed from equation (4.5) and the thermodynamic data in Table 4.1. These total 43.65 kJ for coal, compared to 38.85 kJ for an exergetic equivalent amount of methane, excluding oxygen (which is extracted from rather than wasted to the atmosphere). The sum of the exergies of mixing of the coal exhaust gases are 12% higher than those of methane, indicating that coal combustion is likely to have a greater atmospheric environmental impact (due primarily to increased $CO_2$ emissions from carbon) than methane combustion. (Additional environmental considerations have not been included in this analysis and may make coal even less attractive: For example, the presence of mercury increases the loading of pollutants to the atmosphere; ash disposal may present a problematic land impact; acid rain may contaminate watersheds.)

The dispersion of pollutants throughout the environment is essentially a process that converts the exergy of mixing embodied in the initial state (the concentrated pollutant) into entropy of the final state (the dispersed pollutant). In general, chemical species that appear with the greatest frequency in the environmental sink of interest are those with the lowest exergy of mixing and least potential for harm. Conversely, organisms and ecosystems have little evolutionary experience with chemical species that occur only rarely in nature, and introduction of these (having greater exergy of mixing) may be particularly disturbing.

## Reference State

Because exergy of mixing computations are sensitive to the choice of an ideal reference environment, the pollution potential of some substances might be very high in some reference environments but low in others. For example, the environmental effects of wasting brine to freshwaters may be comparatively much higher than wasting to ocean waters. Identifying an appropriate reference condition is one of the biggest obstacles to implementing a pollution potential approach, and some subjectivity may be introduced in deciding what reference mole fractions are to be employed. In general, it may be advisable to select reference conditions that represent a preindustrial environment in which a particular ecological system has evolved (sometimes over millions of years) such that pollution potentials are measured with respect to a zero-environmental-effect groundstate. However, it may also be reasonable to select a reference state on other bases. For example, the optimal level of carbon dioxide in the atmosphere is unknown but may be greater than the approximately 280 ppm representative of a preindustrial atmosphere. In evaluating the pollution potentials created by various chemical emissions scenarios, it is likely that a matrix-based hierarchy is called for that ranks each substance based on regional or global locus, type of medium, and near-, intermediate-, or long-term concern. Different substances may emerge as the leading environmental concerns under different criteria.

## Pollution Potential

Waste exergy of mixing is an indication of the ideal work of mixing (or separation) required to remove a pollutant from a waste stream (as in an end-of-pipe treatment). Calculation of the environmental effect of released pollutants requires additional sophistication. The *pollution potential* hypothesis proposes that the extent of environmental effects associated with chemical release is proportional to the extent of chemical change in the environment caused by that release. This phenomenon is measured by the change in the *configurational* exergy per mole of the chemical species in the environment. [The thermodynamic potential present in a concentration gradient or distribution has been alternatively termed *configurational energy, composition-dependent chemical exergy,* or the *free energy of mixing.* Although any of these terms may have equally justifiable bases for scientific discourse, mixing is the most concise, if least precise, and is easily interpreted in the context of pollution. For that reason, it was used in the previous section to introduce the general principle. The more sophisticated term *configurational* implies a nonhomogeneous concentration distribution such as that which usually results in the environment from a pollutant point source. Therefore, *configurational* is employed in the context of pollution potential. See Denbigh (1981) for an explanation of configurational entropy.]

Because the environment is not at thermodynamic equilibrium, changes in configurational exergy depend upon several factors in addition to the exergy of mixing of the pollutant stream, namely:

- Degradation of the pollutant over time (e.g., before mixing is complete)
- Migration of the pollutant beyond the geographic boundaries or environmental medium of interest
- Sorption, deposition, volatilization, or partitioning of the pollutant into a different environmental medium of interest
- Catalysis of side reactions or creation of daughter products that further affect the chemical composition of the environment

Provided the chemodynamic details of any particular pollutant stream are known or estimable, the changes in environmental configurational exergy associated with any introduced chemical species are calculable as a function of time (from the changing concentration distribution). This is equivalent to the ideal configurational thermodynamic work required to remediate the environment instantaneously or revoke the chemical perturbation.

Consequently, a new measure may be introduced: *pollution potential*. It is defined as the difference between the configurational exergy per mole of a species $i$ in the polluted environment at some time $t$ subsequent to chemical release and the configurational exergy per mole of that species $i$ in the reference environment. The pollution potential may be thought of as the chemical distance between the reference and polluted environments. It is expressed mathematically for any species in the environment as the product of the temperature, ideal gas constant, and natural log of the concentration ratio:

$$\psi = \frac{\partial B}{\partial n} - \frac{\partial B^0}{\partial n} = -T\left[\frac{\partial S}{\partial n} - \frac{\partial S^0}{\partial n}\right] = -T\left[R\ln\left(\frac{y}{y^0}\right)\right] \quad (4.9)$$

where $\psi$ = pollution potential

$\frac{\partial B}{\partial n}$ = configurational exergy per mole in polluted environment (J/mol)

$\frac{\partial B^0}{\partial n}$ = configurational exergy per mole in reference environment (J/mol)

$y$ = molar concentration in polluted environment (mol/mol)

$y^0$ = molar concentration in reference environment (mol/mol)

$\frac{\partial S}{\partial n}$ = configurational entropy per mole in polluted environment (J/K/mol)

$\frac{\partial S^0}{\partial n}$ = configurational entropy per mole in reference environment (J/K/mol)

$T$ = absolute temperature (K)

$R$ = ideal gas constant (8.314 J/K/mol)

Unlike configurational exergy per mole, which is a property of a state applicable to a single chemical in a thermodynamic system and defined by a partial derivative (such as chemical potential), pollution potential ($\psi$) is defined as the *difference* between configurational exergetic potentials in disturbed and reference environments. Although the physical units of pollution potential (kJ/mol) are identical to those of a thermodynamic potential, the term *potential* should be interpreted poetically, rather than in a strictly thermodynamic sense. It implies a sense of possibility to create harm, just as other environmental "potentials" do (see Chapter 1). The pollution potential approach suggests that pollutants that occur rarely (if at all) in nature, mix quickly, and are long-lived may have high pollution potential (create the greatest change in chemical composition of the environment) and be the most dangerous. Several notorious environmental surprises may fit this category: CFCs, dichlorodiphenyltrichloroethane (DDT), polychlorinated biphenyls (PCBs), and lead (e.g., in gasoline). In these instances, a precautionary or preventive approach may be justified.

As with chemical potentials, it is not proper to talk about the pollution potential of an entire thermodynamic system, only individual species within it. To facilitate life-cycle approaches, however, pollution potentials must be logically amenable to aggregation as *extensive* measures. (*Intensive* measures, like concentrations, may not be added in a simple sum without first being converted to extensive measures.) When pollution potential is multiplied by the total number of excess moles of pollutant in the environment, it represents an estimate of the total ideal thermodynamic work require to revoke the pollution. From here forward, this total is defined as the *pollution exergy* of the pollutant for the system of interest. By considering all pollutants, the mole-weighted sum of the individual pollution potentials may be estimated and reported as a pollution exergy for the entire environmental system:

$$B_{\text{env}} = \sum_i \left(n_i = n_i^0\right)\psi_i \tag{4.10}$$

Environmental concentration distributions are rarely uniform; therefore, aggregation of pollution potential in different spatial compartments requires additional sophistication to compute an average or representative pollution potential for the entire system. The mole-weighted average pollution potential of any species in a large environmental volume may be computed by summing the pollution exergy of the species in smaller volumes (represented by the index $z$) and dividing by the total moles of the pollutant in the entire volume:

$$\overline{\psi_i} = \frac{\sum_z \left(n_i - n_i^0\right)\psi_i}{\sum_z \left(n_i - n_i^0\right)} \tag{4.11}$$

It may also be convenient to aggregate the changes in daughter products, end products, and side reactants to the direct effects attributable to release of a single pollutant and to express these in terms of moles of the original pollutant. In this respect, pollutants may be compared on an intensive basis for total environmental effect over large volumes:

$$\psi_{\text{total}} = \frac{\sum_{z}\sum_{i}^{m}\left(n_i - n_i^0\right)\psi_i}{\sum_{z}\left(n_p - n_p^0\right)} \tag{4.12}$$

where $m$ = total number chemical species in environment perturbed by primary pollutant

$n_i^0$ = molar abundance of species $i$ in reference environment (mol)

$n_i$ = molar abundance of species $i$ in polluted environment (mol)

$p$ = primary pollutant

$\psi_{\text{total}}$ = average pollution potential attributable to release of primary pollutant $p$ (kJ/mol)

The aggregation and averaging of pollution potentials for many related components of a system improves the utility of the measure as an environmental metric, but it should also be clear that equations (4.11) and (4.12) are not appropriate in a strictly thermodynamic sense, as the resulting ratio is no longer a thermodynamic variable of state defined by a single partial differential for a single component. (The authors hope that this limitation does little to impugn the scientific credibility or utility of the measure.)

## Quantitative Example: HFCs and HCFCs

The case of chlorofluorocarbon (CFC) replacements illustrates an application of the pollution potential concept. Both hydrofluorocarbons (HFCs) and hydrochlorofluorocarbons (HCFCs) have been used as replacements for CFCs in a variety of industrial applications (e.g., refrigerants, nonaqueous cleaning solvents, foam blowing agents). HFCs and HCFCs and their eventual degradation products are largely nontoxic, have low chemical reactivities, and have low photochemical smog formation potentials (see Hayman and Derwent, 1997; McCulloch, 1999). The principal adverse environmental effects attributable to CFCs are global warming (the greenhouse effect) and catalytic destruction of stratospheric ozone by release of chlorine to the upper atmosphere.

The most economical and readily available CFC replacements are HCFCs, which contain at least one hydrogen atom in place of a chlorine or fluorine. The carbon–hydrogen bond is subject to hydroxyl radical attack in the troposphere, speeding degradation of the HCFC to less harmful end products. However, phase-out of HCFCs under the Montreal Protocol will likely result in increased use of HFCs, which contain no chlorine. The extent to which chlorine, fluorine, and hydrogen are used in the chemical formula largely dictates the environmental properties of the substance.

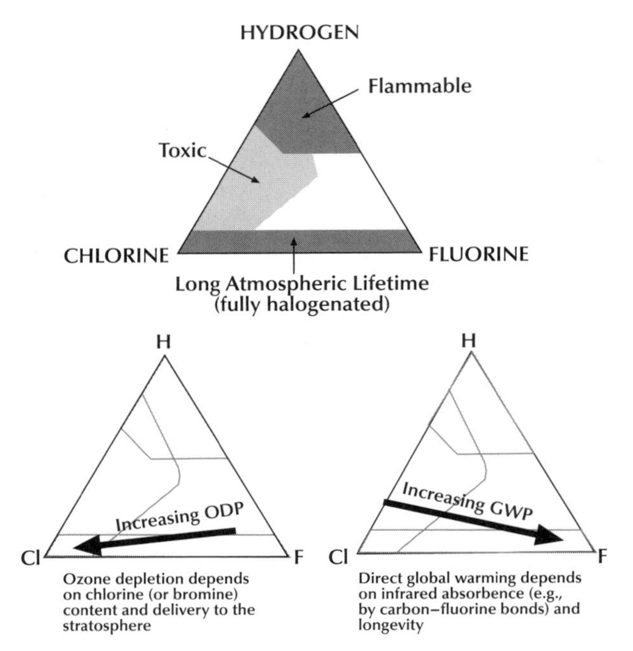


**Figure 4.5 Trade-offs in halogenated hydrocarbon CFC replacements.** The most economical and readily available CFC replacements are HCFCs, which contain at least one hydrogen atom in place of a chlorine or fluorine. The carbon–hydrogen bond is subject to hydroxyl radical attack in the troposphere, speeding degradation of the HCFC to less harmful end products. However, phase-out of HCFCs under the Montreal Protocol will likely result in increased use of HFCs, which do not contain any chlorine. The extent to which chlorine, fluorine, and hydrogen are employed in the chemical formula largely dictates the environmental properties of the substance. (Adapted from Calm and Didion, 1998.)

Figure 4.5 depicts the environmental trade-offs that are involved in selection of CFC replacements. Greater fluoride content typically results in longer atmospheric lifetimes, which increases global warming potential (GWP) due to absorption of infrared radiation by the halocarbons before they are destroyed. Greater chlorine (or bromine) content typically shortens lifetime but exacerbates ozone depletion potential (ODP) by releasing chlorine (or bromine) *after* destruction of the primary halocarbon by oxidation or photolysis in the stratosphere.

Both GWP and ODP are important to assess the environmental implications of any particular CFC substitute. However, they are estimated under slightly different conditions:

- ODP is reported relative to CFC-11; GWP is reported relative to carbon dioxide.
- ODP is evaluated at steady state (which represents the effects of decades, if not centuries, of constant mass release); GWP is reported at multiple time horizons of 20, 100, or 500 years.
- ODP is calculated for different emissions levels at a constant change in ozone; GWP is based on different levels of total radiative forcing over the time horizon employed, for *the same* emissions level.

To understand the pollution potential approach, the chemodynamics of the halocarbons themselves (which are closely related to GWP) and the configurational entropy of the indirect effects of the degradation products (which are closely related to ozone depletion) must be understood more clearly.

Fortunately, the environmental fate and degradation pathways of HCFCs and HFCs have been described in detail. [See McCulloch (1999), World Meteorological Organization (1999), or Prather and Spivakovsky (1990) for good examples.] A general chemodynamic diagram is presented in Figure 4.6. Most HCFCs and HFCs are destroyed by hydroxyl radical attack in the troposphere, although the reaction kinetics vary widely between chemical species. The end products are principally water-soluble halogenated organic acids (e.g., trifluoroacetic acid) and free ions (fluoride and chloride salts) that wash from the troposphere in wet precipitation. (The average tropospheric lifetime of water-soluble end products is estimated at about 9 days.)

A portion of some HCFCs or HFCs may disperse intact through the tropopause and reach the stratosphere, where OH attack continues at a slower rate (due principally to colder temperatures). Once in the stratosphere, the reaction end products are removed more slowly, as they must disperse downward through the tropopause before precipitating. For some HFCs and HCFCs, photolysis by ultraviolet (UV) light (available in or above the ozone layer) can be an important destruction mechanism and may lead to rapid release of all organic forms of fluoride and chloride. Stratospheric photolysis is the only significant destruction mechanism for CFCs. Generally, longer-lived CFCs are photolyzed higher in the atmosphere.

Fluoride is not active in the stratosphere and contributes little to overall pollution potential, but stratospheric chloride is highly reactive. Although partitioned among many forms, total stratospheric chloride abundance is roughly constant with height (Zander et al., 1994). Some

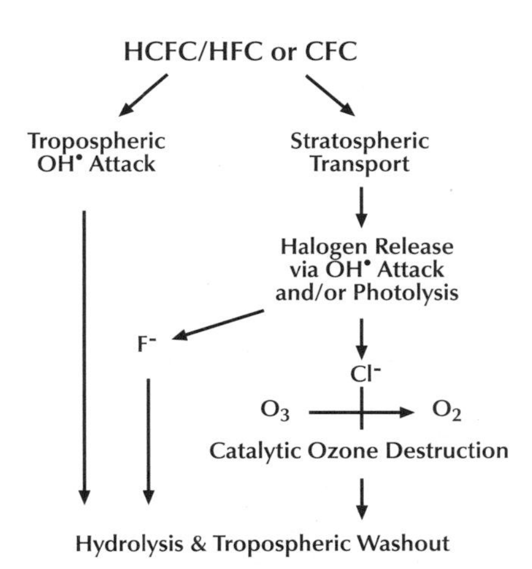


**Figure 4.6 Simplified chemodynamic model of CFC replacement degradation.** (*Source:* Seager and Theis, 2003; used with permission of Springer-Verlag.)

forms catalyze the destruction of stratospheric ozone, converting it to simple oxygen, but even nonreactive forms can later be converted to active forms of chlorine. The carbon and hydrogen contained in HCFCs and HFCs are eventually converted to carbon dioxide and water, but their pollution potential attributable to CFC and substitute release is negligible. Consequently, they may be excluded from the chemodynamic model. For HFCs, which do not contain chlorine or contribute to ozone depletion, it would be simple enough to stop here and compute the pollution potential slowly based on the primary halocarbon concentration distribution, but to estimate the pollution potential of HCFCs, the configurational entropy of ozone column depletion must be added to the results of the HCFCs themselves, and additional detail is warranted.

The chemistry of stratospheric ozone is especially complex. ODP was devised by Wuebbles (1981) to simplify policy decisions. Calculation of ODP relies on sophisticated chemodynamic models of the atmosphere to reduce chloride- (and bromide-) induced changes in stratospheric ozone to a single number reporting the destructive effectiveness of HCFCs relative to CFC-11. By definition, the ODP of CFC-11 is 1; therefore, an equivalent mass release of HCFC with an ODP of 0.1 would be expected to reduce stratospheric ozone levels by only one-tenth as much as CFC-11. In its original formulation, ODP was based on a release of 1 billion pounds per year of a particular halocarbon and determined by the ratio of the expected change in stratospheric ozone divided by the expected change due to 1 billion pounds per year of CFC-11. Subsequent determinations, however, are estimated by adjusting the steady-state mass release required to model a 1% reduction in globally averaged total column ozone, so that ODP is calculated by the mass ratios at equivalent ozone losses (Fisher et al., 1990). Mathematically, this is expressed as:

$$\text{ODP} = \frac{\text{steady-state mass release of HCFC to result in 1\% ozone loss}}{\text{steady-state mass release of CFC-11 to result in 1\% ozone loss}}$$

Globally averaged total ozone losses probably peaked near the end of the 20th century at approximately 5%. (A stratospheric column abundance of 300 Dobson units—or $1.34 \times 10^5$ mol $O_3$—may be assumed as an unperturbed reference condition representative of the global atmosphere.) Calculating ODP at the lower level of 1% depletion reduces distortion in the estimates due the nonlinearities in the relationship between stratospheric chloride levels and ozone destruction—especially at high concentrations.

Because the purpose of pollution potential is to compare all environmental effects on the same scale, under the same emissions scenarios, and over identical time horizons, the differences in the GWP and ODP models must be reconciled. The matter of scale is accommodated in the pollution potential itself, which compares all pollutants in terms of changes in configurational entropy per mole. To reconcile the other differences, it is most sensible to use the 500-year time horizon, effectively representing steady-state conditions for all the ozone-active species, and constant mass release scenarios (because pollution potential is especially sensitive to the total pollutant loadings).

The pollution potential of ozone depletion depends on the absolute (rather than relative) decline in total column ozone—a value that is difficult to calculate with certainty and varies according to season and latitude. Polar regions experience greater ozone losses than temperate or tropical areas. In the Antarctic, reductions as high as 60% are reported on a seasonal basis. Consequently, averaged expressions may not represent regions that experience acute effects. The local pollution potential due to ozone loss could vary substantially from a globally and annually averaged value. It is essential, therefore, to scale ODP to an absolute level of total column ozone loss to determine total pollution potential. Using 1% (as suggested by ODP models) would underestimate the contribution of ODP to the total pollution potential of CFC replacements, given that current ozone levels are 4 to 5% below preindustrial levels. A more conservative approach would be to evaluate at levels higher than current global averages and represent better the possibility of more severe losses at midlatitudes. Using 9.4% (representing annually averaged losses for northern European latitudes), the pollution potential of the ozone depletion attributable to an HCFC results in equation (4.13):

$$\psi_{p,\mathrm{ozone}} = RT\,\frac{[1.34 \times 10^{5}(-0.094)(\mathrm{ODP})]\ln[1-(0.094)(\mathrm{ODP})]}{n_{\mathrm{HCFC}} - n^{0}_{\mathrm{HCFC}}} \tag{4.13}$$

Substitution of 60% depletion (typical of Antarctic spring) for the 9.4% represented in equation (4.13) results in higher pollution potentials (see Figure 4.7). Table 4.3 compares estimates of GWP, ODP, and pollution potential for several CFC substitutes in descending order of globally averaged pollution potential for the entire column height. Here $\psi_p$ refers to the direct effects of the primary pollutant; $\psi_{p,\mathrm{Cl}}$ and $\psi_{P,\mathrm{F}}$ refer to the free chloride and fluoride end products; and $\psi_{p,\mathrm{ozone}}$ refers to the pollution potential of ozone only. Both midlatitude (assuming 9.4% total losses) and extreme Antarctic (assuming 60%) values are reported. Here $\psi_{p,\mathrm{total}}$ refers to the sum of all ozone effects, for all species over the column height.

For CFC-11, the pollution potential of ozone depletion at midlatitudes is on the order of one-sixth that of direct effects, but the pollution potential of the Antarctic ozone hole *exceeds* that of direct global effects almost by a factor of 10. At current levels of stratospheric chloride loading, the pollution potential hypothesis suggests that ozone depletion is a problem that is largely confined to polar regions; whereas the potential effects of global warming, which are correlated with direct pollution potentials and more widespread.

It may be anticipated that extension of the pollution potential methodology to other media (water, land) and other problems of chemical contamination could result in similar regional variances. In many instances, globally and/or annually averaged values may be inappro-

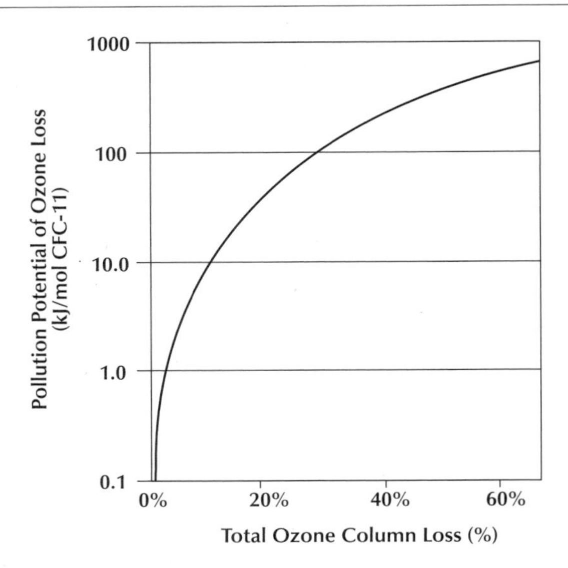


**Figure 4.7 Ozone pollution potential vs. total ozone column loss.** The pollution potential of stratospheric ozone depletion due to CFC-11 varies accordingly to local and seasonal conditions. In sensitive areas such as the polar regions, which have low rates of ozone formation, the impact of CFC-11 results in more severe total ozone column loss, and the ozone portion of pollution potential is higher. Typical midlatitude (45°N) losses probably peaked at the end of the 20th century at slightly less than 10%.

priate, and reference conditions that vary temporally or geographically are likely to be needed. For CFC substitutes, ozone depletion is not a large contributor to total pollution potential, suggesting that ozone effects for the HCFCs are not significant. This result may be surprising, given the fact that all chlorine-containing CFC replacements are regulated by the Montreal Protocol, but it is consistent with other scientific approaches. One study indicates that continued use of low-ODP compounds such as HCFC-123 would result in "indiscernible" effects on the ozone layer (Calm et al., 1999).

## Instantaneous Pollution Potential of Notable Trace Gases

One way to simplify the pollution potential approach is to circumvent the chemodynamic modeling by using actual environmental measurements to estimate instantaneous pollution potentials from equation (4.9). Apportioning ozone losses (attributable to a large number of chlorinated and brominated compounds) to a single compound requires estimates of total ozone losses (estimated here at 3% to represent near-term *projected* globally averaged depletion levels), total stratospheric chlorine and bromine loadings, and an understanding of the relative effects of these on ozone. Reasonable estimates may be made from available data in the scientific literature and aggregated with the results from equation (4.9). These figures are described as instantaneous (rather than

**Table 4.3** Comparison of ODP, GWP, and Pollution Potential for CFC Replacements[a]

| | | | | | | | | Global | Polar | Total |
|---|---|---|---|---|---|---|---|---|---|---|
| Chemical Formula | Trade Name | Lifetime (yr) | ODP | 500-yr GWP | $\psi_{p,Cl}$ (kJ/mol) | $\psi_{p,F}$ (kJ/mol) | $\psi_p$ (kJ/mol) | $\psi_{p,ozone}$ (kJ/mol) | $\psi_{p,ozone}$ (kJ/mol) | $\psi_{p,total}$ (kJ/mol) |
| $CHF_3$ | HFC-23 | 243 | 0 | 11900 | 0 | 0.0006 | 62.3 | 0 | 0 | 62.3 |
| $CO_2$ | Carbon dioxide | 80 | 0 | 1 | 0 | 0 | 0.01 | 0 | 0 | 0.6 |
| $CF_3CH_3$ | HFC-143a | 53.5 | 0 | 2000 | 0 | 0.5 | 58.4 | 0 | 0 | 58.9 |
| $CCl_3F$ | CFC-11 | 45 | 1 | 1600 | 0.49 | 1.02 | 57.2 | 9.2 | 550 | 67.9 |
| $CHF_2CF_3$ | HFC-125 | 32.6 | 0 | 1200 | 0 | 0.43 | 56.6 | 0 | 0 | 57 |
| $CH_3CClF_2$ | HCFC-142b | 18.5 | 0.05 | 720 | 0.007 | 0.5 | 55.8 | 0.04 | 1.7 | 56.4 |
| $CH_2FCF_3$ | HFC-134a | 13.6 | 0 | 500 | 0 | 0.45 | 55.08 | 0 | 0 | 55.5 |
| $CHF_2CHF_2$ | HFC-134 | 10.6 | 0 | 370 | 0 | 0.56 | 54.5 | 0 | 0 | 55.1 |
| $CHClF_2$ | HCFC-22 | 11.8 | 0.04 | 590 | 0.005 | 0.5 | 55.13 | 0.035 | 1.4 | 55.6 |
| $CH_3CCl_2F$ | HCFC-141b | 9.2 | 0.1 | 220 | 0.048 | 0.5 | 53.8 | 0.4 | 16.8 | 54.7 |
| $CH_2F_2$ | HFC-32 | 5.6 | 0 | 270 | 0 | 0.24 | 55.3 | 0 | 0 | 55.5 |
| $CH_3F$ | HFC-41 | 3.7 | 0 | 43 | 0 | 0.5 | 54.8 | 0 | 0 | 55.3 |
| $CHF_2CH_2F$ | HFC-143 | 3.8 | 0 | 120 | 0 | 0.36 | 52.6 | 0 | 0 | 53 |
| $CH_3CHF_2$ | HFC-152a | 1.5 | 0 | 58 | 0 | 0.34 | 51.1 | 0 | 0 | 51.5 |
| $CF_3CHCl_2$ | HCFC-123 | 1.4 | 0.014 | 36 | 0.005 | 1.11 | 48.9 | 0.066 | 2.7 | 50.1 |
| $CH_2FCH_2F$ | HFC-152 | 0.5 | 0 | 13 | 0 | 0.3 | 48.9 | 0 | 0 | 49.2 |
| $CH_3CH_2F$ | HFC-161 | 0.25 | 0 | 3 | 0 | 0.18 | 48 | 0 | 0 | 48.2 |

[a]Atmospheric lifetime, ODP and GWP data are from World Meteorological Organization (1999). 500-yr GWP refers to the total amount of infrared radiation absorbed during 500 yr of constant mass-flux release of the substance from the surface of Earth, relative to a mass-equivalent release of carbon dioxide. ODP refers to the steady-state reduction in total column ozone attributable to a constant mass-flux release relative to CFC-11. Direct pollution potentials are evaluated under pollutant loadings (constant 2.44 $g/km^2/day$ mass-flux release) and time frames (500 yr) comparable to those assumed for ODP and GWP, respectively. In all cases except HFC-23, the 500-yr elapsed model time is sufficient to approximate steady-state conditions. In the absence of naturally occurring sources of a particular HCFC, an arbitrary reference condition of $1 \times 10^{-20}$ mol/mol (column abundance of $3.5 \times 10^{-9}$ mol/$km_2$) was employed. Reference conditions for $Cl^-$ and $F^-$ are assumed to be $1 \times 10^{-9}$ and $1 \times 10^{-12}$ mol/mol, respectively. Other data are from Seager and Theis (2003). (Used with permission of Springer-Verlag.)

steady state) because they are based on atmospheric measurements that may change over time. The advantage of the instantaneous approach is that pollutants may be ranked according to current environmental conditions rather than future scenarios.

The results for 21 globally significant pollutants are summarized in Table 4.4 in descending order of total pollution exergy. They differ significantly from the modeled pollution potentials. All the substances listed are generally recognized as high environmental priorities, even though the range of current pollution potentials varies. Some substances with high pollution potential are already regulated under the Montreal Protocol. Largely, the remainder are under increasing scientific and political scrutiny, e.g., carbon dioxide, methane, hydrofluorocarbon, and perfluorocarbon regulation is proposed under the Kyoto Protocol.

No single figure accurately represents the residence time (or cycling rate) of carbon dioxide in the natural environment, principally because carbon dioxide is stored in all three environmental media: the atmosphere, hydrosphere, and lithosphere. Equilibration among these three takes place through a variety of physical and biological mechanisms at different rates. An estimated atmospheric residence time of 80 years is listed; a figure combining all media may be closer to 200 years. [See Ravishankara and Lovejoy (1994) for a discussion of the importance of atmospheric lifetime as an indicator of environmental insult.] The pollution potential of carbon dioxide is very low, comparable to only methyl bromide and methyl chloride (both of which also occur naturally but have much shorter atmospheric lifetimes). However, it is apparent that the long atmospheric lifetime and massive rate of emission of carbon dioxide allows rapid accumulation of excess molar quantities that are extremely difficult to revoke. Consequently, carbon dioxide has high pollution *exergy*; that is, small amounts of carbon dioxide have little atmospheric effect at the margin, but the combined effect of *all* the carbon dioxide is the single largest chemical perturbation in the atmosphere today.

The data in Table 4.4 represent the sum of historical use patterns rather than hypothetical scenarios, and they ignore the fact that use of some pollutants is expanding while others are diminishing. Some substances (e.g., perfluorocarbon) may have serious environmental consequences but have yet to be employed or accumulate in sufficient quantities to rank among current pollution potential leaders. Reported instantaneous pollution potentials may change as patterns of use (and environmental abundances) vary.

## Sustainability Metrics in Design: Freezer Insulation

Comparing the environmental implications of different technological options is especially problematic. Although the financial and thermodynamic tools are generally available to evaluate dissimilar life-cycle

**Table 4.4** Instantaneous Pollution Potentials and Total Pollutant Exergy for Some Notable Atmospheric Pollutants[a]

| Name | Chemical Formula | Atmospheric Lifetime* (yr) | 500-yr GWP* | ODP* | Current (ppt)* | Ref. (ppt) | $\psi_p$ (kJ/mol) | $\psi_{ozone}$ (kJ/mol) | $\psi_{total}$ (kJ/mol) | Atmospheric Abundance $\Delta n$ (mol) | Pollution Exergy $B_{env}$ (TJ) |
|---|---|---|---|---|---|---|---|---|---|---|---|
| Carbon dioxide | $CO_2$ | 80 | 1 | 0 | $3.65 \times 10^8$ | $2.80 \times 10^8$ | 0.6 | | 0.60 | $1.90 \times 10^{16}$ | $1.15 \times 10^7$ |
| Methane | $CH_4$ | 12.2 | 7.5 | 0 | 1725000 | 700000 | 2 | | 2.0 | $2.29 \times 10^{14}$ | $4.70 \times 10^5$ |
| CFC-12 | $CCl_2F_2$ | 100 | 5200 | 0.82 | 530 | $1.00 \times 10^{-8}$ | 56.1 | 0.54 | 56.6 | $1.19 \times 10^{11}$ | 6714 |
| CFC-11 | $CCl_3F$ | 45 | 1600 | 1 | 265 | $1.00 \times 10^{-8}$ | 54.5 | 0.65 | 55.2 | $5.93 \times 10^{10}$ | 3271 |
| Nitrous oxide | $N_2O$ | 120 | 190 | 0 | 315000 | 275000 | 0.3 | | 0.3 | $8.95 \times 10^{12}$ | 2760 |
| HCFC-22 | $CHClF_2$ | 11.8 | 590 | 0.034 | 123 | $1.00 \times 10^{-8}$ | 52.8 | 0.02 | 52.8 | $2.75 \times 10^{10}$ | 1453 |
| Carbon tetrachloride | $CCl_4$ | 35 | 450 | 1.2 | 101 | $1.00 \times 10^{-8}$ | 52.3 | 0.78 | 53.1 | $2.26 \times 10^{10}$ | 1200 |
| CFC-113 | $CCl_2FCClF_2$ | 85 | 2700 | 0.9 | 83.5 | $1.00 \times 10^{-8}$ | 51.9 | 0.59 | 52.5 | $1.87 \times 10^{10}$ | 980 |
| HFC-23 | $CHF_3$ | 243 | 11900 | 0 | 11 | $1.00 \times 10^{-8}$ | 47.3 | | 47.3 | $2.46 \times 10^9$ | 116 |
| HCFC-142b | $CH_3CClF_2$ | 18.5 | 720 | 0.043 | 7.6 | $1.00 \times 10^{-8}$ | 46.4 | 0.03 | 46.5 | $1.70 \times 10^9$ | 79 |
| Methyl chloride | $CH_3Cl$ | 1.3 | 5 | 0.02 | 550 | 300 | 1.4 | 0.01 | 1.4 | $5.59 \times 10^{10}$ | 77 |
| HCFC-141b | $CH_3CCl_2F$ | 9.2 | 220 | 0.086 | 5.4 | $1.00 \times 10^{-8}$ | 45.7 | 0.06 | 45.7 | $1.21 \times 10^9$ | 55 |
| Methyl chloroform | $CH_3CCl_3$ | 4.8 | 42 | 0.11 | 93 | 25 | 3.0 | 0.05 | 3.0 | $1.52 \times 10^{10}$ | 46 |
| H-1211 | $CBrClF_2$ | 11 | 390 | 5.1 | 3.5 | $1.00 \times 10^{-8}$ | 44.7 | 3.33 | 48 | $7.83 \times 10^8$ | 38 |
| HFC-134a | $CH_2FCF_3$ | 13.6 | 500 | 0 | 3 | $1.00 \times 10^{-8}$ | 44.3 | | 44.3 | $6.71 \times 10^8$ | 30 |
| H-1301 | $CBrF_3$ | 65 | 2700 | 12 | 2.2 | $1.00 \times 10^{-8}$ | 43.6 | 7.85 | 51 | $4.92 \times 10^8$ | 25 |
| CFC-13 | $CClF_3$ | 640 | 16300 | 0.5 | 5 | 0.005 | 15.7 | 0.33 | 16.0 | $1.12 \times 10^9$ | 18 |
| Carbon tetrafluoride | $CF_4$ | 50000 | 8900 | 0 | 76 | 40 | 1.5 | | 1.5 | $8.06 \times 10^9$ | 12 |
| Sulfur hexfluoride | $SF_6$ | 3200 | 32400 | 0 | 3.7 | 0.05 | 9.8 | | 9.8 | $8.17 \times 10^8$ | 8 |
| Methyl bromide | $CH_3Br$ | 0.7 | 1 | 0.36 | 10 | 5 | 1.6 | 0.12 | 1.7 | $1.12 \times 10^9$ | 2 |
| HCFC-123 | $CF_3CHCl_2$ | 1.4 | 36 | 0.012 | No data | No significant expenditure of exergy currently anticipated | | | | | |

[a]Data columns marked with an asterisk (*) are from measurements reported in World Meteorological Organization (1999). Other data are from Seager and Theis (2002). Reference conditions for CFCs, HFCs, and HCFCs are assumed to be $1 \times 10^{-8}$ ppt. A mass of the total atmosphere = $5.14 \times 10^{18}$ kg, avg. molecular weight of 22.97 g/mol and surface area of $5.1 \times 10^8$ km$^2$ are assumed. Reference concentrations for methyl chloride and methyl bromide are estimated from remote ocean or rural observations reported in Seinfeld and Pandis (1998). Direct pollution potentials $\psi_p$ exclude ozone effects. $\psi_{ozone}$ is based upon a globally and annually averaged 3% reduction in stratospheric ozone from a reference column abundance of 300 Dobson units ($1.34 \times 10^5$ mol/km$^2$) with total pollution potential of stratospheric ozone apportioned among individual pollutants according to an ODP-weighted fraction of all ozone-active species in the atmosphere. (Used with permission of Elsevier.)

inventories in terms of dollars or joules, it is a formidable scientific task to compare different environmental priorities. Optimizing design for a specific environmental metric, such as ODP, TEWI, or human toxicity potential (HTP, see Chapter 1 on environmental metrics), inevitably leads to suboptimization of other environmental criteria. As demonstrated by the example of CFC replacements, trade-offs are inherent in the selection of technological alternatives. Materials requirements, types of pollutants, and affected media are likely to vary among different technologies. Qualitatively different environmental risks are likely to be engendered in different time frames and different media. Within the constraints of the Montreal Protocol, designers may optimize for any number of criteria: least capital cost, least operating cost, least total cost of ownership (TCO), and so forth. Application of different metrics often results in different policy recommendations, management strategies, or engineering designs. *The pollution potential hypothesis suggests that optimization for least change in environmental chemical composition may represent the best compromise, resulting in least overall environmental impact.*

Deciding the ideal thickness of foam-blown polyurethane insulation in a freezer provides an illustrative example. A great deal of research effort has been focused on finding new working fluids for refrigerator/freezer compressors, but the problem of creating environmentally acceptable insulation may be even more perverse. [A previous study of the life-cycle impacts of refrigeration showed that selecting a refrigerant for high-energy efficiency during use could be more important to minimizing TEWI than selecting for least GWP (McCulloch, 1994)].

Polyurethane foam insulation is formed by blowing small bubbles (or cells) of gas that expand the volume of the polyurethane, which may be shaped into boards, panels, or other convenient forms. Some of the gas becomes dissolved in the polyurethane itself, but most is trapped in bubbles. Because approximately 75% of the heat that traverses a polyurethane foam board may be carried by conduction through the gas bubbles, the insulating properties of the panel depend on the thermal conductivity of the gas used to blow the foam. Over the lifetime of a typical freezer (approximately 15 years), the foam-blowing agent is released extremely slowly, meaning that the environmental consequences of the blowing agent are muted by the fact that it remains trapped in the panel itself. Even when discarded, the rate of release is small. (By contrast, refrigerants may leak throughout the lifetime of an appliance, may be inadvertently bled to the atmosphere during maintenance, or may be released rapidly when the unit is discarded or damaged.)

When appliances using foam panels are recycled, however, they typically are shred for easy handling and recovery of valuable metals (Lambert and Stoop, 2001). When shredded, the gas bubbles containing blowing agent are ruptured, and as much as 50% of the agent can be released (Kjeldsen and Jensen, 2001). Consequently, the TEWI of refrigerator/freezer foam-insulating panels is the sum of the carbon dioxide equivalents associated with manufacture, with use, *and* with recycle or remanufacture. The manufacturing phase is insignificant compared to use and remanufacture (Campbell and McCulloch, 1998; Yanagitani and Kawahara, 2000). The most important elements of foam insulating panel design, therefore, are the thermal conductivity of the blowing agent, the thickness of the panels employed, and the environmental characteristics of the blowing agent. (Physical attributes such as the optimal size and distribution of cells or blowing agent/polyurethane molar ratio can be similar for all HCFC/HFC substances under consideration.)

The equivalent warming associated with the use phase is proportional to the heat loading on the compressor system. Heat may come from several sources—defrost cycles, leakage around the door seal, internal lights—but perhaps the most important factor is the heat gained through the insulating panels. This heat gain is proportional to the reciprocal of the thickness of the panel, according to the well-known diffusive heat equation.

The equivalent warming of the remanufacture (i.e., shredding) phase is governed by the GWP and mass of blowing agent released when the freezer is shredded. TEWI of remanufacture varies linearly with insulation thickness, therefore. (For mathematical convenience, the total volume of insulation employed may be assumed to vary linearly with insulation thickness, and insulation cross-sectional may be assumed constant.) Assuming the molar ratio of blowing agent to polyurethane is constant regardless of which agent is employed, the equivalent warming of remanufacture may be computed as a function of insulation thickness, GWP, and molecular weight of any halocarbon blowing agent. Assuming 0.67 kg of carbon dioxide is released for every kilowatt-hour of electricity generated, 50 W for the average additional heat load of door openings, air leakage, defroster, and lights, and an average of 25 W for the auxiliary electricity demand of the lights, fans, and defroster, the TEWI is estimated by:

$$\text{TEWI} = 0.67\left[\frac{\left(\frac{D\,\Delta T}{x}50{,}000 + 50\right)}{\text{COP}} + 25\right]\frac{(365.25)(24)}{1000}(15)$$
$$+\left(\frac{1}{2}\right)\frac{50{,}000x}{29{,}000}(M_w)\,\text{GWP} \qquad (4.14)$$

where $D$ represents the one-dimensional diffusive thermal conductivity of the foam in units of W/cm/K (see Table 4.5), and $x$ represents the foam insulation thickness in cm; $\Delta T$ is a constant 47.2 K. The total insulation area is assumed to be 50,000 $cm^2$. Watts are converted to kWh/yr for 15 years. COP represents the coefficient of performance of the compressor/heat exchanger system (including motor), assumed to be 1.4 W/W. The typical volume of foam containing one mole of blowing agent is assumed to be 29,000 $cm^3$/mol, although only half of the agent is assumed to be released; $M_w$ is the molecular weight of the foam-blowing agent in kg/mol. GWP represents the global warming potential of the foam-blowing agent in equivalent kg of $CO_2$/kg.

**Table 4.5** Chemical Data and Optimal Freezer Insulation for TEWI, $B_{env}$[a]

| | Molar Wt (g/mol) | $D$ (W/cm/K) | Min. TEWI (eq. k $CO_2$) | Optimum TEWI × (cm) | Min. $B_{env}$ (kJ) | Optimum $B_{env}$ Thickness × (cm) | 8 cm $B_{env}$ (kJ) |
|---|---|---|---|---|---|---|---|
| CFC-11 | 137.5 | $9.00 \times 10^{-5}$ | 8500 | 8 | 830 | 2 | 1700 |
| HCFC-142b | 100.5 | $1.50 \times 10^{-4}$ | 7700 | 19 | 700 | 4 | 800 |
| HFC-134a | 102 | $1.40 \times 10^{-4}$ | 7300 | 22 | 600 | 4.5 | 650 |
| HCFC-141b | 117 | $1.30 \times 10^{-4}$ | 6700 | 29 | 500 | 5 | 550 |
| HCFC-123 | 153 | $1.20 \times 10^{-4}$ | 5900 | 61 | 320 | 13 | 340 |
| Carbon dioxide | 44 | $2.20 \times 10^{-4}$ | 5500 | ~550 | 330 | 22 | 400 |

[a]Thermal conductivities $D$ of insulating foams estimated from data in Perkins et al. (2001). [Other data from Seager and Theis (2004).]

Figure 4.8 contrasts the relative importance of the use and recycle phases and depicts TEWI as a function of insulation thickness for a CFC-11 blown foam (which has excellent thermal properties but high GWP). The optimum insulation thickness might be selected at the minimal TEWI, found between 8 and 9 cm.

Table 4.5 lists chemical and environmental characteristics of foam-blowing agents recently in use and shows the insulation thickness that would result in minimal TEWI. The most common agent employed currently is HCFC-141b. In contrast to CFC-11, foam blown with HCFC-141b has thermal properties that are almost as desirable, but much lower GWP. It yields minimal TEWI at an insulation thickness of close to

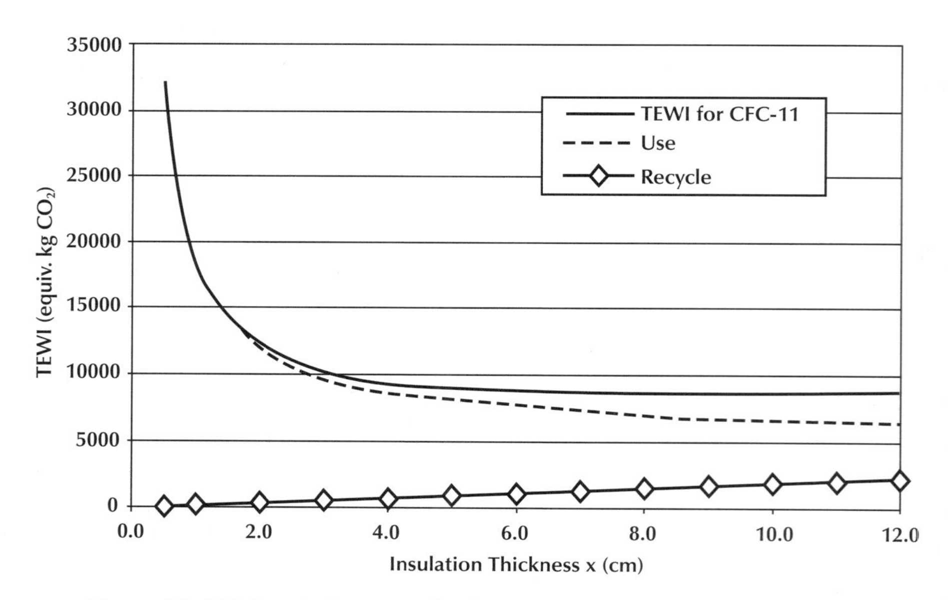


**Figure 4.8 Total equivalent warming impact (TEWI) vs. CFC-11 foam blown insulation.** Increasing freezer insulation reduces energy consumption and associated carbon dioxide but increases carbon equivalents released during destruction of foam insulation upon recycle.

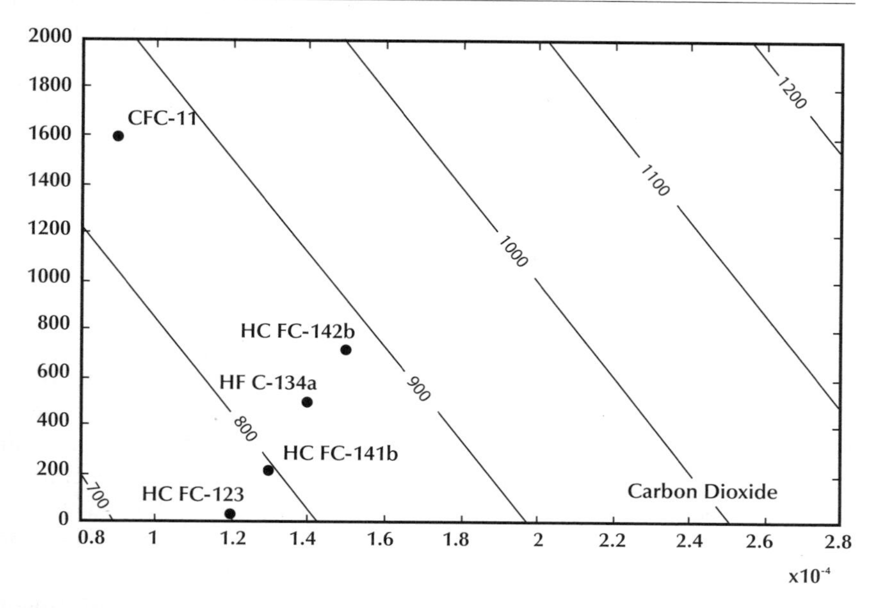


**Figure 4.9 TEWI vs. GWP and thermal conductivity.** Contour lines depict the total equivalent warming impact (TEWI, equiv. kg $CO_2$) over the lifetime of a typical North American freezer (with 8 cm insulation) as a function of the thermal conductivity of the foam insulation (*x* axis) and the global warming potential (GWP) of the agent used to blow the foam.

29 cm, much greater than the 7 to 8 cm typically found in modern North American freezers. This suggests that current design practice is to sacrifice climate-forcing impacts for economic (or other) benefits.

Figure 4.9 is a contour plot that depicts the relationship between TEWI, GWP, and panel thermal conductivity for a constant thickness of 8 cm. Black dots indicate values that are typical of the GWP and thermal conductivity of panels blown with a variety of halocarbon agents. The closer to the lower-left-hand corner of the plot, the lower the TEWI (and the more environmentally advantageous the represented blowing agent from a global warming perspective). Substitution of HFC-134a (which may be most readily substitutable among the HFCs studied) for HCFC-141b would result in higher TEWI if insulation thickness and other design elements remain unchanged.

The TEWI completely neglects ozone effects, but the pollution potential approach can determine whether the increased global warming effects of HFC-blowing agents are justified by elimination of ozone depletion effects. The life-cycle environmental effect of the freezer that is related to the global warming and ozone characteristics of the insulating foam is best represented by the total change in atmospheric configurational exergy $B_{\text{env}}$ introduced in equation (4.10). (For the sake of simplicity, environmental effects associated with environmental criteria other than global warming or ozone depletion have been neglected. For example, sulfur dioxide, nitrous oxides, or mercury vapors may result from the electricity consumption associated with the use phase, but these pollution potentials are not included.) For the assumptions already listed above, $B_{\text{env}}$ is represented by:

$$B_{\text{env}} = 0.67\left[\frac{\left(\frac{D\,\Delta T}{x}50{,}000 + 50\right)}{\text{COP}} + 25\right]\frac{(365.25)(24)}{1000}\psi_{\text{CO}_2}$$

$$+\left(\frac{1}{2}\right)\frac{50{,}000x}{29{,}000}\frac{\tau_{\text{HCFC}}}{15}\psi_{\text{total}} \tag{4.15}$$

[Equation (4.15) differs slightly from (4.14) in treatment of the lifetime of the freezer and the blowing agent. Because TEWI is normalized to carbon dioxide via GWP, it may be reported as either a single pulse release as in (4.14). However, pollution exergy is not normalized and is estimated in (4.15) as a steady state. It should be clear from (4.15) that the longer the freezer remains in service, the less the contribution of the remanufacture term to overall pollution exergy].

Figure 4.10 depicts $B_{\text{env}}$ as a function of insulation thickness for a CFC-11 blown foam. The optimal CFC-11 blown insulation thickness

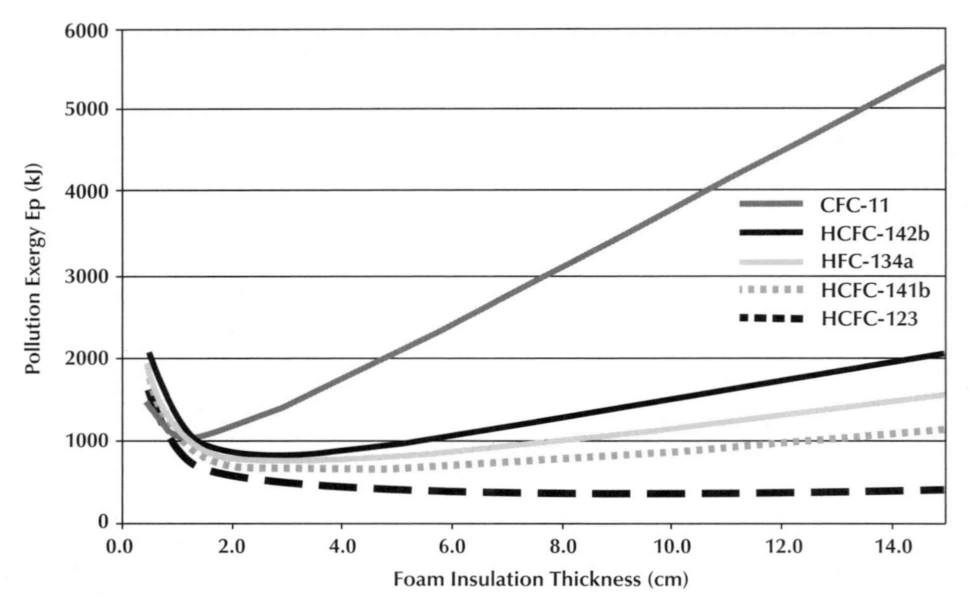


**Figure 4.10 Pollution exergy $B_{\text{env}}$ vs. freezer insulation thickness.** Increased insulation reduces the energy consumption and concomitant carbon dioxide release but increases the amount of blowing agent released upon recycle. Compared with the TEWI approach (Figure 4.8), the minimal exergetic impact of CFC-11 foams is found a much thinner dimension of 1 to 2 cm. This is because the $B_{\text{env}}$ approach is a mole-weighted aggregation of pollution potentials that accounts for global warming and ozone effects, whereas TEWI is a GWP-weighted aggregate focused only on global warming. (*Source:* Seager and Theis, 2004.)

is 1 to 2 cm. In contrast with Figure 4.8, which depicts only climate concerns, the pollution potential perspective suggests that much less CFC-11 insulation should be used.

Table 4.5 lists the minimal TEWI, minimal $B_{env}$, and optimal insulation thickness $x$ from global warming and pollution potential perspectives as well as the $B_{env}$ expected to result from the 8 cm of insulation that is current design practice. A TEWI perspective favors use of additional insulation, but the pollution potential perspective weighs the effect of recycling more heavily and suggests thinner designs. According to *both* measures, however, substitution of HFC-134a for HCFC-141b results in greater environmental impact. In this case, the pollution potential and TEWI approaches result in a consistent evaluation of the relative environmental merits of each halocarbon alternative but disagree on the optimal amounts of insulation to employ. The proper conclusion may be that ozone effects for HCFC foam blown insulation have less effect on the environment (and are less important as a design criteria) than global warming effects. Table 4.5 also identifies HCFC-123 as potentially environmentally preferable to HCFC-141b, which is the current standard. If other factors (e.g., ease of manufacture) are equal, failure to use HCFC-123 may result in unnecessary environmental opportunity costs.

## Limitations

### Pollution Potential Approach

Although pollution potential represents a novel approach to assess the adverse environmental effects of chemicals released to the environment, it is not a solution for management of all sustainability problems. Pollution potential is limited to problems of chemical, not biological, concern. It is not an ecological metric because it is not related to the health of populations or functioning of ecosystems. As a thermodynamic tool, it must draw boundaries to exclude the organisms themselves, measuring only the departure between their present chemical environment and that under which they are likely to be most successful, namely the reference state. Metaphorically speaking, pollution potential is a measure of the soil, not the seed. Nevertheless, several of the environmental surprises of the last few decades might have been identified earlier had a reliable method for ranking chemicals by potential environmental effects been available. The example of CFC replacements discussed in this chapter illustrates the utility of pollution potential as a metric that compares pollutants across different environmental effects and provides supporting evidence for $\psi$ as a meaningful environmental evaluation tool that may reconcile global warming and ozone policies.

## Sustainability Engineering Metrics

Both life-cycle analysis and systems approaches suffer from the same shortcoming: the lack of a uniform basis for comparing or expressing disparate material and energy requirements, emissions, or environmental effects. LCA is severely limited in its applicability to complex problems (such as technology replacement) because the resources and pollutants probably cannot be compared, and environmental systems analysis to date has primarily been used to identify cost-savings opportunities rather than environmental improvements. Unfortunately, no scientific methodology is available to answer the most critical question of industrial ecology: *Which of the technological alternatives is preferable from a sustainability perspective?*

Objectively assessing the comparative sustainability of any technological design may be impossible when trade-offs among different design criteria are involved. Consequently, there is a clear need for further understanding of the underlying relationships between different quantitative metrics, particularly with regard to integrating ecological and sociopolitical metrics with the framework of those more familiar to engineers and scientists. The challenge facing practitioners of industrial ecology is to synthesize different forms of knowledge so that policy makers, designers, and decision makers can produce conscious choices about the financial, thermodynamic, environmental, ecological, and sociopolitical ramifications of alternatives. It is always preferable to be aware of the different dimensions related to a decision, even if the ultimate decision is made subjectively. To participate in such a discussion, engineering students and practitioners must be able to communicate with scholars and professionals of other disciplines, with politicians and government regulators, and with business people and consumers. In sustainability, there may be no "right" answers, only different viable alternatives. Selecting among them is a collective decision-making process that requires broad participation.

### Neoclassical Economic Analog

Pollutants that are associated with high financial costs (e.g., greatest EPC) are those that either are the most expensive to abate or remediate or those that create environmental damage that is most expensive to repair. Similarly, pollutants with high pollution potential are those that require the greatest ideal thermodynamic work (of mixing) to remove. The underlying argument is that the proper expenditure of exergy (money) could restore or preserve an acceptable environment state, and is consequently a valid indication of the environmental impact.

The analogy is hardly coincidental, but requires further explanation. It is revealed in the foundations of quantitative neoclassical economics, which rest on a model of economic activity borrowed from the physical sciences—thermodynamics in particular (Mirowski, 1992; Brody, 1994). This is especially evident in use of the terms *equilibrium* and *reversibility*, which may have different meanings to different audiences, but identical origins. In economics, equilibrium describes a state of maximum economic utility (minimum free energy) in which the price a buyer pays is exactly equal to the benefit the buyer derives, and exactly equal to the benefit the seller has forgone. In thermodynamics, equilibrium refers to a condition in which forward reactions release exactly the same amount of exergy as backward reactions capture and proceed at exactly the same rate, resulting in no net change in the system. Equilibrium should not be confused with steady state in which the system does not change but forward reactions outstrip backwards and a constant input of exergy is required to maintain the system. Nor should equilibrium be confused with systems that do change, but only very slowly.

At the equilibrium price, the buyer and seller are indifferent to the transaction (since neither captures a net benefit) and would presumably be willing to revoke it without net economic impact (assuming transaction costs are zero). Such a transaction (chemical reaction) is called *reversible.* However, transactions may occur at nonequilibrium prices for many reasons; for example, monopoly power, incomplete knowledge, or market externalities. These transactions, especially those far from the equilibrium price, result in a net benefit for at least one party, who would presumably be reluctant to revoke it and surrender the excess economic utility captured. Such a transaction is *irreversible*, but not *irrevocable* (because, presumably, the benefited party would be willing to trade back for some other, greater consideration outside the original transaction). [A concise explanation of the concept of reversibility in chemical thermodynamics may be found in Denbigh (1981). For an explanation of revocability compared with reversibility in economics see Georgescu-Roegen (1971)]. Similarly, chemical reactions can be revoked in the sense that the products can almost always be used to reconstitute the reactants, provided that the system is invested with whatever exergy was lost upon reaction.

However, real processes do not take place at 100% efficiency. All economic systems contain irreversibilities in the form of transaction costs and dissipation of economic utility through misallocation of resources at nonequilibrium prices. In real economic and ecological systems the transaction costs are never zero, equilibrium is unachievable, and there is no *prima facie* reason to believe that economic or ecological systems should tend in this direction. Just as "true" prices in economics are estimable but unattainable, the pollution potential approach represents only an idealized basis for comparison of different alternatives. The actual exergy required to remove any pollutant from the environment would be much higher than that estimated by the pollution potential approach due to inefficiencies in the removal process.

Some of the same criticisms that have been levied against environmental monetization schemes may be applicable by analogy to pollution potential. That is, both monetary and thermodynamic measures lend themselves to the suggestion that environmental damage, like financial transactions and chemical reactions, may be undone. However, evolution may not. The most fundamental difference is the fact that evolutionary systems are path dependent, which is to say that the variables of state at any time may depend upon the previous state and consequently the path by which the current state was achieved. By contrast, thermodynamic variables of state are path *in*dependent (as are prices), and are too simplistic to model economic systems faithfully. Evolutionary pathways are selected both on the basis of boundary conditions such as climate or chemical environment and initial conditions such as population distributions. Even a temporary departure from a specific pathway may result in irrevocable ecological damage, such as the extinction of a species, which cannot be restored even when boundary conditions return to normal. For more on thermodynamics, economics and evolution see Ayres (1994) and Ayres and Martinas (1994).

## Exercises

### Discussion Questions

1. The Montreal Protocol will eventually prohibit all chlorine-containing CFC replacements. (HCFC production restrictions began in 2002.) The readily available replacements are HFCs, which are not regulated by the Montreal Protocol and are listed in the Kyoto Protocol because they generally have high GWP. Should the Montreal Protocol be amended to allow production of low-ODP, low-GWP CFC replacements if it can help meet the objectives of the Kyoto Protocol?

2. Suppose an economical zero-ODP, zero-GWP alternative substance became available. How can we investigate whether this material will have other unforeseen adverse environmental effects? If we can never be *certain*, how should the new substance be regulated?

3. The three factors that contribute to create high pollution potential are rarity, longevity, and mobility in the natural environment. Evaluate as many of the substances listed below as you can for each factor. (You may need to do some independent research.) Use a scale of 0 to 3, with 0 meaning that the substance does not display any aspect of that characteristic and 3 meaning that the substance displays that characteristic to the greatest possible extent. For example, CFC-11 might be ranked a 3 for rarity because it does not occur at all in the natural environment. It could be ranked a 2 for longevity as it is long-lived at approximately 45 years but not as long-lived as sulfur hexafluoride (10,000 years) or some radioactive materials. Lastly, CFC-11 could be ranked 3 for mobility; as a gas it is dispersed as quickly as any substance could be.

   Polychlorinated biphenyls (PCBs)
   Dichlorodiphenyltrichloroethane (DDT)
   Tetra ethyl lead
   Mercury
   Methyl mercury
   Polyvinyl chloride (PVC)
   Carbon dioxide
   Carbon monoxide
   Glass
   Ethanol

Dioxin
Cadmium
Methyl tetra butyl ether (MTBE)
Atrazine
Aluminum metal
Water vapor (or steam)
Methane
Nitrous oxide
Crushed rock
Incinerator ash

Experiment with your assessment. First, rank the substances by the sum of all three ratings. For example, CFC-11 would be $3 + 2 + 3 = 8$. Then, rerank the substances by the product of the three numbers; for CFC-11: $3 \times 2 \times 3 = 18$. Which system do you think is more informative?

### Mathematical Exercises

Using equation (4.14) perform a sensitivity analysis showing how changes in different parameters affect the TEWI of a typical North American freezer. Which parameters are most important? How would a 10% improvement in the coefficient of performance (COP) affect the TEWI compared with a 10% reduction in the thermal conductivity $D$ or 10% reduction in GWP (assuming all other variables remain constant)? What if the mix fuel sources in the power grid changed so that only 0.60 kg $CO_2$ were released for every kilowatt-hour generated instead of 0.67? Use the results of your analysis to suggest strategies for reducing the global warming effects of refrigeration.

Repeat the sensitivity analysis described above, but use equation (4.15) for pollution exergy. Are the results and recommendations similar? Are the sensitivities affected by your choice of blowing agent?

## References


Ahrendts J. 1980. Reference states. *Energy*, **5**:667–677.

Ayres, R. U. 1998. Ecothermodynamics: Economics and the second law. *Ecological Economics*, **26**(2):189–210.

Ayres, R. U. 1994. *Information, Entropy and Progress: A New Evolutionary Paradigm*. American Institute of Physics Press: New York.

Ayres, R .U. and K. Martinas. 1994. A non-equilibrium evolutionary economic theory. In *Economics and Thermodynamics: New Perspectives on Economic Analysis*. P. Burley and J. Foster, eds. Kluwer Academic: Norwell, MA.

Ayres, R. U., L. W. Ayres, and B. Warr. 2003. Exergy, power, and work in the US economy 1990–1998. *Energy*, **28**(3):219–273.

Bejan, A. 1996. *Entropy Generation Minimization: The Method of Thermodynamic Optimization of Finite-Size Systems and Finite-Time Processes*. CRC Press: Boca Raton, FL.

Brody, A. 1994. The use of thermodynamic models in economics. In *Economics and Thermodynamics: New Perspectives on Economic Analysis*. P. Burley and J. Foster, eds. Kluwer Academic: Norwell, MA.

Calm, J. M. and D. A. Didion. 1998. Trade-offs in refrigerant selections: Past, present, and future. *International Journal of Refrigeration—Revue Internationale Du Froid*, **21**(4):308–321.

Calm, J. M., D. J. Wuebbles, and A. K. Jain. 1999. Impacts on global ozone and climate from use and emission of 2,2-dichloro-1,1,1-trifluoroethane (HCFC-123). *Climatic Change*, **42**(2):439–474.

Campbell, N. J. and A. McCulloch. 1998. The climate change implications of manufacturing refrigerants: A calculation of 'production' energy contents of some common refrigerants. *Transactions of the Institution of Chemical Engineers*, **76**(Part B):239–244.

Denbigh, K. G. 1981. *The Principles of Chemical Equilibrium*, 4th ed. Cambridge University Press: New York.

Fisher, D. A., C. H. Hales, D. L. Filkin, M. K. W. Ko, N. D. Sze, P. S. Connell, D. J. Wuebbles, I. S. A. Isaksen, and Stordall. 1990. Model calculations of the relative effects of CFCs and their replacements on stratospheric ozone. *Nature*, **344**(5):508–512.

Georgescu-Roegen, N. 1971. *The Entropy Law and the Economic Process.* Harvard University Press: Cambridge, MA.

Gunnewick, L. H. and M. H. Rosen. 1998. On exergy and environmental impact. *International Journal of Energy Research*, **10**:261–272.

Hayman, G. D. and R. G. Derwent. 1997. Atmospheric chemical reactivity and ozone-forming potentials of potential CFC replacements. *Environmental Science and Technology*, **31**(2):327–337.

Kjeldsen, P. and M. H. Jensen. 2001. Release of CFC-11 from polyurethane foam waste. *Environmental Science and Technology*, **35**(14):3055–3063.

Lambert, A. J. D. and M. L. M. Stoop. 2001. Processing of discarded household refrigerators: Lessons from the Dutch example. *Journal of Cleaner Production*, **9**(3):243–252.

Lindley, D. 2001. *Boltzmann's Atom: The Great Debate That Launched a Revolution in Physics*. Free Press (a division of Simon and Schuster): New York.

McCulloch, A. 1999. CFC and halon replacements in the environment. *Journal of Fluorine Chemistry*, **100**(1–2):163–173.

McCulloch, A. 1994. Lifecycle analysis to minimize global warming impact. *Renewable Energy*, **5**:1262–1269.

Mirowski, P. 1992. *More Heat Than Light: Economics as Social Physics, Physics as Nature's Economics*. Cambridge University Press: Cambridge, UK.

Perkins, R., L. Cusco, J. Howley, A. Laesecke, S. Matthes, and M. L. V. Ramires. 2001. Thermal conductivities of alternatives to CFC-11 for foam insulation. *Journal of Chemical & Engineering Data*, **42**:428–432.

Prather, M. and C. M. Spivakovsky. 1990. Tropospheric OH and the lifetimes of hydrochlorofluorocarbons. *Journal of Geophysical Research*, **95**(D11):18723–18729.

Preston, B. L. 2002. Indirect effects in aquatic ecotoxicology: Implications for ecological risk assessment. *Environmental Management*, **29**(3):311–323.

Rant, Z. 1956. Exergie: Ein neues Wort für "technische Arbeitfähigkeit." *Forsch. Ing. Wis.*, **22**:36–37 [in German].

Ravishankara, A. R. and E. R. Lovejoy. 1994. Atmospheric lifetime, its application and its determination: CFC substitutes as a case study. *Journal of the Chemical Society Faraday Transactions*, **90**:2159–2169.

Seager, T. P. and T. L. Theis. 2004. A taxonomy of metrics for testing the industrial ecology hypotheses with application to freezer insulation. *Journal of Cleaner Production*, accepted for publication.

Seager, T. P. and T. L. Theis. 2003. A thermodynamic basis for evaluating environmental policy trade-offs. *Clean Technologies and Environmental Policy*, **4**:217–226.

Seager, T. P. and T. L. Theis. 2002. Exergetic pollution potential: Estimating the revocability of chemical pollution. *Exergy: An International Journal*, **2**:273–282.

Seinfeld, J. H. and S. N. Pandis. 1998. *Atmospheric Chemistry and Physics: From Air Pollution to Climate Change*. Wiley: New York.

Solomon, S., M. Mills, L. E. Heidt, W. H. Pollock, and A. F. Tuck. 1992. On the evaluation of ozone depletion potentials. *Journal of Geophysical Research*, **97**:825–842.

Stumm, W. and J. J. Morgan. 1996. *Aquatic Chemistry: Chemical Equilibria and Rates in Natural Waters*, 3rd ed. Wiley: New York.

Szargut, J., D. R. Morris, and F. R. Steward. 1988. *Exergy Analysis of Thermal, Chemical and Metallurgical Processes*. Hemisphere: New York.

Von Baeyer, H. C. 1998. *Maxwell's Demon: Why Warmth Disperse and Time Passes*. Random House: New York.

Wall, G. 1997. Energy, society and morals. *Journal of Human Values*, **3**(2):193–206.

World Meteorological Organization. 1999. *Scientific Assessment of Ozone Depletion: 1998 (Vols. 1 and 2)*, Global Ozone Research and Monitoring Project—Report No. 44. Geneva, Switzerland.

Wuebbles, D. J. 1981. *The Relative Efficiency of Several Halocarbons for Destroying Stratospheric Ozone*. Lawrence Livermore National Laboratory: Livermore, CA.

Yanagitani, K. and K. Kawahara. 2000. LCA study of air conditioners with an alternative refrigerant. *International Journal of Lifecycle Assessment*, **5**(5): 287–290.

Zander, R., E. Mahieu, M. R. Gunsin, M. C. Abrams, A. Y. Chang, M. Abbas, C. Aellig, A. Goldman, F. W. Irion, N. Camper, H. A. Michelson, M. J. Newchurch, C. P. Rhineland, R. J. Salwitch, G. P. Stiller, and G. C. Toon. 1994. The northern midlatitude budget of stratospheric chlorine derived from ATMOS/ATLAS-3 observations. *Geophysical Research Letters*, **23**:2357–2360.

CHAPTER 5

# Implementation of an ISO 14001 Environmental Management System

PETER S. SEGRETTO AND MARC J. DENT

Many social observers have long considered industry to be the single greatest threat to the integrity of the environment. This notoriety may have been well earned from the early days of the Industrial Revolution to the time of Love Canal in the mid-1970s, but over the last few decades industry has made impressive progress both in understanding its influence on the environment and in reducing its adverse environmental effects. In the 1950s and early 1960s, industry began to realize that clean water and air, along with all other natural resources, were not infinite. At the same time, science advanced and developed new substances that had never previously been part of any ecosystem. With the development of sophisticated analytical techniques, some of these substances, such as dichlorodiphenyltrichloroethane (DDT) and polychlorinated biphenyls (PCBs), were found to have migrated into ecosystems around the globe.

Industry was first prompted to examine its effect on the environment by government regulations. Industry was required to reduce pollutant loading to air and water and was forced to remediate hazardous waste sites—sometimes decades old. Corporate leaders began to understand that pollution prevention, not just pollution treatment and remediation, actually paid dividends. Moreover, pollution prevention techniques that were prompted because of legal requirements actually boosted efficiencies in product production and made good business sense.

As the understanding of environmental systems matured, especially in Europe and North America, much of industry also matured in its understanding of environmental responsibility. Society began to look for environment-friendly ("green") products to buy and "green" companies with which to do business and in which to invest. By the end of the 1990s, some leaders in industry began to understand that ultimately only the best environmental performers would prosper. Terms such as *environmental responsibility* and *sustainability* began to identify society's expectations of industry. The new understandings among industrial leaders were consistent with the 1987 Brundtland Commission's statement on sustainability that "humanity has the ability to make development sustainable—to ensure that it meets the needs of the present without compromising the ability of future generations to meet their own needs."

Figure 5.1 illustrates the Alcan, Inc., management strategy toward sustainability. Alcan, Inc., has operations in 41 countries in primary aluminum, fabricated aluminum, as well as flexible and specialty packaging. In 2002, it had revenues of $12.5 billion and by 2003 had grown to 53,000 employees. The pyramid shows the evolution of an environmental management strategy toward sustainable development. The company's basic values are the foundation of the pyramid; the pyramid is further supported by the needs of customers, other significant stakeholders, and the requirements of selected management systems. On the left side, the basic strategy of survival and company image tends to protect value, not create or enhance it. Through the lower levels of the pyramid, distinct environmental, economic, and social factors are addressed to meet the needs of the management system. Toward the top of the pyramid, these factors begin to meld as the company's strategy begins to maximize value and as the business grows and develops toward sustainability. At the peak of the pyramid, the business has moved to a value-based management strategy that will sustain its growth and development. This move toward sustainability cannot be attained without recognition of and attention to the elements of the foundation.

**Figure 5.1 Alcan Inc. Management Strategy—toward sustainability.**

Sustainable development cannot be achieved unless the business process incorporates an environmental management system (EMS). ISO 14001 is a sophisticated method for, first, fostering the integration of environment into the business and, second, for managing the process of environment. This chapter will help you to understand the purpose for an ISO 14001 EMS and how it can be implemented. Such an EMS has the potential to mitigate adverse environmental effects that otherwise might be a consequence of an engineer's work. This chapter illustrates the process by showing how specific elements of ISO 14001 were implemented by facilities in Alcan Aluminum Corporation, a subsidiary of Alcan, Inc.

## Multiple Criteria for Success

Industry has learned to respond to the demands of the marketplace to survive and grow. Many leaders of industry now realize that they must make not only capital investments in the traditional sense with dollars, but they must also invest in environmental and social capital as well. In the metals and packaging business, for instance, the marketplace presents many competing materials for a customer's choice.

Figure 5.2 illustrates some of these choices in light of present consumer values. In the choice process, consumers (clients and customers) screen their choices with their established values (environment, health concerns, ethics). These feed the roots of the tree in the illustration.

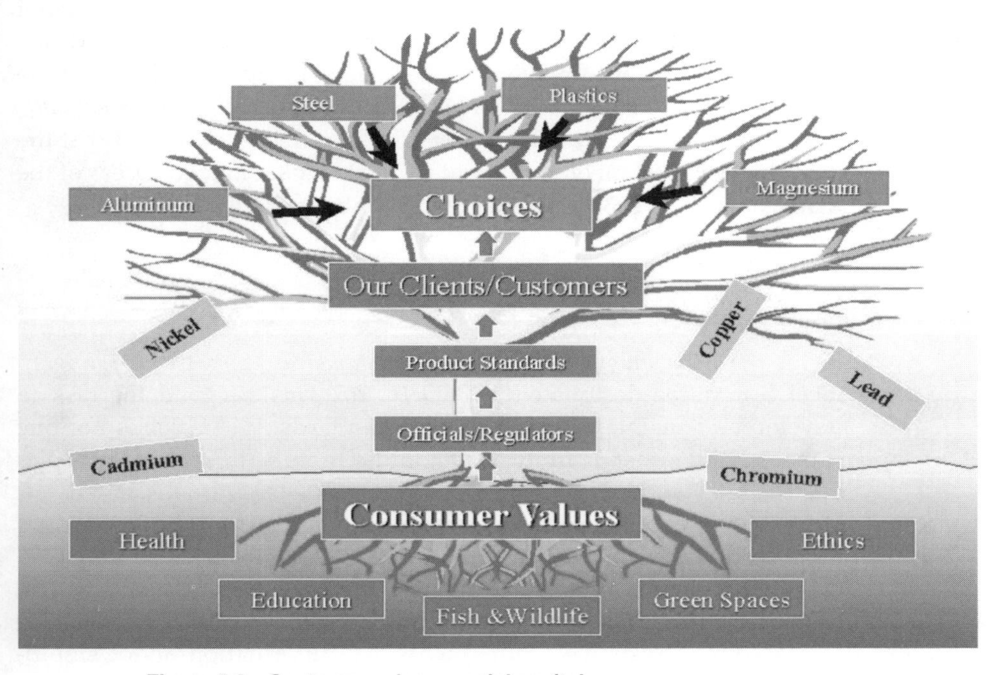


**Figure 5.2 Customer value = social capital.**

Government regulation and other relevant standards support the trunk of the tree. The fruits of the tree represent competing materials (aluminum, steel, plastic, nickel). Some of the competing materials have fallen off the tree because they are no longer nourished or supported. Some are in decline and in danger of falling off. Others are still healthy and thrive, but to maintain this condition they must continue to gain strength from the tree through support and consumer value. This illustration of the marketplace clearly demonstrates the merit of understanding social values and the importance of investments in social capital.

Investments in social capital help the marketplace to make the choices that enhance sustainability. Many companies recognize this fact. In 1998, Alcan Aluminum Corporation made a commitment that its aluminum fabrication facilities would implement an EMS and seek ISO 14001 registration. In May 2001, Richard Evans, executive vice president of Alcan Inc, stated in his address to the International Aluminum Forum in New York City that the aluminum industry and its associations must "clearly position aluminum on the 'sustainability map' in the minds of investors, customers, employees, politicians, regulators and other stakeholders." Alcan followed up on its commitment in May 2002 with the publication of its first corporate sustainability report. Alcan is one of numerous businesses that has responded to the call toward sustainability.

Industry and society have made significant progress in understanding the effect of their operations and their demands on the environment. With this progress comes the following question: Is Earth better off today than it was over three decades ago when the first Earth Day (April 22, 1970) was celebrated? The debate rages. Countless environmental groups continue to warn of impending environmental doom and disaster. Recently, however, new voices have been heard that document the progress that has been made. In his book *The Skeptical Environmentalist, Measuring the Real State of the World,* Bjorn Lomborg points out that the data show significant environmental improvement in many areas of the world over the last 30 or more years (Lomborg, 2001).

## What Is ISO 14000?

ISO is the *International Organization for Standardization*. In 1993, a technical committee designated as *ISO/TC 207* was assigned the task to develop and maintain the ISO 14000 standards. ISO 14000 is not one standard, but rather a family of standards (Table 5.1). Some of the topics and related standards that are included in this series are discussed:

In 1996, ISO published its international standard for environmental management systems and labeled it ISO 14001:1996(E). ISO 14001:1996 (E) "specifies requirements for an environmental management system, to enable an organization to formulate a policy and objectives taking into account legislative requirements and information about significant environmental impacts. It applies to those environmental aspects,

**Table 5.1** Components of ISO 14000 Standard

| ISO Standard | Content |
|---|---|
| ISO 14001 and 14004 | Environmental management system |
| ISO 14010, 14011, and 14012 | Environmental auditing |
| ISO 14021, 14024, and 14025 | Environmental labeling |
| ISO 14031 | Environmental performance evaluation |
| ISO 14060 | Environmental aspects in product standards |
| ISO 14041, 14042, and 14043 | Life-cycle analysis |

*Source:* International Organization for Standardization 2002 (http://www.iso.ch/iso/en/iso9000-14000/iso14000/iso14000index.html).

which the organization can control and over which it can be expected to have influence. It does not itself state specific environmental performance criteria."

At the time of this writing, ISO TC 207 was in the process of revising a number of the standards in the ISO 14000 family, and the ISO 14001:1996(E) revision was in the form of a committee draft. The introduction of ISO/CD 14001.2 indicates that the revision will focus on clarification of the 1996 version and will take in consideration the provisions of the ISO 9001:2000 quality standard to enhance the compatibility of the two standards for the benefit of the user community. The fundamentals of implementation discussed in this chapter will remain relevant with publication of the revised standard.

An organization may choose to implement ISO 14001 for the entire organization or for specific activities or facilities. It should be noted that ISO 14001 contains only requirements that may be objectively audited for registration or certification or for self-auditing. ISO 14001 does not include requirements for occupational health and safety management; however, it does not discourage an organization from developing these management system tools.

The structure of ISO 14001 follows the Deming PDCA Cycle (Plan, Do, Check, Act; Table 5.2). It is designed to support an environmental policy and promote continual improvement in environmental performance. It is important to understand that *ISO 14001 is designed as a business process that is meant to facilitate integration of environment into the*

**Table 5.2** Relation of ISO 14000 to Deming PDCA Cycle

| Plan | Do | Check | Act |
|---|---|---|---|
| Section 4.3: Planning | Section 4.4: Implementation and operation | Section 4.5: Checking and corrective action | Section 4.6: Management review |

*business*. The business cycle starts with a business plan. When environment is truly integrated, it is just another aspect that is considered when developing and carrying out the business plan.

It is also important to note that ISO 14001 defines a standard to which a corporation's EMS must conform. It provides a detailed blueprint and a set of guidelines. Each corporation adapts ISO 14001 to its own culture and circumstances in its implementation. At the end of the plan–do–check–act cycle, individuals in management review the system they have devised based on ISO 14001 and determine how effectively they were in achieving their goals. Their review determines what was effective and what needs to be improved. The cycle is repeated year after year, and this system is designed to improve not only business performance but also environmental performance. Figure 5.3 shows that the system continually improves through repeated cycles of the process.

Implementation of the ISO 14001 EMS is no small task, but the benefits do justify the effort. One major benefit is the increased effectiveness and efficiency of managing complex environmental issues in the context of a functional management system (ISO 14001) integrated with the business. The challenge is reaching this functional integration. It is best achieved at the onset of the implementation process. If the management system is developed in its own functional silo, solely by functional professionals such as the corporate environmental manager, integration with the business management system may never be achieved.

Several functional management system standards, such as ISO 9001: 2000, ISO 16949 (quality), and Occupational Health and Safety Assessment Series (OHSAS) 18001 (safety) foster a organizational structure similar ISO 14001. When a corporation intends to implement two or

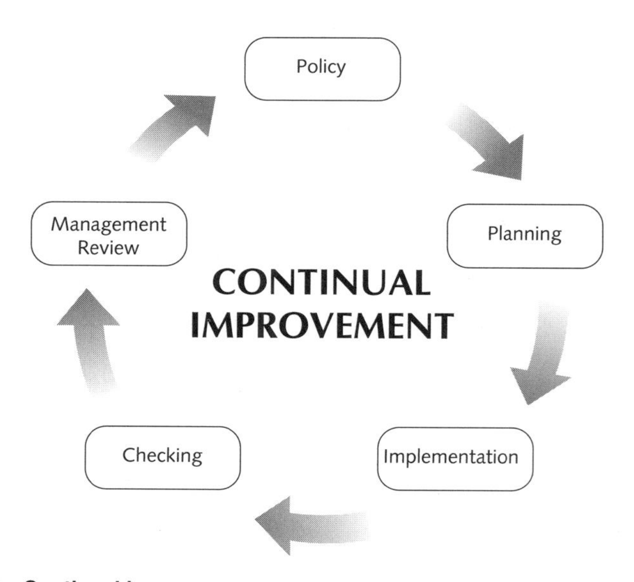


**Figure 5.3 Continual improvement.**

more of these standards, complementary development of common elements from each system makes good business sense because such an approach facilitates integration of these functions with the business process. Whether an organization intends to put into practice a functionally integrated management system or only ISO 14001, integration with the business process will only be achieved if it is initiated at all relevant facets of the organization.

## Why Implement ISO 14001?

Implementation of ISO 14001 should not be taken lightly. Applying the standard to a company is a labor-intensive effort and should not be undertaken merely because it is a "nice thing to do." Anyone who has implemented a quality management system such as ISO 9001 understands the effort required to put into action and then maintain such a system. So the question is: Why implement ISO 14001? The most significant reasons to implement an ISO 14001 management system include:

**Customer Request** This marketplace driver is one of the most powerful encouragements for implementation of ISO 14001, and it is one of the easiest to justify. Many times customers require that their suppliers be registered to the ISO 14001, or equivalent, management system standard. This is the case, for example, with General Motors Corporation and Ford Motor Company.

**Other Marketplace Drivers** Another driver directly related to the marketplace, but not as straightforward as a customer's request, is society's move toward sustainability. The move toward sustainability cannot be achieved easily without a comprehensive EMS.

**Increasing Complexity of Environmental Issues** Drivers related to the operations of a facility—having nothing to do with the marketplace—can motivate use of an EMS. Increased regulation and social expectations of companies as well as self-perceived corporate responsibility can motivate businesses to address environmental issues. The complexity of these issues and the traditional ways of managing them are proving to be inefficient. Integration of environmental matters into the business process is believed to increase the effectiveness and efficiency of managing these complex issues. Increased efficiency can translate to reduced cost and, therefore, an addition to the bottom line.

**Other Cost–Benefits** Implementation of an EMS can also help a business to improve its environmental performance by targeting or focusing its efforts on selected issues such as waste generation, pollution prevention, or energy use. It can, thus, further reduce its operating costs. Additions to the bottom line from these activities can more than compensate for the effort spent on implementation.

**Corporate Responsibility and Image** The increased attention to detail from implementation of an EMS can enhance compliance with environmental regulation. Benefits derived from the resulting good corporate image are hard to quantify but are tangible.

Once a company makes a decision to implement ISO 14001, it is critical to understand what kind of a commitment it is making.

## Elements of the ISO 14001 Management System

The requirements of the various elements of ISO 14001 are presented in this section. No set way has been established to implement the elements of this international standard. ISO TC 207 has developed a set of guidelines on principles and supporting techniques for environmental management systems designated as ISO 14004:1996(E). This document is helpful in understanding the requirements of the standard. In addition, to facilitate implementation, this section discusses the various requirements of the standard and uses examples of how various facilities in Alcan Aluminum Corporation developed their management system to meet these requirements. (Note: For the remainder of this discussion, reference to Alcan will indicate facilities in the Alcan Aluminum Corporation organization. Reference to other organizations, such as Alcan, Inc., will be indicated by their full title.)

ISO 14001 comprises the following six elements:

- 4.1 General requirements
- 4.2 Environmental policy
- 4.3 Planning
- 4.4 Implementation and planning
- 4.5 Checking and corrective action
- 4.6 Management review

**Important Terms and Definitions Used in ISO 14001**

**Continual improvement:** Process of enhancing the EMS to achieve improvements in overall environmental performance in line with the organization's environmental policy. *Note:* The process need not take place in all areas of activity simultaneously.

**Environment:** Surroundings in which an organization operates, including air, water, land, natural resources, flora, fauna, humans, and their interrelation. *Note:* Surroundings in this context extend from within an organization to the global system.

**Environmental aspect:** Element of an organization's activities, products, or services that can interact with the environment. *Note:* A significant environmental aspect is an environmental aspect that has or can have a significant environmental impact.

**Environmental impact:** Any change to the environment, whether adverse or beneficial, wholly or partially resulting from an organization's activities, products, or services.

**Environmental management system:** The part of the overall management system that includes organizational structure, planning activities, responsibilities, practices, procedures, processes, and resources for developing, implementing, achieving, reviewing, and maintaining the environmental policy.

**Environmental management system audit:** A systematic and documented verification process of objectively obtaining and evaluating evidence to determine whether an organization's EMS conforms to the EMS audit criteria set by the organization and for communication of the results of this process to management.

**Environmental objective:** Overall environmental goal, arising from the environmental policy, that an organization sets itself to achieve, and which is quantified where practicable.

**Environmental performance:** Measurable results of the EMS, related to an organization's control of its environmental aspects, based on its environmental policy, objectives, and targets.

**Environmental policy:** Statement by the organization of its intentions and principles in relation to its overall environmental performance, which provides a framework for action and for the setting of its environmental objectives and targets.

**Environmental target:** Detailed performance requirement, quantified where practicable, applicable to the organization or "parts" thereof, that arises from the environmental objectives and that needs to be set and met in order to achieve those objectives.

**Interested party:** Individual or group concerned with or affected by the environmental performance of an organization.

**Organization:** Company, corporation, firm, enterprise, authority or institution, or part or combination thereof, whether incorporated or not, public or private, that has its own functions and administration. *Note:* For organizations with more than one operating unit, a single operating unit may be defined as an organization.

**Prevention of pollution:** Use of processes, practices, materials, or products that avoid, reduce, or control pollution, which may include recycling, treatment, process changes, control mechanisms, efficient use of resources, and material substitution. *Note:* The potential benefits of prevention of pollution include the reduction of adverse environmental impacts, improved efficiency, and reduced costs.

*Source:* Section 3, ISO 14001:1996(E)

## General Requirements

The organization shall establish and maintain an EMS, the requirements of which are described in ISO 14001 [4.1, ISO 14001:1996(E)].

*Environmental Policy*

> Top management shall define the organization's environmental policy and ensure that it: (a) is appropriate to the nature, scale and environmental impacts of its activities, products or services; (b) includes a commitment to continual improvement and prevention of pollution; (c) includes a commitment to comply with relevant environmental legislation and regulations, and with other requirements to which the organization subscribes; (d) provides the framework for setting and reviewing environmental objectives and targets; (e) is documented, implemented and maintained and communicated to all employees; (f) is available to the public. [4.2, ISO 14001:1996(E)]

An organization's environmental policy must address these points. If any of the requirements is missing, the management system supporting and implementing that policy cannot conform to the international standard.

The environmental policy establishes an overall sense of direction and forms the basis upon which the corporation sets its objectives and targets. The policy should be clear so that individuals inside and outside the company understand it. The policy should be periodically reviewed and revised to reflect changes. An environmental policy for an individual facility within a company should reflect the facility's role in the environmental policy of the corporation. The environmental policy must be documented, implemented and maintained, and communicated to all employees. It must be made available to the public.

Top managers of the corporation set the direction for the corporation. Top managers at an individual facility of the corporation set the direction for the facility, but they are guided by corporate management. When it comes to the environmental policy, the facility or plant must support the corporate policy in its role as one facility in the organization. Some roles set by the policy are purely corporate in nature; others are better suited to the facility. Once this is understood at a plant, implementation of the policy at the plant is much easier.

Figure 5.4 illustrates the Alcan, Inc., corporate environmental health and safety (EHS) policy. It meets the requirements of ISO 14001. During implementation of ISO 14001, some facilities within Alcan had difficulty with the broad statements of that policy. The statements became more understandable when the role of each organization was put into the context of the total organization. Literal interpretation of the corporate policy for implementation purposes can cause confusion at a plant level until the role of the plant in the overall organization is considered. Some plants in Alcan opted to develop site-specific policy statements that supported or complemented the corporate policy. Where this was done, the individual policy statements were crafted to conform to the requirements of the appropriate standards. Other facilities chose to implement the Alcan, Inc. policy.

# ALCAN POLICY
## ENVIRONMENT, HEALTH AND SAFETY

Alcan is committed to excellence in environment, health and safety (EHS) through continual improvement of our awareness, understanding, and performance. Our goal is to protect and promote the environment, and the health and safety of all employees and communities where we operate. This will contribute to greater sustainability thereby benefiting all employees, communities, customers, suppliers and shareholders. Every Alcan employee is expected to actively support this policy and to implement the following guiding principles:

**GUIDING PRINCIPLES**

- Integrate EHS as an essential part of Alcan's management and decision-making process. Our shared objective is to demonstrate leadership through performance that contributes to maximization of value.
- Cooperate with customers to understand their needs and support their use of best EHS practices in the design and manufacture of safe and reliable products that take full advantage of our materials' properties throughout their life cycles.
- Demonstrate leadership in EHS to reflect the superior and life-enhancing characteristics of our products for the benefit of all society.
- Ensure a working environment that motivates and supports all employees in their efforts to achieve zero work-related injuries and illnesses.
- Minimize any adverse environmental impact from operations and business practices, and use natural resources and energy more efficiently through the effective use of management systems that continually improve EHS performance.
- Consider and establish appropriate EHS requirements when selecting business partners and contractors.
- Audit operations and business practices at regular intervals to assess EHS performance and compliance.
- Comply with legal requirements and Alcan's internal standards.
- Engage in open and transparent communication with stakeholders to achieve greater environmental, health and safety understanding and to improve performance.

Travis Engen

Travis Engen,
President and CEO
January 2002

**Figure 5.4 Alcan's EHS Policy Statement.**

Communication of the environmental policy to all employees is relatively straightforward and can be achieved in a number of ways. The point to distributing information about the environmental policy is that employees must internalize the information and apply it in their daily activities. Alcan, Inc., developed a formal procedure for rollout of its newly integrated EHS policy issued in September of 2001. It put posters in strategic locations throughout each facility; pocket-sized brochures were distributed to all employees and other relevant stakeholders as defined by the plants. In addition, a compact disc describing the EHS policy was created. The disc included a slide presentation and guide that was intended for use by each facility's management team in the policy rollout process. It was recommended that each facility place the presentation on its computer network to facilitate employee access.

The method used to communicate the policy will necessarily vary from location to location depending on the size and availability of the workforce. Postings and handouts are common; their distribution should be customized to the location. In most cases, the plant manager should disseminate the policy, or at least an introduction to the policy, to employees in person. Most often, the environmental policy can be communicated in the form of a general environmental awareness training session. It can take place as part of the agenda at regularly scheduled employee meetings. The goal of communication is to have each employee realizes that an environmental policy has been promulgated and that he or she has a role to play in its implementation.

Communication of the environmental policy to the public is not mandated by the standard. ISO 14001 only requires that the policy be made available to the public. Some facilities in Alcan have completed this requirement by annually publishing the corporate environmental policy in the local newspaper. Others have determined whom they consider relevant stakeholders and sent special mailings to them. Still others have merely put a procedure in place that states that if requested by the public, a copy of the environmental policy statement will be provided. Implementation of this ISO requirement should be customized to the culture of the location and the management of each facility.

## *Plan* Do Check Act: Planning

Most organizations develop a business plan to set and achieve business objectives. Part of this business planning process should include implementation of the organization's environmental policy. During the development of the business planning process, ISO 14001 requires that a company consider certain environmental factors. In addition to commitments set forth in its environmental policy, the organization must consider the significant environmental aspects of its activities, products, and services as well as the legal requirements of its operations as it develops its environmental objectives and targets. In addition, the standard requires that the organization establish appropriate management programs (action plans) to achieve these goals and targets with appropriate internal performance criteria established. This planning process sets the foundation for the facility's entire environmental management system.

*Environmental Aspects*

> The organization shall establish and maintain (a) procedure(s) to identify the environmental aspects of its activities, products or services that it can control and over which it can be expected to have an influence, in order to determine those which have or can have significant impacts on the environment. The organization shall ensure that the aspects related to these significant impacts are considered in setting its environmental objectives. The organization shall keep this information up-to-date. [4.3.1, ISO 14001:1996(E)]

Although these requirements are straightforward, implementation is challenging. ISO defines *environmental aspect* as an "element of an organization's activities, products or services that can interact with the environment" [Section 3, ISO 14001:1996(E)]. This is a broad, sweeping statement. The organization or facility must scrutinize its operations and its potential environmental impacts. In doing so, it must establish and maintain a procedure and follow that procedure in identifying its environmental aspects and their environmental impacts. Significant environmental aspects are those that can have a significant environmental impact.

Environmental aspects must be inventoried; those that can have a significant effect on the environment must be highlighted. This is the development of a *significant aspect list*. The process does not end there because the standard requires that the inventory and significant aspect list be kept up to date. That requires periodic review and assurance that new projects or process changes are reviewed for their impact on the aspect inventory or significant aspect list. The level of detail in developing the environmental aspect inventory can vary from plant to plant and depends somewhat on the complexity of the operation and the individuals involved. Because the significant aspect list is central to other elements of the standard, this step is critical in establishing the foundation for the rest of the EMS.

A lone individual should not develop the aspects inventory. The process should be completed as a team effort with participation from all parts of the company, including marketing, operations, maintenance, engineering, and management to encompass all components of the process. This approach will help to establish the "buy in" from the organization so that the EMS becomes part of the business process. If the process is carried out by the functional professionals, it will forever be seen as just another environmental program.

Figure 5.5 is a simplified illustration that provides a systematic methodology to develop an environmental aspect inventory for the aluminum rolling process. This example starts with recycling and proceeds through cold rolling. Any portion of the diagram can be adapted to develop the inventory for any given segment of the process. In this illustration, the inventory is developed by listing the inputs and outputs for each step in the aluminum rolling process.

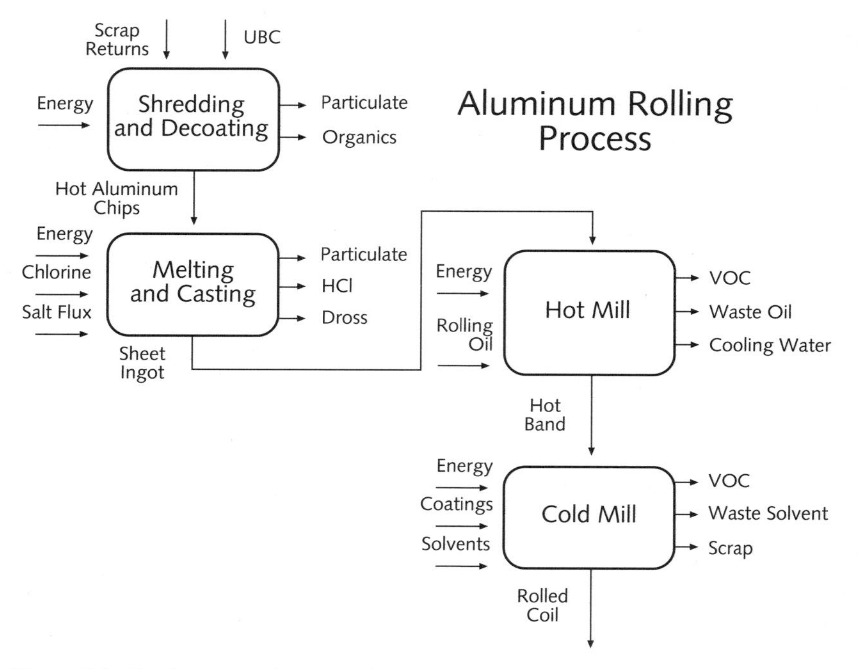


**Figure 5.5 Environmental aspects inventory, rolling.**

Starting with the recycling step of shredding and decoating aluminum scrap, the inputs to the process include scrap returns, used beverage containers, and energy. Outputs include air emissions including particles and organic compounds and hot aluminum chips. The portion of the diagram that includes the hot mill shows inputs of sheet ingot from the melting and casting process, energy, and rolling oil. Outputs include volatile organic compounds (VOC), waste oil, cooling water, and hot band. The inventory can and will become much more complex as the various components of each input and output are factored in. The satisfactory level of complexity is determined by each facility.

After identifying the elements of the inventory, their potential effects on the environment must be identified. Many Alcan facilities took guidance from ISO 14004 and developed a spreadsheet much like that in Table 5.3. A multidisciplinary team considered each process at the facility so that no viewpoint or consideration would be missed. The resulting inventories from the various facilities ranged from less than 100 aspects to over 1000.

Once the aspects inventory is complete, a method must be developed to identify the aspects that have a significant effect on the environment. Thus, a rating system along with a subjective threshold of significance must be developed. The procedure can be unique and based on a value set by the facility. It may be influenced by local issues or parameters that the evaluation team considers relevant for the business, process, and location. For example, if a plant is located in a region where ozone alerts are common because of climatic or other physical

**Table 5.3** Environmental Aspects Inventory

| Date | Plant Process | Process, Activity, or Service | Environmental Aspect | Impact | Significance Ranking Score |
|---|---|---|---|---|---|
| 6/6/2002 | Hot rolling | Handling of rolling oils | Potential for spill | Contamination of soils and/or ground water | × |
| 6/6/2002 | Recycling | Melting scrap | Particulate emissions | Air pollution | × |
| 6/6/2002 | Remelt | Melting scrap | Energy use | Resource depletion | × |

conditions, VOC emissions will probably make the list of significant environmental aspects. Another facility with a similar process but in a different setting—say, fewer industries in the area—may not consider VOC emissions significant. The critical point to this process is that a methodology must be developed to identify significance at each location. The methodology must be applied across the aspect inventory to identify and address significant environmental aspects.

Many facilities have developed an aspect scoring system based on a matrix to identify significant aspects. Table 5.4 illustrates one such matrix developed from the criteria suggested in Section 4.2.2 ISO 14004: 1996(E). Each impact identified would be evaluated using the seven criteria. Depending on the probability of occurrence, high, medium, or low points would be assessed for each criterion. This can be modified to suit the needs of the evaluation team. Each element in the aspect inventory should be evaluated using the matrix and be assigned a resulting score. Once all the aspects have been scored, it is necessary to identify which aspects have the potential to affect the environment substantially.

Some organizations have a predetermined limit. They identify the top $x\%$ of their scores as "significant." In other cases, natural breaks in scores are apparent, and the top group is easily identified. In any case, the management system must *describe the process*. Must a minimum or maximum number of significant aspects be identified? No. The process chosen by the system should be reasonable to the operation and organization, and it must be justifiable. In addition, the process should not be so complex that it is difficult for the evaluation team to understand or apply.

The EMS, by design, promotes continual improvement. Thus, as time passes, as objectives are set, and as programs are developed to mitigate adverse environmental impacts, environmental aspects whose related effects were once considered minimal may become more significant. What is not captured in the first cycle may well be captured in the future. ISO requires that the aspect inventory be current. Any physical change to the facility must be reviewed to determine if the aspect inventory requires updating. Should a new aspect be added, or has the significance of an existing aspect been influenced? The review must be

**Table 5.4** Evaluation of Significance of Impacts

| Criteria | High = 3 Points | Medium = 2 Points | Low = 1 Point |
|---|---|---|---|
| A. Severity scale of impact | Very hazardous. Material, effect will be severe. | Medium potential for environmental effect. | Release to the environment will be negligible. |
| B. Potential regulatory and legal exposure | Process is not capable of continued compliance. | Requirements that are more stringent will not be met with present controls. | Compliance is not an issue. |
| C. Cost of changing impact | Substitute material greatly increases raw material cost. | Substitute material moderately increases raw material costs. | Many substitutes available with little increase in cost. |
| D. Effect of change on other activities and processes | Process or material change will affect entire operation. | Some changes will be necessary in operations with changes in material or process. | Process or material change will not affect other operations or activities. |
| E. Concerns of interested parties | Process or materials are the subject of continual public concern. | Material exposures are being questioned by scientific community. | Little or no concern by media or public. |
| F. Effect on public image of the organization | Very hazardous material. Any release will cause significant public criticism. | Moderate hazard. Release may cause public criticism. | Little or no effect on public image. |
| G. Probability of occurrence | Historically, process controls have failed on a regular basis. | With proper operational controls, process is relatively stable. | Releases or failures are uncommon. |

performed if new equipment is installed, existing equipment is modified, or if materials used in the process are substituted (such as change the roll coolant base oil). If an increase in the magnitude or effect of some environmental aspect is caused by a project or action, the change will have to be reflected in the management system. Since the list of significant aspects establishes the foundation of the EMS, a change in this list causes a domino effect. If the significant aspect list is changed, its effect on the related elements of the EMS must be ascertained.

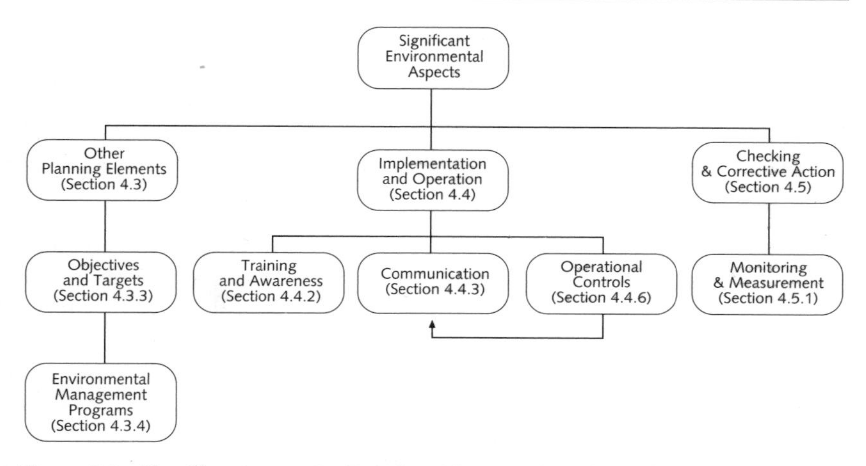


**Figure 5.6 Significant aspects: Relationships to other elements.**

Figure 5.6 illustrates the relationship of significant environmental aspects to the other elements of the management system structure. As this chapter reviews the requirements of the remaining elements of ISO 14001, note the relationship and influence of significant aspects on each of the elements in the illustration. Significant aspects have a direct impact on three sections of the management system; Section 4.3 Planning, Section 4.4 Implementation and Operation, and Section 4.5 Checking and Corrective Action.

In Section 4.3 Planning, significant aspects must be considered when the organization sets its environmental objectives and targets (4.3.3). In addition, once the objectives and targets have been established, the standard goes on to require that environmental management programs be developed for achieving those environmental objectives and targets (4.3.4). In Section 4.4, Implementation and Operation, the organization must identify its training needs (4.4.2). In doing so, it must assure that all personnel whose work may create a significant impact on the environment (significant aspect) receive appropriate training. *Significant aspects* are also addressed in Communication (4.4.3). Here the organization must consider processes for external communication regarding its significant aspects. The organization must also establish operational controls (4.4.6) for those operations and activities that have been identified as having significant impacts on the environment. Figure 5.6 shows that there is another requirement of communication related to operational controls. Specifically, relevant procedures and requirements spelled out in operational controls related to significant aspects must be communicated to suppliers and contractors.

In Section 4.5, Checking and Corrective Action, the organization must establish and maintain documented procedures in element 4.5.1 to monitor and measure, on a regular basis, the key characteristics of its operations and activities that can have a significant impact on the environment.

*Legal and Other Requirements*

> The organization shall establish and maintain a procedure to identify and have access to legal and other requirements to which the organization subscribes, that are applicable to the environmental aspects of its activities, products or services. [4.3.2, ISO 14001:1996(E)]

The organization must have access to and be able to identify its legal obligations. It must not lose sight of its additional obligations set forth by its corporate policies, guidelines, and other requirements. Many Alcan facilities established a registry of legal and other requirements and defined how that registry would be kept current. Many information services will, for a fee, provide this service customized for an organization. The advantage of such a service is that it highlights only the regulations pertinent to the operation as specified. The federal government [U.S. Environmental Protection Agency (USEPA)] and most state governments make their environmental regulations available on the Internet at no cost. In addition the U.S. Government Printing Office makes the *Federal Register* available online at no cost and will email the table of contents on a daily basis upon request.

The stated procedure must define how the registry will be kept up-to-date and how changes will be tracked. In addition, the procedure must also define how the legal and other requirements will be communicated to the rest of the organization.

*Objectives and Targets*

> The organization shall establish and maintain documented environmental objectives and targets, at each relevant function and level within the organization. When establishing and reviewing its objectives, an organization shall consider the legal and other requirements, its significant environmental aspects, its technological options and its financial, operational and business requirements, and the views of interested parties.The objectives and targets shall be consistent with the environmental policy, including the commitment to prevention of pollution. [4.3.3, ISO 14001:1996(E)]

In the previous elements of ISO 14001, procedures had to be established and maintained. For Section 4.3.3, environmental objectives and targets must be established, maintained, and *documented*. The standard is clear on where hard documentation is required. A company must be able to produce documented evidence of the organization's environmental objectives and targets. In addition, it must be able to show that, in setting the objectives and targets, it has considered significant environmental aspects, legal, and other requirements as well as technological options and financial and operational requirements and the views of interested parties.

If objectives clearly address these considerations, there is no problem in making the connection. Where environmental objectives do not clearly tie to these considerations, however, how can the company establish that the considerations clearly took place? One way is to document a procedure for setting objectives and targets that requires that these considerations take place. Thus, they are not forgotten. In addition, when meetings take place to discuss and set objectives and targets, they should also be documented in the form of minutes to produce the hard evidence that proper consideration was given in the process.

*Environmental Management Programs*

> The organization shall establish and maintain (a) program(s) for achieving its objectives and targets. It shall include: (a) designation of responsibility for achieving objectives and targets at each relevant function and level of the organization; (b) the means and time frame by which they are to be achieved.
>
> If a project relates to new developments and new or modified activities, products or services, program(s) shall be amended where relevant to ensure that environmental management applies to such projects. [4.3.4, ISO 14001:1996(E)]

The requirements of this element are straightforward. It requires that an action plan be established for each objective and target. The components that need to be identified in the action plan are clearly laid out; they include identification of accountability, milestones, and timetable. Periodic review of the programs should take place. Progress reports and/or minutes of project meetings provide evidence that this activity has taken place. This documentation should include amendments to management programs when relevant.

## Plan *Do* Check Act: Implementation and Operation

To execute and maintain an effective EMS, all employees in the company must be committed to it, from top management to staff personnel. Management must commit human, technology, and financial resources. Training needs must also be identified and provided to appropriate personnel to carry out specific environmental management activities. Specific procedures must be in place to address communications, both internally and externally, and to maintain sufficient documentation to implement the EMS. In addition, procedures to identify potential accidents and how to respond to them must be established and maintained. None of these requirements is impossible or even complex, but a few tips are worth mentioning that can facilitate implementation.

*Structure and Responsibility*

> Roles, responsibility and authorities shall be defined, documented and communicated in order to facilitate effective environmental management.
>
> Management shall provide resources essential to the implementation and control of the EMS. Resources include human resources and specialized skills, technology and financial resources.

> The organization's top management shall appoint (a) specific management representative(s) who, irrespective of other responsibilities, shall have defined roles, responsibilities and authority for: (a) ensuring that EMS requirements are established, implemented and maintained in accordance with this International Standard; (b) reporting on the performance of the EMS to top management for review and as a basis for improvement of the EMS. [4.4.1, ISO 14001:1996(E)]

Section 4.4.1 of ISO 14001 is another area where documentation is essential for meeting the requirements of the standard. One common way to meet this requirement is to have current organization charts and job descriptions in place. The most common problem in this area is to keep the documents current. In addition, the documents cannot be silent on the role of the management representative and the roles and responsibilities of each relevant position in respect to the EMS. With respect to the environmental management representative, it is critical that the individual not only have the defined role and responsibility but also have the authority to establish, implement, and maintain the management system. This cannot be delegated to an individual in the organization that does not have the authority to get the job done.

*Training, Awareness, and Competence*

> The organization shall identify training needs. It shall require that all personnel whose work may create a significant impact upon the environment have received appropriate training.
>
> It shall establish and maintain procedures to make its employees or members at each relevant function and level aware of: (a) the importance of conformance with the environmental policy and procedures and with the requirements of the EMS; (b) the significant environmental impacts, actual or potential, of their work activities and the environmental benefits of improved personal performance; (c) their roles and responsibilities in achieving conformance with the environmental policy and procedures and with the requirements of the EMS, including emergency preparedness and response requirements; (f) the potential consequences of departure from specified operating procedures.
>
> Personnel performing the tasks which can cause significant environmental impacts shall be competent on the basis of appropriate education, training, and/or experience. [Section 4.4.2, ISO 14001:1996(E)]

The scope of the organization's training program is somewhat defined by the significant aspects identified by the EMS. Before ISO 14001 is implemented, training is still required either by management or by existing regulations. Once the training process is formalized under a 14001-style management system, however, far more attention to detail is required to verify that the ISO training conditions are met. The training obligation became a topic of discussion at many environmental network meetings in Alcan. Most plants created a detailed training matrix that defined who needed what specific training and how often it had to be delivered or refreshed to meet this requirement.

The matrix was generally developed on a spreadsheet. The vertical axis of the matrix lists the job titles that required training. (Job titles were used in place of individual names to keep the matrix up to date when individuals changed jobs or left the organization or new individuals were added.) The horizontal axis defines the training topic and frequency, such as training for spill prevention control and countermeasures (SPCC). If certain job listings had specific requirements for training frequency that were different from others, that need was noted on the matrix. Table 5.5 illustrates one such training matrix that was developed at Alcan Rolled Products in Warren, Ohio.

Each facility had to develop a procedure to identify how atypical events would be addressed. For example, the procedure would define how and when new hires would be trained and how makeup sessions would be held for those trainees that missed regularly scheduled sessions. In that way, the EMS would assure that all the required training was completed.

Most facilities developed materials and completed training with in-house resources. Some of the larger facilities were able to dedicate training staff and physical resources to this task, and they provided individual training materials and computerized training modules. Others outsourced parts or this entire task. Regardless how the task is accomplished, the training matrix has become an essential tool.

#### *Communication*

> With regard to its environmental aspects and EMS, the organization shall establish and maintain procedures for: (a) internal communication between the various levels and functions of the organization; (b) receiving, documenting and responding to relevant communication from external interested parties.
>
> The organization shall consider processes for external communication on its significant environmental aspects and record its decision. [4.4.3, ISO 14001:1996(E)]

The organization needs to determine how it will communicate relevant environmental information to internal as well as external stakeholders. What is the process for receiving and responding to employee concerns relative to environment or the EMS? What is the process for addressing similar concerns from parties outside the organization such as neighbors or regulatory agencies? In addition to responding to requests, how does the organization communicate information on environmental performance? Is critical information derived from environmental audits moved through the organization to the appropriate level so that deficiencies are fixed? These are just a few of the questions that need to be

**Table 5.5** Environmental Training Plan

| Legend | Course Title | Frequency | Authority |
|---|---|---|---|
| A | RCRA Hazardous Waste | 1 year | 40 CFR 2656.16 |
| A′ | RCRA Haz. Waste Management Certification Course | 1 year | 40 CFR 2656.16 |
| B | (1) ISO 14001 Policy, Guidelines and EMS (job specific) | As needed | ISO 14001 |
| D | DOT HM-18 Overview (1-2h) | 3 years | 49 CFR 172 Sub H |
| D′ | DOT Hazardous Materials 126F (8h) | 3 years | 49 CFR 172 704 |
| F | SPCC General Overview | 3 years | 40 CFR 112 7(e)100 |
| K | Rain Water Discharge Procedures | 3 years | SPCC |
| N | (2) Environmental Incident Reporting | 3 years | ISO 14001 |
| P | Operational Control Pollution Control Systems SOPs/PMs | 3 years | ISO 14001 |
| Q | Coating Line Permit Conditions (Awareness) | 3 years | ISO 14001 |
| R | Operational Controls Pits | 3 years | ISO 14001 |
| S | Environmental Permitting in Ohio (8-12h) | Initial | ISO 14001 |
| T | ISO 14001 Internal Auditing, 3rd Party (5,2 or 1 day) | Initial | ISO 14001 |
| U | Plant Emergency Coordinator Training (3h) | Initial | 1910, 120, 38(a)(5), 1665(b)(4) |
| V | ISO 14001 Internal Auditor Course (6 hours in-house) | Initial | ISO 14001 standard |
| X | External Communications for Security Personnel | Initial | ISO 14001 standard |
| Y | EC Competency or EPRP Overview | Initial | ISO 14001 standard |

● = Training required this year

| Job Titles | A | A′ | B | D | D′ | F | K | N | P | Q | R | S | T | U | V | X | Y |
|---|---|---|---|---|---|---|---|---|---|---|---|---|---|---|---|---|---|
| C/L Operator | ● | | ● | ● | | ● | ● | ● | | | | | | | | | |
| Electrician | | | ● | | | ● | ● | ● | ● | ● | | | | | | | |
| Finish Helper | | | ● | | | ● | ● | ● | | | | | | | | | |
| Machinist | | | ● | | | ● | ● | ● | | | | | | | | | |
| Mechanic | | | ● | ● | | ● | ● | ● | ● | | ● | | | | | | |
| Relief Helper | ● | | ● | | | ● | ● | ● | ● | | ● | | | | | | |
| Utility | | | ● | ● | | ● | ● | ● | | | | | | | | | |
| Accountant | ● | | ● | | | ● | ● | ● | | | | | | | | | |
| BU Manager | | | ● | | | | | | | | | | | | ● | | |
| Clerk | | | ● | ● | | | | ● | | | | | | | | | |
| Controller | | | ● | | | | | | ● | | | | | | | | |
| Drafting | | | | | | | | | | | | | | | | | |
| Electrical Rel Cd | | | ● | | | | | | ● | ● | | | | | | | |
| Environmental Coordinator | | ● | ● | | ● | | | ● | ● | ● | | ● | | ● | ● | | ● |
| Lab Technician | ● | | ● | | | ● | ● | ● | | | | | | | | | ● |
| Maintenance Superintendent | | | ● | | | ● | ● | ● | ● | | | | | | | | |
| Material Coordinator | | | ● | | | | | | | | | | | | | | |
| Mechanical Engineer | | | ● | | | | | | ● | ● | | ● | | | | | |
| Mechanical Rel Cd | ● | | ● | | | ● | ● | ● | ● | ● | ● | | | | | | |
| Nurse | | | | | | | | | | | | | | | | | |
| Personnel | | | | | | | | | | | | | | | | | |
| Process Engineer | | | ● | | | ● | | ● | ● | | | | ● | | | | |
| Process Technician | ● | | ● | | | | | | | ● | | | | | | | |
| Purchasing | | | ● | | | | | | | | | | | | | | |
| QA Manager | | | ● | ● | | | | | ● | | | ● | | | | | |
| SH&E Manager | ● | | ● | ● | | ● | ● | ● | ● | ● | | | | | | | |
| Security | | | | | | | | | | | | | | | | | ● |
| Shipping Coordinator | | | ● | ● | | | | | | | | | | | | | |
| Site Manager | | | ● | | | | | | ● | ● | | | | | | | |
| Systems | | | ● | | | | | | | | | | | | | | |
| Team Leader | ● | | ● | ● | | ● | ● | ● | ● | ● | | | | | | | ● |

considered when developing the procedures to meet this element of the standard. Other pertinent information and practical help on this topic is found in Section 4.3.3.1 ISO 14004:1996(E).

*EMS Documentation*

> The organization shall establish and maintain information, in paper or electronic form, to: (a) describe the core elements of the management system and their interaction; (b) provide direction to related documentation. [4.4.4, ISO 14001:1996(E)]

The nature and extent of the environmental management system documentation is dictated by the size and complexity of the organization. Independent of this issue, however, the management system documentation at a minimum must describe the core elements of the management system and their interaction; it must also provide direction to the support documentation. Generally, this is demonstrated in the organization of the EMS manual and the related procedures. (Further relevant information concerning EMS documentation is found in the next section on document control.)

*Document Control*

> The organization shall establish and maintain procedures for controlling all documents required by this International Standard to ensure that: (a) they can be located; (b) they are periodically reviewed, revised as necessary and approved for adequacy by authorized personnel; (c) the current versions of relevant documents are available at all locations where operations essential to the effective functioning of the EMS are performed; (d) obsolete documents are promptly removed from all points of issue and points of use, or otherwise assured against unintended use; (e) any obsolete documents retained for legal and/or knowledge preservation purposes are suitably identified.
>
> Documentation shall be legible, dated (with dates of revision) and readily identifiable, maintained in an orderly manner and retained for a specified period. Procedures and responsibilities shall be established and maintained concerning the creation and modification of the various types of document. [4.4.5, ISO 14001: 1996(E)]

ISO 14001 requires that system documentation must be readily retrievable, and it must be reviewed and revised as necessary so that it is current. In addition, obsolete documents must be removed from the system or be clearly labeled as "obsolete" if they are to be retained for specific purposes such as legal documentation. The EMS must assure that when critical procedures or work instructions are being conducted, current procedures are available. Documentation and document control are discussed extensively in the ISO guidance because these support functions are critical to an operating EMS. Record-keeping elements define how the management system should be recorded and how required records should be controlled. A company's document control procedure affects the ease with which implementation of this element can be verified.

Specifications for document control can be managed on paper in a small organization with relatively few processes. Once the organization or facility becomes complex and the number of operations and procedures escalates, a paper system becomes a significant challenge. Many Alcan facilities selected to manage documentation in electronic format. This decision takes a commitment from the organization and could require making electronic workstations available throughout the workplace, something that is not always practical. Some facilities have selected to use a hybrid system where the documentation is in electronic format, but some is also maintained in paper format. Where this practice has been adopted, the paper phase is very selective and limited. Once the decision is made to go an electronic format, printed documentation generally becomes "uncontrolled" and should be identified as such. Where paper or printed documentation is necessary, its location must be identified in the EMS. When revisions are made, the obsolete material must be manually replaced promptly.

Several commercially available software packages can assist in maintaining and managing EMS documentation. Once documents are added to the database, they are available to the general workforce in a read-only format. Change can only be made by individuals with the appropriate clearance and only in a manner described by the approved procedures. This approach ensures that unauthorized changes cannot be made. In addition, when changes are made, reason for the change is recorded, and the changes can be traced.

### *Operational Control*

> The organization shall identify those operations and activities that are associated with the identified significant environmental aspects in line with its policy, objectives and targets. The organization shall plan these activities, including maintenance, in order to ensure that they are carried out under specified conditions by: (a) establishing and maintaining documented procedures to cover situations where their absence could lead to deviations from the environmental policy and the objectives and targets; (b) stipulating operating criteria in the procedures; (c) establishing and maintaining procedures related to the identifiable significant environmental aspects of goods and services used by the organization and communicating relevant procedures and requirements to suppliers and contractors. [4.4.6, ISO 14001:1996(E)]

Figure 5.6 shows the relationship of operational controls to significant aspects. Where operational activities are associated with an organization's significant aspects as well as its policy, objectives, and targets, the organization must have detailed documented procedures in place to address these issues. The organization should consider its different operations and the related activities that contribute to significant environmental impacts when developing operational controls.

*Emergency Preparedness and Response*

> The organization shall establish and maintain procedures to identify potential for and respond to accidents and emergency situations, and for preventing and mitigating the environmental impacts that may be associated with them.
>
> The organization shall review and revise, where necessary, its emergency preparedness and response procedures, in particular, after the occurrence of accidents or emergency situations.
>
> The organization shall also periodically test such procedures where practicable. [4.4.7, ISO 14001:1996(E)]

Although the requirements of Section 4.4.7 appear to be relatively straightforward, two areas can be problematic for implementation. Because of other regulatory requirements, most organizations have emergency preparedness procedures and plans in place—and generally stop there. ISO 14001:1996(E) requires that these plans or procedures be reviewed, especially after the occurrence of an accident or emergency. After the review, the plan must be revised if it was deemed inadequate. If a plan is not tested by the occurrence of an actual emergency or accident, the organization has an obligation to test the plan. This can be accomplished through periodic drills. The standard does not address the frequency of these drills because this should be prescribed in the organization's procedure. Once it is specified, it becomes a mandate. Once a plan has been tested though implementation or drill, the evaluation of its adequacy should be documented. Any subsequent revisions should be noted through the EMS document control system.

## Plan Do *Check* Act: Checking and Corrective Action

The procedures of the EMS are conventions to guide action in a predictable and organized manner. Once the EMS is in place, it cannot be allowed simply to operate. An essential step is to check its operation and utility. Checking and corrective action of the EMS identifies opportunities for improvement. A continuous improvement process betters overall environmental performance.

*Monitoring and Measurement*

> The organization shall establish and maintain documented procedures to monitor and measure, on a regular basis, the key characteristics of its operations and activities that can have a significant impact on the environment. This shall include the recording of information to track performance, relevant operational controls and conformance with the organization's environmental objectives and targets.

> Monitoring equipment shall be calibrated and maintained and records of this process shall be retained according to the organization's procedures.
>
> The organization shall establish and maintain a documented procedure for periodically evaluating compliance with relevant environmental legislation and regulations. [4.5.1, ISO 14001:1996 (E)]

When developing procedures to monitor and measure elements of the EMS, a realistic schedule should be established to monitor environmental performance, to calibrate and maintain monitoring equipment, and to evaluate compliance with relevant environmental legislation and regulations. This can be a significant task in a large complex organization. Use of electronic document control has truly facilitated implementation of this component of the standard. When monitoring or measurement tasks are carried out, the only written work instruction available is in an electronic format. This assures that the work instruction followed is the current one and that an obsolete work instruction will not be followed. In certain circumstances, such as when the work will be performed in a remote location, personnel would be allowed to print out the work instruction for reference with an "uncontrolled document" label printed on the document. Once the work was completed, the document would be destroyed.

*Nonconformance and Corrective and Preventative Action*

> The organization shall establish and maintain procedures for defining responsibility and authority for handling and investigating nonconformance, taking action to mitigate any impacts caused and for initiating and completing corrective and preventive action.
>
> Any corrective or preventive action taken to eliminate the causes of actual and potential nonconformance shall be appropriate to the magnitude of problems and commensurate with the environmental impact encountered.
>
> The organization shall implement and record any changes in the documented procedures resulting from corrective and preventive action. [4.5.2, ISO 14001:1996(E)]
>
> Considerations for corrective and preventive action include:
>
> - What process does the organization have to identify corrective and preventive action and improvement?
> - How does the organization verify that corrective and preventive actions and improvements are effective and timely? [4.5.3 in ISO 14004:1996(E)]

A process should be established to identify corrective and preventative action and to verify that corrective and preventative actions and improvements are effective and timely. Maintenance of these procedures along with retrieval of related records is facilitated by use of an electronic format. Alcan's software systems also have links to various Lotus

Notes databases. The software packages track progress and completion of corrective and preventative actions and facilitate implementation of Sections 4.5.2 and 4.5.3 of the management system standard.

#### *Records*

> The organization shall establish and maintain procedures for the identification, maintenance and disposition of environmental records. These records shall include training records and the results of audits and reviews.
>
> Environmental records shall be legible, identifiable and traceable to the activity, product or service involved. Environmental records shall be stored and maintained in such a way that they are readily retrievable and protected against damage, deterioration or loss. Their retention times shall be established and recorded.
>
> Records shall be maintained, as appropriate to the system and to the organization, to demonstrate conformance to the requirements of this International Standard. [4.5.3, ISO 14001:1996(E)]

Documentation has already been shown to be instrumental to the EMS. When developing procedures for record keeping it is essential that the records are readily retrievable by any individual in need of specific information and that key indicators of performance and appropriate environmental data are recorded so that the organization can track its progress in achieving its objectives.

#### *EMS Audit*

> The organization shall establish and maintain (a) program(s) and procedures for periodic EMS audits to be carried out, in order to: (a) determine whether or not the EMS conforms to planned arrangements for environmental management including the requirements of this International Standard; and has been properly implemented and maintained; and (b) provide information on the results of audits to management.
>
> The organization's audit program, including any schedule, shall be based on the environmental importance of the activity concerned and the results of previous audits. In order to be comprehensive, the audit procedures shall cover the audit scope, frequency and methodologies, as well as the responsibilities and requirements for conducting audits and reporting results. [4.5.4, ISO 14001:1996(E)]

The periodic audits of the EMS assess conformance with the ISO standard and proper implementation of the various elements. Moreover, they furnish critical information to management. The definition of *periodic* or the frequency of these audits should be spelled out in the procedure. Generally, this period will coincide with development and evaluation of the business plan. In any event, completion of this internal audit of the management system must be completed before the management review process described below.

Use of an electronic format can streamline the management system audit process. When documentation is easily retrievable and when related and relevant documents are linked, the entire process moves smoothly. Alcan has taken the next step and developed an electronic management system audit tool based on Lotus Notes to assess conformance with the standard. This arrangement facilitates analysis of results and, as multiple cycles are completed, provides a mechanism to begin to assess continuous improvement. The management system audit tool has been developed on an exception format. When conformance to a protocol question cannot be verified, a finding is recorded. Thus, at the end of the process nonconformances or other findings are identified.

According to the standard, the internal audit should be designed to answer two questions:

- Does the management system conform to the requirements of the international standard?
- Has is been properly implemented and maintained?

If the results of the internal audit only record deficiencies or exceptions, the record does not provide the information necessary to verify why or how conformance was demonstrated. The key word is *verify*. Conformance to the standard and evidence of implementation must be demonstrated during the audit process. Whatever tool or process is used to audit the system, the results must record this evidence. To supplement the information recorded in electronic format on an exception basis, auditors' notes must record the evidence that demonstrated conformance to each of the elements. The evidence may come from documents reviewed, observations made, or interviews conducted and recorded. This audit trail should be recorded and maintained to demonstrate that the audit process was complete. Exception reporting is still the focus of the Alcan tool because it provides direction to the elements that require improvement. Alcan's focus is believed to add value to the process. The results of the internal audit are made available to management for review and action.

## Plan Do Check *Act*: Management Review

The organization's top management shall, at intervals that it determines, review the EMS, to ensure its continuing suitability, adequacy and effectiveness. The management review process shall ensure that the necessary information is collected to allow management to carry out this evaluation. This review shall be documented.

The management review shall address the possible need for changes to policy, objectives and other elements of the EMS, in the light of EMS audit results, changing circumstances and the commitment to continual improvement. [4.6, ISO 14001:1996(E)]

Most organizations conduct management reviews of their business plan, and it is logical to include review of the EMS in this process. Several key items should be considered when developing the management review process:

- What mechanism/system is in place to review the EMS periodically?
- Are the appropriate individuals involved in the evaluation and possible revisions to the EMS?
- How does the system address input from interested parties during the management review process?

The record or documentation of this review is critical to verification. One requirement that troubled many plant managers was the reference to "changes to the policy. . . ." Where facilities have developed their own environmental policies in support of the corporate policy, review and appropriate modification identified by the management review raised no issue. In the facilities that did not develop their own separate policy, there was discomfort created by this requirement because of their perceived inability to change the corporate policy. At this point, plant management must once again reflect on the role of corporate management to set the direction for the corporation and define the role of the facility in the organization. Although unlikely, if management review discovers a conflict between the role of the plant and the corporate policy, the disagreement will have to be brought to and resolved with corporate management.

All facets of the management review including decisions, conclusions, and actions related to the adequacy and effectiveness of the EMS must be recorded and maintained as evidence of conformance with this element of the standard. Once the management review is complete, and required actions are taken, the process begins again with the planning of the next business cycle.

## Conclusions

Congratulations, by developing and implementing the EMS components described above, the management system is in place. The process is not over, however. This is just the end of the first cycle. The cycle is beginning again, and now the goal is continuous improvement. How is the system modified to make it better? To answer that, first answer these questions.

- What are the performance indicators?
- How is the success of the EMS measured?
- What should be measured and how should it be measured?
- Is the management system effective? This is a key question to be assessed during the management review. What will be reviewed to determine the answer?

The answers to these questions may not be obvious. Answers will be found in several places. First, consider the reason the management system was established. Then ask whether it accomplished its goal. If ISO 14001 was implemented in response to a customer requirement to get registration, that goal may have been satisfied and business with that customer has been maintained. Look more deeply, however, for additional advantages or benefits from implementation. Consider the corporate goals and targets. Recall that ISO 14001 requires that significant environmental aspects as well as legal and other requirements be considered in setting corporate goals and related management programs. Were goals actually set around those significant aspects? How did the company perform against these goals?

A formal EMS is not necessary to set environmental goals, but the EMS does help to focus and target the areas where the company may have the greatest outcome for the effort. When setting the business plan, ISO 14001 forces a look at environmental considerations. It does not permit setting business goals in an environmental vacuum.

The set of environmental performance indicators (EPIs) that are selected to measure environmental progress must be quantifiable. Goals that are quantifiable can be tracked and progress toward achieving them can be measured. It is much more difficult to estimate the value of improved performance if goals are subjective. An example is an energy-related goal. If energy consumption is a significant cost of operation, energy usage may be an appropriate EPI. To set an objective to "improve energy usage" may seem logical, but how will performance be measured against that objective? A more appropriate statement of the objective may be to "reduce energy consumption in the next year by 2%." This objective has set a quantifiable criterion: to reduce energy usage by 2% over a 1-year time frame. It may be appropriate to define this objective further by stating a baseline year from which progress will be measured. Progress can be tracked throughout the year so that appropriate adjustments can be made to achieve the goal within the required time. If the company has just completed a major expansion and production is expected to increase by 25%, an absolute reduction in energy use may be an impossible challenge. Energy can remain an EPI, but the goal may have to be restated in a way that will reduce the environmental effect but still allow the business to grow. The restated goal could be based on efficiency. An example of this type of goal may be to "reduce the number of gigajoules (GJ) per ton of production." The goal could also be stated as to "increase energy efficiency per ton of production by 2%." If in 2003 the operation used 100,000 GJ to generate one ton of production, the goal in 2004 could be to use only 98,000 GJ to produce that ton of product. Achieving the target will require improvements to the operation and may mean investments in burners or motors that are more energy efficient.

It is very important in the next step to assess all of the economic benefits achieved in addition to those reduced environmental impacts. In this case, the cost of a gigajoule of energy can be calculated. After the costs have been accounted for and energy savings measured, the result may be that total production costs have gone down. These savings need to be quantified to determine the overall economic benefit of the management system. This type of analysis becomes easier as the EMS matures. In the beginning, corporate resources may have to be invested to establish baseline data. Data systems required to measure some of these indicators may not be in place and will have to be developed. It may take several cycles of plan–do–check–act to determine the benefit.

In 2002, Alcan facilities developed over 50 projects designed to reduce environmental impacts of significant aspects. At the end of that year, the benefits in reduced environmental effects were tabulated and the most significant were reported in the company's updated sustainability report. This was the first time that this information was generated as a result of reduced impacts targeted and realized from implementation of ISO 14001.

This chapter began by saying that toward the end of the 20th century, industry quickly learned how pollution prevention paid benefits. That fact is still true today, and implementation of an ISO 14001 management system can help to realize the entire benefit. This is done by targeting the proper areas, setting sound environmental objectives, and measuring appropriate parameters. Thus, this quickly becomes a win/win situation and both the environment and business benefit.

## Exercises

1. In 1987, the Brundtland Commission reported that "humanity has the ability to make development sustainable—to ensure that it meets the needs of the present without compromising the ability of future generations to meet their own needs." In Alcan, Inc.'s, management strategy toward sustainability, what three factors meld together as the company moves up the sustainability path? Explain.

   *Note:* As a company moves up the sustainability path toward a value-based company, environmental, economic, and social factors become integrated into the business management system and are no longer identifiable as separate standalone activities.

2. The authors suggest that for industry to thrive, it must make investments not only in economic terms (capital) but in environmental and social capital as well. Explain.

   *Note:* See discussion of Figure 5.2. This illustration of the marketplace clearly demonstrates the value of understanding social values and the importance of investments in social capital. To answer this question successfully, the student will have to expand the answer to include investments in environmental capital such as investments in pollution prevention, environmental management systems, reduction in environmental impacts or "footprint," better efficiencies in using raw materials and energy, and so forth.

3. What is ISO 14000? Give Examples.

   *Note:* ISO 14000 is a not one standard but rather a family of standards relating to environment established by the International Organization for Standardization. Some examples were found in Table 5.1 (reproduced here):

| ISO Standard | Content |
| --- | --- |
| ISO 14001 and 14004 | Environmental management system |
| ISO 14010, 14011, and 14012 | Environmental auditing |
| ISO 14021, 14024, and 14025 | Environmental labeling |
| ISO 14031 | Environmental performance evaluation |
| ISO 14060 | Environmental aspects in product standards |
| ISO 14041, 14042, and 14043 | Life-cycle analysis |

4. Explain the Deming PDCA cycle, and relate it to ISO 14001.

   *Note:* Table 5.2 (reproduced here) shows the relationship of various elements of Deming's Plan Do Check Act Cycle to the ISO 14001 Standard. To be most complete the student must indicate that this is a repeating cycle. Once through the cycle is not enough. To stop at that point in the explanation would not be enough. The student should explain that by repeating the cycle one could be expected to achieve continual improvement. That is the ultimate goal.

| Plan | Do | Check | Act |
| --- | --- | --- | --- |
| Section 4.3: Planning | Section 4.4: Implementation and operation | Section 4.5: Checking and corrective action | Section 4.6: Management review |

5. Explain in detail at least four reasons why an organization would want to implement an ISO 14001 environmental management system.

   *Note:* Five reasons are explained in detail in the text. Students should be able to use reasons not given in the text if they develop sound logic in their explanations.

6. An *environmental aspect* is an element of an organization's activities, products, or services that can interact with the environment.
   a. How would you go about identifying an organization's significant aspects? Be specific and use examples.
   b. Explain the importance of identifying *significant environmental aspects* to the functioning of the rest of the ISO 14001 environmental management system.

   *Note:*
   a. The student should start by explaining the development of the *aspects inventory*. Reference should be made to the process exhibited in Figure 5.5; the process of considering the various inputs and outputs of the activities carried out at the facility. Once the aspects inventory has been developed, a mechanism must

be identified to determine which aspects have the potential to have significant effect on the environment. Table 5.4 represents one possible example. Once applied, the list of significant environmental aspects is complete for the first cycle.

b. See Figure 5.6 and the related discussion in the text.

**7.** ISO 14001 Section 4.4.2, Training, Awareness and Competence, can require significant effort to implement effectively. List at least five requirements of this section and describe how they could be implemented effectively.

*Note:* The student may list any five requirements from the requirements of ISO 14000 Section 4.4.2, as presented in the text. The student should identify that the effort for implementation will vary with the size and complexity of the organization. To meet the numerous requirements of the standard, a company could develop an approach that uses a training matrix. Students may develop an alternative approach, but they need to be mindful of the requirements of the standard and the approach demonstrate conformance.

**8.** If you were a site manager who had just implemented an ISO 14001 environmental management system, how would you assess the effectiveness of your effort?

*Note:* Answers to this question can be developed from the Conclusions section of the chapter. Some of the important concepts developed there were:

- Consider the reason the management system was established. Then ask whether it accomplished its goal.
- In the development of the business plan, were significant aspects considered?
- If goals and targets were set around significant aspects, were their effects reduced?
- Goals and targets should be quantifiable.
- As the management system matures, measuring performance should get easier.

The student may widen the discussion around additional measures of assessment if they are logically developed.

## References

Brumtland, O., ed. 1987. *Our Common Future: The World Commission on Environment and Development.* Oxford University Press: Oxford.

International Organization for Standardization. ISO 14000 standards. http://www.iso.ch/iso/en/iso9000-14000/iso14000/iso14000index.html. A fee was paid to the American National Standards Institute for a license to use passages from the ISO 14000 standards.

reproduced in any form, electronic retrieval system or otherwise or made available on the Internet, a public network, by satellite or otherwise without prior written consent of the American National Standards Institute, 25 West 43rd Street, New York, NY 10036.

Lomborg, B. 2001. *The Skeptical Environmentalist, Measuring the Real State of the World.* Cambridge University Press: Cambridge.

# CHAPTER 6

# Zero Liquid Discharge—The Ultimate Design Challenge

TIMOTHY J. BARRY AND MICHAEL J. TRACY

## Water Use and Automobile Manufacturing

Industries, cities, and farms across the globe have voracious claims on water. Products must be manufactured, thirsts slaked, and living plants irrigated. With populations and economies galloping ever forward, the demand for water might outpace the supply—all things being equal. "All things," however, are never equal. The ability of humans to apply the principles of sustainability to water can extend supplies or even lessen the demand. Water is fungible; that is, one form of water, say a polluted one, can sometimes be substituted for another, say freshwater from a stream—once adequately treated. Even New York City's seemingly endless appetite for additional water will likely be met in the future with creative and sustainable application of the existing supply rather than the construction of new reservoirs (Endreny, 2001). This chapter presents the intricate but feasible complexity of implementing sustainable principles to extend an industrial water supply at the DaimlerChrysler automobile assembly plant in Toluca, Mexico. The water system is a closed-loop system with only about 20% of the water lost to evaporation. This approach has permitted the plant to reduce its demand for groundwater by almost 1 million liters/day (264,000 gal/day). Treatment and reuse of the water have saved the company more than $1 million/year (DaimlerChrysler, 2003).

Every grade-school pupil is familiar with the hydrologic cycle (evaporation–transpiration, condensation, precipitation, runoff–percolation). The *use* of water also follows a cycle: withdrawal, transport, application, discharge, and reuse. Water is central to any economy. Manufacturing enterprises use water in vast quantities to produce individual products—such as 5000 gal for a single ton of steel or 100,000 gal for a single automobile—including its steel, plastics, glass, and fabric (National Academy of Sciences, 2002, citing Thompson, 1999). Economic growth, population growth, and climatic conditions in specific locales call for efficient water management in both quantity and quality. (See pattern of population growth and water use in Figure 6.1.) Water is also a bellwether for the environment since it is the ultimate sink for

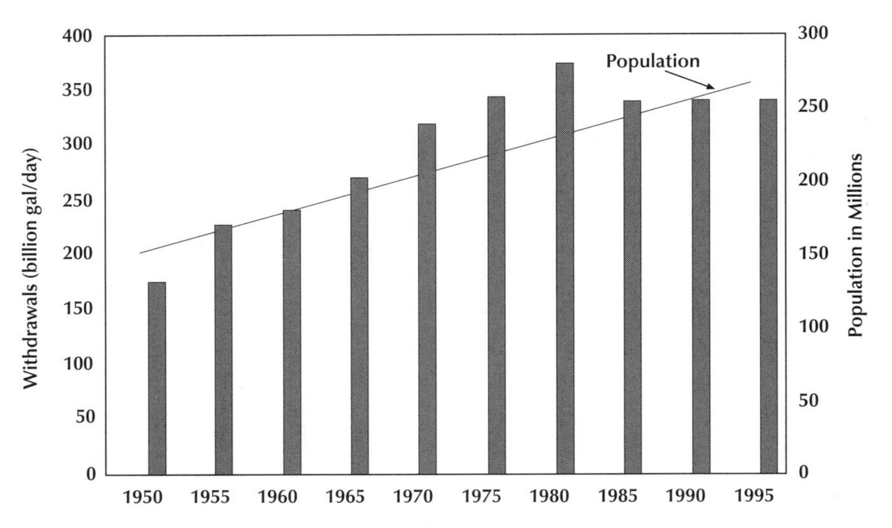


**Figure 6.1 Trend of water use and population in United States.** (*Source*: Adapted from National Academy of Sciences, 2002.)

pollutants. Even chemicals on the ground surface will reach groundwater; air pollutants will eventually wash down to the oceans, lakes, or streams and can reach aquifers.

Human uses of water interface with the natural system first at the point of withdrawal and again at the point of discharge. These uses usually begin with withdrawals from the natural water system of streams, water bodies, and aquifers; some uses of water, however, such as hydroelectricity generation, do not require that the water be withdrawn. Humans use a variety of water facilities to transport water for use—intakes, conveyance pipes, pumps, and treatment facilities. The discharged water is generally different in chemical characteristics (such as salinity) or physical characteristics (such as temperature) than was the withdrawn water. In addition, the quantity of the water returned to the natural system is often less than that withdrawn because of consumption, loss in the manufacturing process, evaporation, or other use. Furthermore, the discharged and altered water that is returned to the natural system may introduce contaminants, by-products of the water's use, to the wider water resources. These contaminants may be returned to the original source or may be sent to a new watershed in the case of an interbasin transfer of water (*use* in one watershed—*discharge* in another).

Water is hardly just a natural commodity. It is also an economic commodity and a political commodity. Water's scarcity—even in water-rich regions it is not always in the place where it will be used—gives it a value clearly recognizable in the enormous costs of water resource projects such as dams and treatment facilities. The political factor is clear in the pages of laws that govern water takings and uses. Thus, not only is water the central element of any ecosystem, it is enmeshed with the social and political values of the nation where it flows. Most cultures of the world have even imputed spiritual attributes to water, and it is routinely used with symbolic meaning in religious ceremonies. Conflicts over the allocation of water are most pronounced in arid regions. Arizona long resented California's use of Colorado River water (a water quantity issue), and that same resource has led to arguments between the United States and Mexico (a water quality issue). The United States and Canada have debated more than once the possible interbasin transfer of water from the Great Lakes to the arid southwestern states.

## Industrial Water Use

Since and because of passage of the U.S. Clean Water Act in 1972, industry's use of water in manufacturing has become more efficient, leading to far more recycling of water. The wastewater treatment requirements imposed by the Clean Water Act effectively increased the cost of water, and reuse rather than outright discharge became more economical. If, as is expected, the cost of water continues to rise as scarcity and quality issues limit the quantity of usable water, reuse should continue to increase. Compare rates of water recycling in 1954—before active federal enforcement of water quality standards—and 2000 (Table 6.1) for an indication.

In discussions of sustainable development, water is a central issue. It is addressed in two ways: conservation and integrity. Among the 12 principles of green engineering (see Introduction), the sustainability of resources (inputs) including water is addressed:

- *Reduction of water input:* It is better to prevent waste than to treat or clean up waste after it is formed (principle 2).
- *Reuse of water:* Separation and purification operations should be designed to minimize . . . materials use (principle 3).
- *Closed-loop water use:* Products, processes, and systems should be designed to maximize mass, energy, space, and time efficiency (principle 4). Material and energy inputs should be renewable rather than depleting (principle 12).
- *Attuned to ambient water supply:* Design of products, processes, and systems must include integration and interconnectivity with available energy and materials flows (principle 10) (Anastas and Zimmerman, 2003).

**Table 6.1** Water Recycling Rates (%)

| Industry | 1954 | 2000 |
|---|---|---|
| Paper | 2.4 | 11.8 |
| Chemicals | 1.6 | 28.0 |
| Petroleum | 3.3 | 32.7 |
| Primary metals | 1.3 | 12.3 |
| Manufacturing | 1.8 | 17.1 |

*Source:* National Academy of Science (2002) from Thompson (1999). Reprinted with permission of Elsevier.

How much water does it take to make an automobile? How much water is too much? Neither question is easily answered. Accounting for water—unless performed by the company actually making the automobile—is fraught with the opportunity for errors, errors that multiply upon themselves. Moreover, different companies typically account for water differently, too, making comparisons across companies challenging, if not impossible. And, the volume of water used depends on the supply available. Surely, the water-rich state of Michigan is less concerned with water consumption than is the perennially dry Georgia—except in times of drought.

The German automaker BMW is concerned with the use of water, although its primary environmental concerns are fuel efficiency and the recycling of used automobiles (BMW, 2003b). In 2001, BMW used

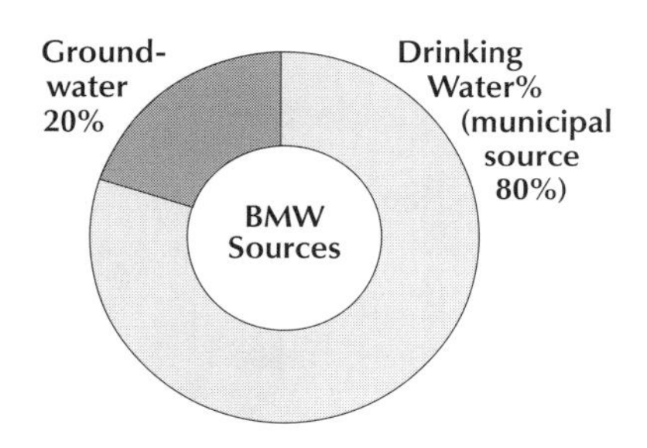


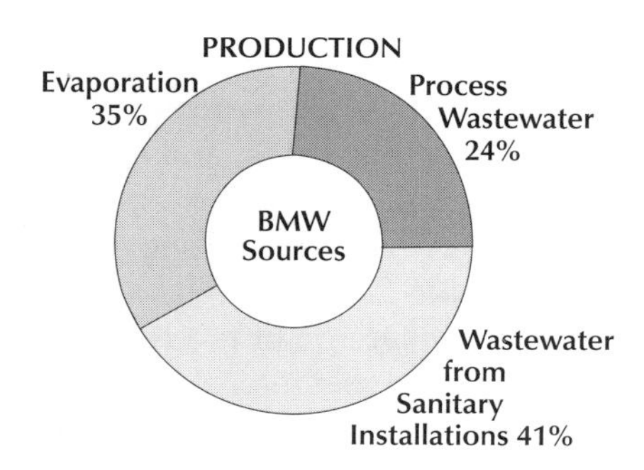


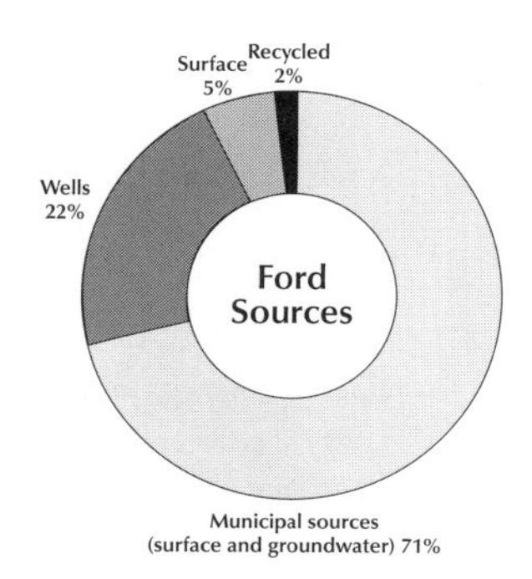


**Figure 6.2 Sources and uses of water.**

3.5 million cubic meters (924 million gallons) worldwide. In 1991, the comparable figure was 4.7 million cubic meters [1242 million gallons, about the amount New York City used in a single day in 2002 (NYC Dept. of Environmental Protection, 2003)]. Figure 6.2 (top) shows that most of the water was taken from a municipal source. The middle graph of Figure 6.2 shows that approximately 60% of the water was used in BMW automobile production (process and evaporation), and the remainder was used in sanitary facilities. The fact that sanitation is a major use of water in a manufacturing facility comes sometimes as a surprise. In many industries that require labor-intensive manufacturing and assembly, the sanitary wastewater from on-campus cafeterias, restrooms, and showers can contribute a significant portion of the facility's overall wastewater. In the DaimlerChrysler facility that is described in the second part of this chapter, the sanitary wastewater stream was specifically accounted for in its recycling strategy.

BMW identifies water conservation as a goal and the use of "recycling and circulation systems based on natural models" as ways to meet its goal (BMW, 2002). Ford Motor Company (Figure 6.2, bottom) explicitly identifies recycled water as a significant source (Ford Motor Company, 2003).

Automobile manufacturers worldwide are looking to reduce the volume of water used—and consumed. In some cases, they wish to reduce wastewater treatment costs; in other cases, they have plants located in arid regions and have restricted supplies—as with DaimlerChrysler in Toluca, Mexico; and in still other cases, they see a reduction in water use as a "green goal," albeit one that helps to increase profits. The automaker BMW touts that "process water requirements dropped by over one-third, wastewater disposal by 16 percent. . . ." Its figures are presented in Table 6.2. The company's concerted effort to conserve water use resulted in a decline of 0.21 $m^3$ (65 gal) per automobile by 2000, using 1996 as the base year (see Figure 6.3).

Ford Motor Company said it used 42.2 million cubic meters (11.1 billion gallons) in 2001 (Ford Motor Company, 2003). Between 1996 and 1998, Ford manufacturing plants worldwide reduced water usage by 11.4 million cubic meters (>3 billion gallons). In 2000, the company set a global goal to reduce water use by 3% by 2003. Because of the complexities of water supply and use, individual Ford facilities have set individual targets to participate in the company-wide goal. From 2000 to 2001, it measured, as part of its ISO 14001 environmental management system, a reduction in water use of 8.9% (2.8% on a per-vehicle basis). That is 4.1 million cubic meters (>1 billion gallons). Ford continues to develop tools to measure and manage facility-specific water conservation and management plans to assist it to meet its goals. Toyota North America has set as a goal a reduction of total water use by 15% per unit; it uses 2000 as its base year and 2005 as its target year (Toyota, 2002). This goal builds on Toyota's reductions in water use that started in the 1990s (Table 6.3).

**Table 6.2** Water and Wastewater Use Reported by BMW

| | 1996 | 1997 | 1998 | 1999 | 2000 |
|---|---|---|---|---|---|
| Vehicle production (units) | 639,433 | 672,238 | 706,426 | 755,547 | 834,519 |
| Wastewater total | 2,231,194 $m^3$<br>589,419,000 gal | 1,985,842 $m^3$<br>524,604,000 gal | 2,340,409 $m^3$<br>618,271,000 gal | 2,131,837 $m^3$<br>563,172,000 gal | 2,206,733 $m^3$<br>582,957,000 gal |
| Water consumption/water input | 3,736,900 $m^3$<br>987,184,000 gal | 3,468,948 $m^3$<br>916,399,000 gal | 3,423,820 $m^3$<br>904,478,000 gal | 3,403,209 $m^3$<br>899,033,000 gal | 3,344,939 $m^3$<br>883,639,000 gal |
| Process water input | 2,917,520 $m^3$<br>770,727,000 gal | 2,717,549 $m^3$<br>717,900,000 gal | 2,737,398 $m^3$<br>723,144,000 gal | 2,650,677 $m^3$<br>700,235,000 gal | 2,481,127 $m^3$<br>655,444,000 gal |
| Process wastewater per unit produced | 1.27 $m^3$<br>335 gal | 1.27 $m^3$<br>335 gal | 1.23 $m^3$<br>325 gal | 1.15 $m^3$<br>304 gal | 1.06 $m^3$<br>280 gal |

*Source:* Compiled from BMW (2002).

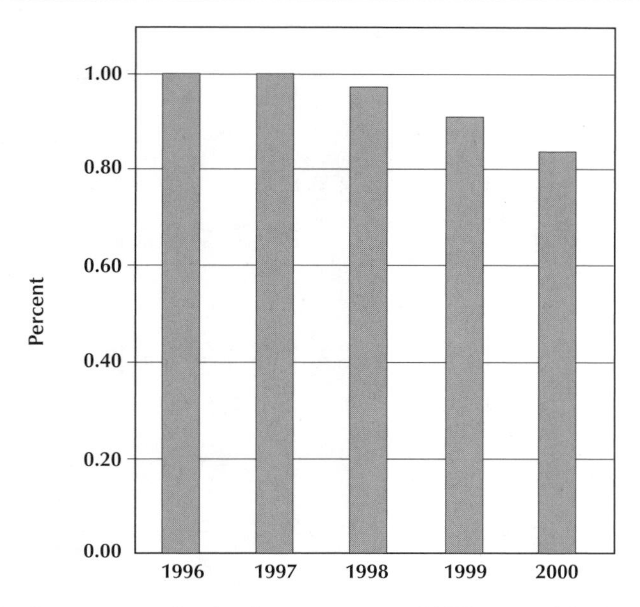


**Figure 6.3 Percentage reduction in water use per BMW automobile, 1996–2000.**

**Table 6.3** Pattern of Water Use at Toyota[a]

| Year | 1997 | 1998 | 1999 | 2000 | 2001 |
|---|---|---|---|---|---|
| Water ($m^3$/vehicle) | 4.6 | 4.6 | 4.3 | 4.2 | 4.3 |

*Source:* Compiled from Toyota (2002).
[a]Fiscal year 2002: April 2001–March 2002.

## Pollution Generation

In 2000, the U.S. Environmental Protection Agency (USEPA) surveyed 54 automobile assembly facilities under its Sector Facility Indexing Project. Each released "an average of 675,393 pounds" of chemicals recorded in the Toxics Release Inventory (TRI; see Chapter 1 and box below). Of these, 49,620 lb were carcinogens (USEPA Sector Facility Indexing Project, 2003a); see Table 6.4. Generally, solvents and heavy metals are the pollutants most reported to the USEPA from automobile assembly plants. See Table 6.5 (USEPA, 2003b); note that the columns do not sum to the total row because only chemicals that were released to surface water are listed.

**Table 6.4** USEPA SFIP Data Averages for 54 Automobile Assembly Facilities

| | |
|---|---|
| Production capacity (vehicles/yr) | 214,707 |
| TRI releases | 675,393 lb |
| TRI off-site transfers | 660,555 lb |
| TRI releases, carcinogens | 49,620 lb |
| TRI releases and transfers, metals | 33,301 lb |
| Total waste generated | 1,666,654 lb |

**Table 6.5** Chemicals Released in 2001 by Transportation Equipment Industry[a]

| Chemical | Surface Water Discharges | Total Air Emissions | Underground Injections | Releases to Land | Total On-site Releases | Total Off-site Releases | Total On- and Off-site Releases |
|---|---|---|---|---|---|---|---|
| Total | 198,256 | 66,691,884 | 750 | 773,497 | 67,664,387 | 13,388,968 | 81,053,354 |
| Nitrate compounds | 136,838 | 1,868 | 0 | 0 | 138,706 | 158,156 | 296,862 |
| *n*-Butyl alcohol | 18,833 | 5,032,351 | 0 | 0 | 5,051,184 | 11,716 | 5,062,900 |
| Zinc compounds | 12,199 | 71,869 | 0 | 14,881 | 98,949 | 1,609,652 | 1,708,601 |
| Ammonia | 8,497 | 605,421 | 0 | 0 | 613,918 | 117,580 | 731,498 |
| Xylene (mixed isomers) | 4,113 | 15,579,263 | 0 | 260 | 15,583,636 | 18,025 | 15,601,661 |
| Copper compounds | 2,520 | 114,412 | 0 | 9 | 116,941 | 309,550 | 426,491 |
| Nickel | 2,364 | 118,751 | 0 | 4,221 | 125,336 | 307,016 | 432,352 |
| Copper | 2,043 | 42,419 | 0 | 191,087 | 235,549 | 720,851 | 956,400 |
| Manganese | 1,821 | 279,894 | 0 | 378,232 | 659,947 | 336,463 | 996,410 |
| Chromium | 1,599 | 102,391 | 0 | 49,505 | 153,495 | 2,500,952 | 2,654,447 |
| Manganese compounds | 897 | 33,432 | 0 | 0 | 34,329 | 903,317 | 937,646 |
| Chromium compounds[b] | 807 | 12,232 | 0 | 10 | 13,049 | 1,721,902 | 1,734,951 |
| Toluene | 766 | 4,409,351 | 0 | 9,240 | 4,419,357 | 9,676 | 4,429,033 |
| Zinc (fume or dust) | 750 | 14,397 | 0 | 23,501 | 38,648 | 109,710 | 148,358 |
| Nickel compounds | 613 | 5,574 | 0 | 10 | 6,197 | 322,925 | 329,122 |
| Lead | 579 | 12,216 | 0 | 2,743 | 15,537 | 173,066 | 188,604 |
| Barium compounds | 492 | 188,860 | 0 | 29,683 | 219,035 | 1,344,317 | 1,563,352 |
| Lead compounds | 330 | 19,011 | 0 | 219 | 19,559 | 423,104 | 442,664 |
| Styrene | 286 | 14,394,547 | 0 | 500 | 14,395,333 | 332,751 | 14,728,084 |
| Methyl ethyl ketone | 263 | 2,765,148 | 0 | 4 | 2,765,415 | 57,380 | 2,822,795 |
| Ethylene glycol | 251 | 126,488 | 750 | 250 | 127,739 | 66,188 | 193,927 |
| Certain glycol ethers | 120 | 5,737,224 | 0 | 0 | 5,737,344 | 285,068 | 6,022,412 |

| | | | | | | | |
|---|---|---|---|---|---|---|---|
| Phenol | 39 | 190,500 | 0 | 34,265 | 224,804 | 20,198 | 245,002 |
| Dichloromethane | 33 | 294,600 | 0 | 0 | 294,633 | 4,162 | 298,795 |
| Aluminum oxide (fibrous forms) | 13 | 237 | 0 | 0 | 250 | 107,633 | 107,883 |
| Methyl isobutyl ketone | 10 | 3,228,743 | 0 | 0 | 3,228,753 | 3,501 | 3,232,254 |
| Ethylbenzene | 8 | 2,344,960 | 0 | 10 | 2,344,978 | 1,570 | 2,346,548 |
| Trichloroethylene | 7 | 501,425 | 0 | 0 | 501,432 | 16,027 | 517,459 |
| Aluminum (fume or dust) | 5 | 46,625 | 0 | 27,015 | 73,645 | 239,643 | 313,288 |
| Nitric acid | 5 | 73,349 | 0 | 13 | 73,367 | 35,919 | 109,286 |
| Methanol | 1 | 1,707,068 | 0 | 0 | 1,707,069 | 728 | 1,707,797 |
| *n*-Hexane | 1 | 1,239,348 | 0 | 0 | 1,239,349 | 262 | 1,239,611 |
| Methyl methacrylate | 1 | 587,113 | 0 | 0 | 587,114 | 45,095 | 632,209 |
| Tetrachoroethylene | 1 | 294,868 | 0 | 0 | 294,869 | 2,181 | 297,050 |
| Sodium nitrite | 1 | 7,580 | 0 | 0 | 7,581 | 134,601 | 142,182 |

*Source:* USEPA, 2003b.
[a]SIC 37. Only chemicals found in surface water discharges are listed here.
[b]Except chromite ore mined in Transvaal region.

**Toxics Release Inventory**

The U.S. Emergency Planning and Community Right-to-know Act (EPCRA) was passed following a fatal chemical-release accident in Bhopal, India, to promote emergency planning, to minimize the effects of accidental chemical releases, and to give the public information on releases of toxic chemicals in their communities. §313 of EPCRA established the Toxics Release Inventory (TRI) Program. TRI is a national database that identifies:

Facilities

Chemicals manufactured, processed, and used at the facilities

Amounts of these chemicals released in routine operations *including spills and accidents* otherwise managed on- and off-site in waste

The 1990 U.S. Pollution Prevention Act included a mandate to expand TRI to address additional information on toxic chemicals in waste and on source reduction methods. Beginning in 1991, the affected facilities were required to report quantities of TRI chemicals treated on-site, recycled, and combusted for energy recovery. This waste management data has strengthened TRI as a tool for providing information on facilities' handling of TRI chemicals in waste as well as for analyzing progress in reducing releases.

The TRI Program has been a notably successful program. The industries that have reported to TRI since its inception have reduced their on- and off-site releases of TRI chemicals by 54.5% or 1.72 billion pounds (for chemicals reportable in all years). For the reporting year 2000, TRI was expanded to include certain new persistent bioaccumulative toxic (PBT) chemicals. In addition, reporting thresholds were lowered for both the newly added PBT chemicals and certain PBT chemicals already on the TRI list.

*Source:* USEPA n.d.

The USEPA data from 2001 (Table 6.5) show that chemicals were released to environmental media thus: 82% to air, 24% to water, <1% to land. The releases on the premises of the vehicle equipment industry were 83%, and off-site releases were 17%. The largest volume of wastewater from vehicle manufacturing is from metal plating and contains metal wastes—cadmium, chromium, copper, lead, mercury, nickel, and cyanide (Paterson, 1985). Coolant from metal-working processes becomes contaminated and spent over time through extended use and reuse. These oils can contain metals such as lead, cadmium, and chromium. Chlorine, sulfur, phosphorous compounds, phenols, cresols, and alkalines can also be found in the spent oils (USEPA, 1995). Liquid and solid wastes may contain metals from pigments and organic solvents. Releases of the chemicals listed in Table 6.6 must be accounted

**Table 6.6** Common Sources of Common Pollutants in Wastewater from Automobile Manufacturing

| | |
|---|---|
| Hexavalent chromium | Used in paint pigments, metal cleaning, and electroplating. Automotive manufacturing is the largest use of chromium plate metal parts. Frequently, major source of chromium is in chromic acid bath and rinse water used in metal plating operations. |
| Barium | White pigment for paints and vulcanized rubber. |
| Aluminum | Cleaning and etching of aluminum parts produces spent baths and rinse waters with aluminum. |
| Cadmium | Inorganic pigments, electroplating, and metallurgical alloying. |
| Iron | Used in paint pigments, metal processing industry, and polishing powder. Iron waste frequently is from acid pickling of iron and steel to remove surface oxides. The presence of cadmium and lead in iron may also contaminate the wastewater (USEPA, 1995). |
| Selenium | Paint pigment |
| Zinc | Painting operation |
| Phosphorus | May be present because of phosphating operations. (Iron phosphating is used to provide steel surfaces with good adhesion to paints; zinc phosphating is used against corrosion and as pretreatment for steel and aluminum surfaces in advance of painting or oiling for cold flow pressing; mangano phosphating impedes the adhesive wear of mechanical parts of internal combustion engines or bearings. |

for when designing wastewater treatment for a vehicle manufacturing facility.

Table 6.7 presents important information about pollution control in the manufacture of automobiles from the perspective of green engineering and green chemistry. Data arrayed as inputs and outputs can be analyzed for components that are the most harmful to the environment and that can be reduced most beneficially. For the purposes of this chapter,

**Table 6.7** Material Inputs and Pollution Outputs

| Process | Material Input | Process Wastes (Wastewater and Liquids) |
|---|---|---|
| Metal cutting and/or forming | Cutting oils, degreasing, and cleaning solvents, acids, and metals | Acid/alkaline wastes and waste oils. Metals copper, chromium, and nickel as well as BTEX (benzene, toluene, ethylbenzene, and xylenes) |
| Heat treating | Acid/alkaline solutions, cyanide salts, and oils | Acid/alkaline wastes, cyanide wastes, and waste oils. Metals copper, chromium, nickel |
| Solvent cleaning | Acid/alkaline cleaners and solvents | Acid/alkaline wastes and BTEX |
| Pickling | Acid/alkaline solutions | Acid/alkaline wastes and metals |
| Electroplating | Acid/alkaline solutions, metal bearing, and cyanide bearing solutions | Acid/alkaline wastes, cyanide wastes, plating wastes, metals, reactive wastes, and solvent wastes |

*Source:* USEPA (1995).

the information in Tables 6.6 and 6.7 give an indication of the chemicals that required treatment on the way to the goal of achieving zero liquid discharge in the Toluca manufacturing facility.

## Vehicle Production in Toluca, Mexico

Toluca, Mexico, has benefited from the movement of industry from Mexico City (Figure 6.4). Toluca is situated just west of the Distrito Federal where Mexico's capital city, Mexico City, has long been the country's primary economic and cultural center. Crowding and air pollution were two factors in the migration of industry to Toluca. In Toluca, businesses found plentiful labor and space. They did not, however, find plentiful water. Located in south-central Mexico on the southwestern border of the Anáhuac plateau, it is the capital of Mexico state. The city is home to more than 325,000 people, and Mexico state has over 11.7 million people in an area of 8200 mi$^2$ (21,200 km$^2$) with 122 municipalities. At 9000 ft (2700 m), Toluca is the highest city in Mexico. Toluca receives about 600 to 1000 mm of precipitation annually (24 to 40 inches), and two small rivers flow through the city. The rainfall is insufficient to recharge the groundwater that is withdrawn for industrial use (DaimlerChrysler, 2003).

Mexico is the tenth largest auto producer in the world. Beginning in 1995, the three major U.S. automobile manufacturers—General Motors, Ford, and DaimlerChrysler—along with Volkswagen AG made major investments in Mexico. One impetus was the 1994 North American Free

**Figure 6.4 Mexico state and Toluca.**

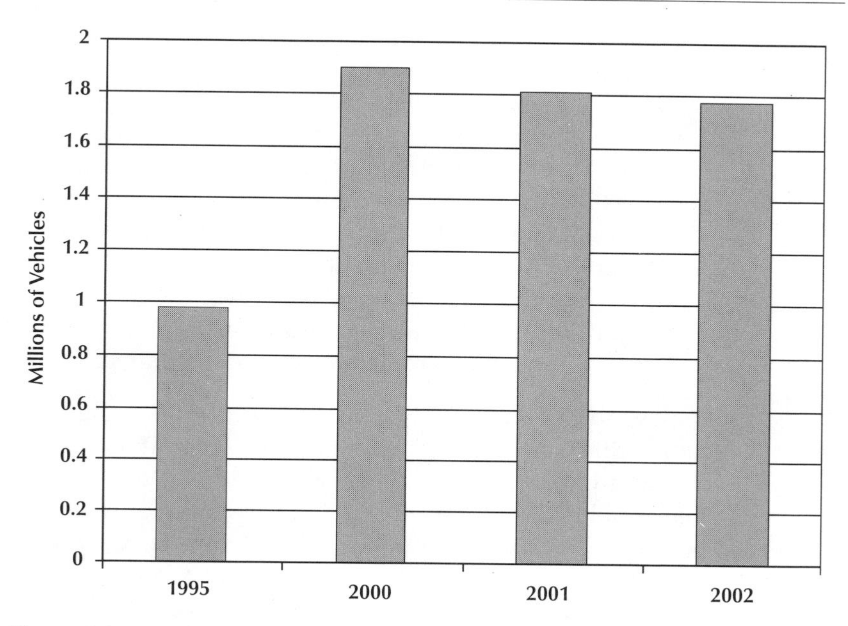


**Figure 6.5 Auto production in Mexico.** (*Sources*: Intelimex, 2001; AutoWorld.com, 2003.)

Trade Agreement (NAFTA), which reduced tariffs between Mexico and the United States and Canada; tariffs were scheduled to be eliminated on January 1, 2004. See Figure 6.5 for vehicle production in Mexico; the bar for 1995 represents output before the effect of NAFTA.

## Water Regulation in Mexico

Mexican government control of water resources has important characteristics that bear on the use of water by industry in Mexico (North American Commission for Environmental Cooperation). The Mexican Constitution states that the nation owns all waters. Before a private party may use any water resource, it must acquire a water-use concession from the Comisión Nacional del Agua (CNA, National Water Commission). Two national laws provide the basis for water resources regulation:

**Ley General del Equilibrio Ecológico y la Protección al Ambiente (General Law of Ecological Balance and Environmental Protection, the Ecology Law)** Sets out criteria to prevent and control water pollution. When a party uses water, the Ecology Law requires that it avoid contaminating the receiving water; interfering with water decontamination processes; physical alterations to streambanks, basins, or groundwater aquifers.

**Ley de Aguas Nacionales (National Water Law) and Its Regulations, Reglamento de la Ley de Aguas Nacionales** Outline the legal institutions and procedures for carrying out the Ecology Law. It established the Comisión Nacional del Agua (CNA, National Water Commission). CNA generally has authority over water hydraulics (takings, diversions, etc.). State and municipal agents are responsible to control wastewater. The law's regulations established the official Mexican water quality standards, the Normas Oficiales Mexicanas (NOMs). CNA has wide-ranging authority, including water resources planning, permitting, management, and enforcement. It also sets standards for water quality and for wastewater discharges, with input from the secretariats of the Navy, Health, and Environment, Natural Resources, and Fisheries. Private parties may use a water resource only when granted a concession or allocation by CNA.

Beyond the NOMs, CNA may institute water quality requirements for specific water resource uses through a permit. Industrial discharges are controlled through industry-specific NOMs. CNA records and makes public a record of all water concessions granted and all discharge permits in the Registro Público de Derechos de Agua (Public Registry of Water Rights). CNA also is authorized to create reserve zones to preserve drinking water sources and to safeguard them from sources of possible contamination. Whenever a water resource is allocated to a private party, the Ecology Law directs CNA to consider sustainable development policies (article 3). CNA's enforcement powers include plant closures for parties that violate provisions of the law. The DaimlerChrysler plant in Toluca has a zero liquid discharge; therefore, it required no discharge permit.

It is within this regulatory matrix and within the natural climate of Mexico state with its scarcity of water that DaimlerChrysler acted to construct a vehicle assembly plant in Toluca. It chose a closed-loop system for water use and treatment to respond to these conditions. DaimlerChrysler, its design engineer, O'Brien & Gere Engineers, Inc., and its vendor, USFilter, addressed many treatment challenges.

As identified in Figure 6.6, the treatment train consisted of chromium reduction, heavy-metals removal using physical/chemical treatment, organics removal using biological treatment, filtration, microfiltration, and reverse osmosis. The permeate (treated water) from the reverse-osmosis process was recycled for beneficial reuse in the manufacturing process. The concentrate (wastewater) was transferred to a distillation and crystallization process to recover water and manage the waste as a dry solid. These processes are described in greater detail in the subsequent sections.

## DaimlerChrysler de Mexico—Treatment Overview

In late 1997, O'Brien & Gere Engineers was retained to provide design, construction, and startup of a new 1 million gallon/day (MGD) "zero-discharge" industrial wastewater treatment facility. The new facility would treat industrial process wastewater and sanitary wastewater

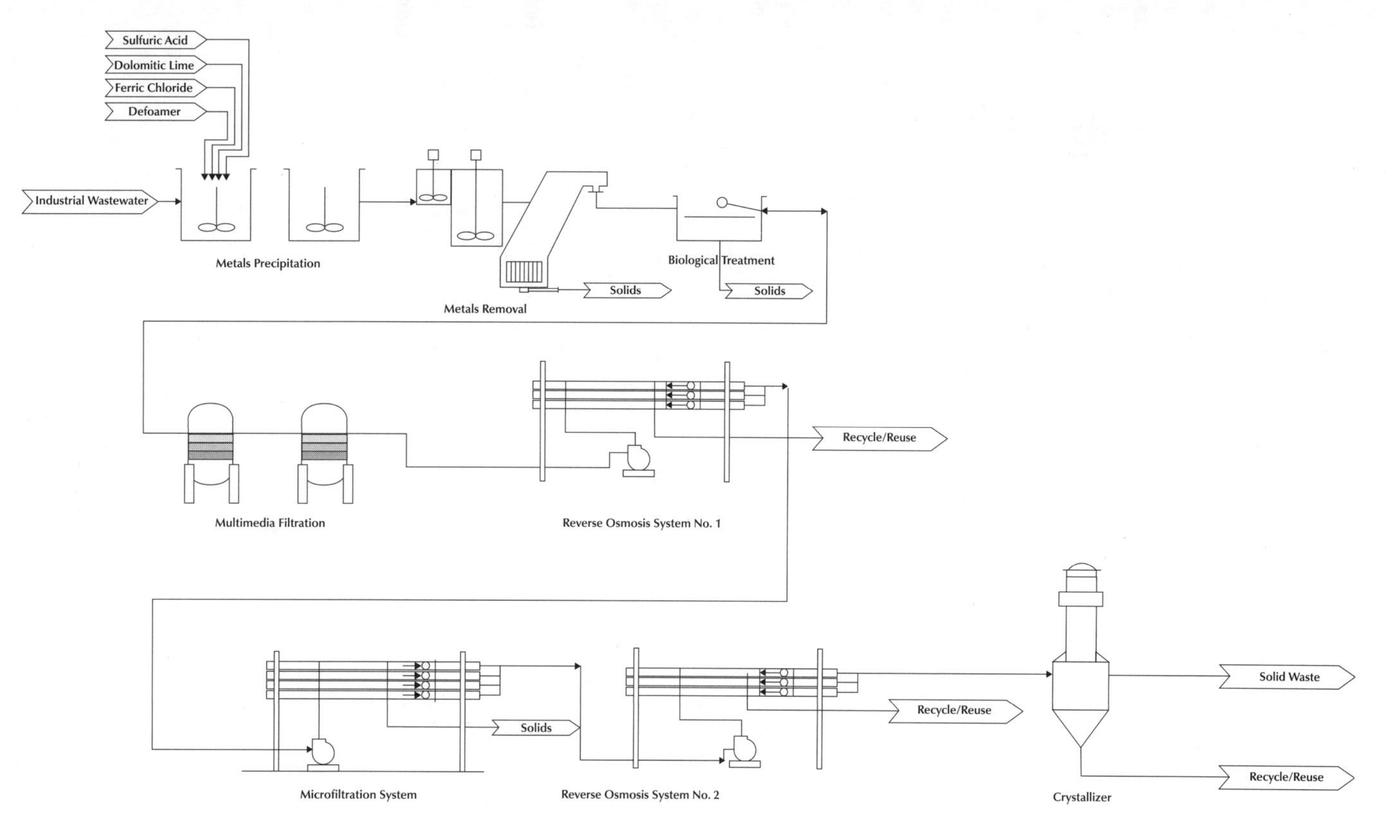


**Figure 6.6 Process flow diagram for DaimlerChrysler de Mexico plant in Toluca.**

generated at DaimlerChrysler's automobile components manufacturing plant in Toluca, Mexico. A fast-track design-build project plan was developed and executed. Construction was completed in late 1998. The system was commissioned in early 1999. The new treatment plant occupies an area of approximately 1.7 acres and required erection of over 3.2 million gallons of process tanks and a 17,000 $ft^2$ process building. Approximately 42,000 linear feet of piping and 20 pumping stations were installed to collect and convey wastewater from various locations of the manufacturing facility to the new wastewater treatment plant and to recycle the treated water back to the automotive facility for reuse. Major treatment processes at the new facility consist of oil separation, chrome reduction, chemical precipitation, clarification, sequencing batch reactors (biological treatment), microfiltration, reverse osmosis, a crystallizer, and three separate solids-handling systems. The plant is designed for zero discharge. Treated process water is returned to the manufacturing facility for use as cooling tower makeup water. Treated sanitary wastewater is used to irrigate the manufacturing campus that includes soccer fields, basketball/tennis courts, and playgrounds for employees and their families.

Tables 6.5 through 6.7 demonstrated that the diverse organic solvents, inorganic pigments, and metals used in the automotive manufacturing process end up in the wastewater streams discharged from the manufacturing facility. Completely removing this array of contaminants such that the resulting liquid effluent is suitable for recycling back to the manufacturing process requires a long series of independent yet integrated unit processes.

It is important to note that a complex design program such as that of the Toluca facility, must begin with a systematic approach to the identification of system inputs (such as flow and quality) and system outputs (such as quality of water required for recycle, permit requirements, and solids generation); in this way, the design process is similar to many other techniques affecting the life cycle of resources. The influent characterization should be validated by collecting physical data. In the case of Toluca, a majority of the wastewater to be generated did not yet exist because the manufacturing plant was under construction. In this case, data from similar operating plants were used. In addition, bench-scale testing was conducted on wastewater samples (or surrogates based on recipes from other operations) to identify plant-specific variables such as chemical dosages required (such as the acid and coagulant).

For a system of this nature and complexity, it is always prudent to develop a materials balance for the plant to predict performance, to size equipment, and to estimate residuals generation. Specifically for Toluca, the importance of the material balance focused on water, metals, salts, and silica. These parameters were deemed most important given the selection of unit operations for the zero-discharge system. The wastewater treatment plant's materials balance was developed using a combination of analytical data, testing data, published engineering data, and experience from similar operations at other plants. It is here where the knowledge of the engineer comes in. Understanding the constituents in the wastewater, how to best treat them, and what unit operations to select is key to developing a technically effective and economical process. This knowledge can be obtained through published information in textbooks and technical articles, equipment vendor information, and, of course, experience.

The process flow diagram (Figure 6.6) depicts the unit processes that the engineer, O'Brien & Gere, used at the DaimlerChrysler facility in Toluca, Mexico. The remainder of this chapter describes the functionality of each unit process and discusses how the unit processes were integrated to produce zero liquid discharge.

## Removing Heavy Metals

Conventional metals removal occurs with several unit operations: rapid mix, coagulation and flocculation followed by sedimentation, and filtration. For the Toluca plant, since the wastewater stream contains organic materials as well as metals, a biological treatment process was included to address the organic material. The biological treatment takes place before filtration and ultimate discharge to the zero liquid discharge subsystem. See the box for definitions of some specialized terms used to describe the removal of metals.

### Definitions

**Caustic** Refers to sodium hydroxide (NaOH). When caustic is added to water, the water becomes strongly alkaline ($pH \gg 7$).

**Concentration** The quantity of a material in a given volume of solution.

**Dilute solution** A weak solution; a relatively small quantity of a material in a large volume of solution (e.g., water).

**Heavy metals** Metals that may pose detrimental health effects when present in water in significant concentrations. Heavy metals include lead, silver, mercury, copper, nickel, chromium, zinc, cadmium, and tin. They must be removed to certain levels to meet discharge requirements.

**Metal hydroxides** When caustic is added to water containing heavy metals, a metal hydroxide solid or precipitate is formed.

**mg/L** Milligrams per liter, a representation of the quantity of material present in a solution. This is equivalent to the units parts per million (ppm).

**pH** A term used to describe the acid–base characteristics of water; it is typically measured by a pH meter. Specifically, pH is the concentration of $H^+$ concentration; that is, $pH = -\log[H^+]$. The following values indicate the classification of a water:

pH $< 7$ refers to acid solutions
pH $> 7$ refers to basic solutions
pH $= 7$ refers to neutral solutions

**Precipitation** Precipitation is the process of producing solids in a solution. In metals removal, it is desirable to precipitate as much metal solid as possible so that the metal can be removed from the water.

**Precipitation region** The region on a solubility diagram that indicates the appropriate concentration and pH value for a metal to form a solid precipitate.

**Solubility** Solubility defines a material's ability to go into solution (dissolve). Materials that are soluble readily dissolve in solution and do not precipitate. Substances that are insoluble do not easily dissolve in solution and stay in their solid form. The goal of metals removal in wastewater is to produce conditions so that metals are insoluble.

**Solubility diagram** A graph that reveals the solubility of metals (through the formation of metal hydroxides) at specific pH values.

The first step in the metals removal process consisted of large pH adjustment tanks where chromium reduction and hydroxide precipitation take place. These pH adjustment tanks operate on a "batch" basis to provide adequate reaction time for the reduction and precipitation processes to occur.

*Chromium Reduction*

Chromium is a common surface coating and is present in many automotive manufacturing and assembly plant wastewater streams. The most prevalent form of chromium in these processes is the hexavalent state (Cr VI). Water containing hexavalent chromium is treated with a chemical reduction process to reduce the oxidation state to trivalent form (Cr III). This conversion is necessary because trivalent chromium reacts with hydroxide at an elevated pH (>8.5) to form a hydroxide precipitate; hexavalent chromium is soluble and remains in solution. The general reduction–oxidation (red–ox) reaction is as follows:

$$Cr^{6+} + \text{reducing agent } (Na_2S_2O_5) + H^+ \rightarrow Cr^{3+}$$
$$+ \text{by-product } (Na_2SO_4) + OH^-$$

Sulfur dioxide, sodium bisulfate, or ferrous sulfate is added to the wastewater as reducing agents, and the pH is lowered to 3.0 or less using acid (typically sulfuric acid). A retention time of 45 minutes is usually maintained to ensure adequate mixing and reaction with the reducing agent. The trivalent form is treated similar to other metals, and the effluent from this process is treated with the other metals' wastewater.

*Hydroxide Precipitation*

As metals enter the treatment process, they are in a stable dissolved aqueous form. The goal of metals treatment by hydroxide precipitation is to adjust the pH (hydroxide ion concentration) of the water so that the metals will form insoluble precipitates. In this state, the metals can easily be removed, and the water, with low metal concentrations, can be transferred to the next wastewater treatment unit process (Ayres et al., 1994).

Metal precipitation primarily depends on two factors: the concentration of the metal and the pH of the water. Heavy metals are usually present in wastewaters in dilute quantities (<0.1% or <1000 mg/L) and at neutral or acidic pH values (< 7.0). Both of these factors are disadvantageous with regard to metals removal, but when caustic is added to water that contains dissolved metals, the metals (indicated as $M^{2+}$ in the following reaction) react with hydroxide ions ($OH^-$) to form metal hydroxide solids:

$$M^{2+} + (OH^-)_2 \rightarrow M(OH)_2 \downarrow$$

The pH corresponds to elevated hydroxide concentrations. Individual metals precipitate at specific pH values, and the solubility of a particular metal is directly controlled by pH. See theoretical solubility graphs for typical metals encountered in wastewater treatment in Figure 6.7. The $y$ axis displays the concentration of dissolved metal in the wastewater, in mg/L, and the $x$ axis displays the pH of the solution. These solubility graphs display regions where the metals are soluble or insoluble. The region within the V formation for each metal signifies that the metals should precipitate as metal hydroxides. This is referred to as the *precipitation region*. The region below or outside the V formation illustrates where the metals are dissolved in solution, no precipitation occurs, and no metal removal takes place. For example, in Figure 6.7, the lowest possible dissolved concentration of copper is approximately 0.0007 mg/L, which occurs at a pH value of 8. Figure 6.7 can be used to evaluate how pH affects the concentration of a metal—say nickel—in water. Suppose a wastewater contains dissolved silver at 4 mg/L and is at pH = 6.8. Since this point is outside the V, which indicates solids formation, the nickel is only present as a dissolved metal. It is not in a solid form, and under these conditions, it will not precipitate. For the nickel example, therefore, caustic is added to adjust the pH of the water to 10.2. This value is selected because the minimum solubility of the metal is based on the V-shaped curve. At this new pH value, for example, most of the nickel forms nickel hydroxide and precipitates from solution.

Thus, simply adjusting the pH has effectively precipitated most of the dissolved metal from the water. Since all metals display similar effects, it is clear that the adjustment of pH is critical when a metal is to be removed from wastewater. The metals, however, are now in another phase or state; they are small solid particles. Metal removal is not complete until these solids are physically removed from the wastewater.

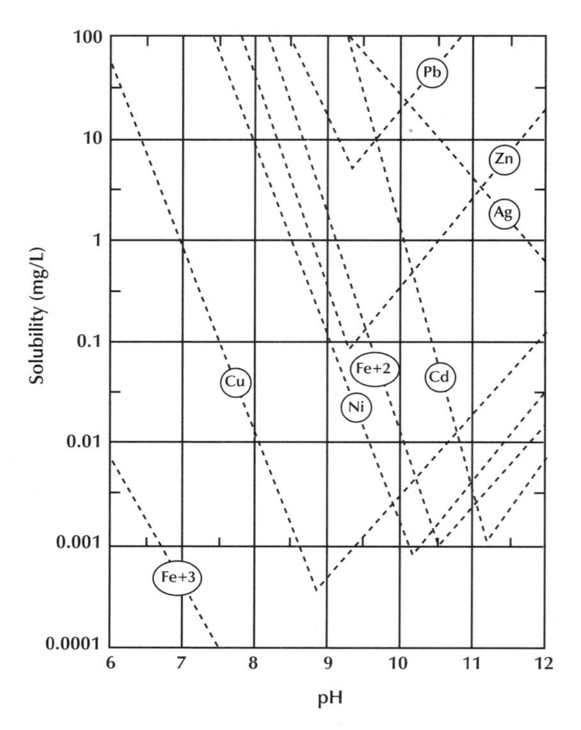


**Figure 6.7 Theoretical solubilities of metal hydroxides.** (*Source*: Adapted from Ayres et al., 1994.)

Their removal is typically accomplished with sedimentation and filtration processes, as discussed below.

The metal solubilities presented in Figure 6.7 are based on theoretical calculations. Some variations in the exact values of the metal concentrations occur because of the presence of other substances in the wastewater. Compounds such as cyanide or ammonia can inhibit precipitation of metals—to the point where discharge limits can be exceeded. In many cases, a phenomenon called co-precipitation has been observed; this causes metals that are not at their respective minimum solubility pH to be precipitated by attachment to other metal hydroxides. Moreover, as can be seen, not all metals have the same minimum solubility. As a general rule, treatability jar testing is conducted to identify the minimum solubility pH for all metals of particular concern and select the optimum pH value to meet the design criteria.

*Silica Removal*

In arid regions, such as Toluca, there is typically a high concentration of silica in the source water. Silica is the primary element present in sand and is present in both the elemental and colloidal form in the source water for manufacturing facilities. The silica is very difficult to remove from the water for process use or wastewater for recycle, and it prematurely fouls a reverse-osmosis membrane resulting in costly operation. The elemental silica can be removed through co-precipitation with magnesium hydroxide. The elemental silica does not form a hydroxide precipitate; however, it binds to the surface charge of the magnesium hydroxide particle and is removed from solution using a sedimentation process. In Toluca, dolomitic lime was used as the reagent to raise the pH of the wastewater. Dolomitic lime has a high percentage of magnesium, the component required for the co-precipitation process. The colloidal silica is removed using standard filtration or microfiltration.

*Solids Removal*

The pH adjustment step is followed by a rapid mix–flocculation–clarification process. For most automotive wastewaters, this process can be implemented with an off-the-shelf inclined plate clarifier system that combines rapid mix, flocculation, and clarification steps in a single unit process (Figure 6.8). The goal of the rapid mix operation is to disperse the coagulant and flocculent chemicals required to aid in the gravity settling of the metal hydroxide precipitate. Coagulant (typically cationic solutions—aluminum, iron salts, or organic polymers) is added to modify the electrical charge of the precipitate and destabilize the particle; that is, to increase the particle's propensity to settle from the aqueous solution.

Flocculent is also added to the water—typically an anionic polymer solution—that attaches to the destabilized metal hydroxide particles. The small metal hydroxide particles become entangled in these polymers, causing the particle size to increase (to form a floc) and promotes settling. The settling effect induced by the inclined plates can be described by an adaptation of Stokes's law; that is, the particle settling

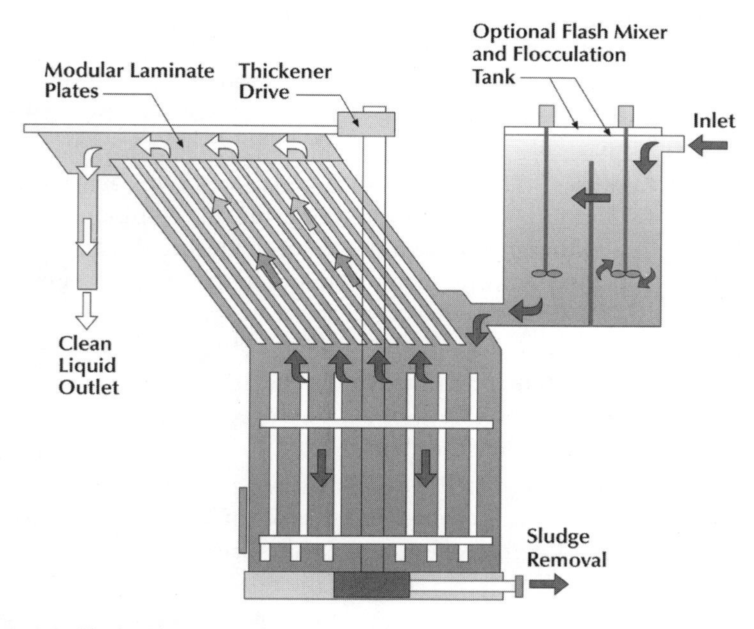


**Figure 6.8 Typical inclined plate clarifier system.**

velocity, vessel size, and flow rate dictate the surface area required for plate separation. The typical surface overflow rate for selecting the surface area for metal hydroxide sedimentation is 0.25 gal/min/ft$^2$. The settling process is enhanced without increasing the floor space by providing multiple parallel plates to create settling chambers in the vessel. For proper settling, an appropriate angle of repose must be selected.

Following clarification, organic biodegradable materials present in the wastewater (such as residual quantities of cutting or lubricating oils) are treated in a biological treatment process.

## Sequencing Batch Reactor

The biological treatment process that the engineers selected to treat the organic materials was a sequencing batch reactor, or SBR. An SBR is a semicontinuous activated sludge process that uses flow equalization, aeration (mechanical means of adding oxygen to wastewater), sedimentation (allowing particles to settle), and clarification to treat wastewater. This is a conventional biological treatment, similar to that used at most municipal wastewater treatment plants. Unlike conventional continuous activated sludge processes, however, the SBR's aeration, sedimentation, and clarification are accomplished in a single bioreactor. The key to effective treatment with the SBR is allowing sufficient time for each phase of the SBR process. The operational sequence and phase time durations are manually controlled or automatically controlled. The operation sequence of the SBR system is as follows:

**Fill** Wastewater is pumped into the SBR tank for treatment.

**Anoxic (Unaerated) Time** This step begins during the fill step and continues for a set time as determined by process conditions. Typically, the anoxic time expires before the end of the fill time, such that there is a period of anoxic fill followed by a period of oxic (aerated) fill.

**Aeration** The aeration time consists of two separate steps: (1) aerated fill and (2) react. Once the fill cycle is complete, the SBR spends the remaining aeration time in react. During the react mode, wastewater is mechanically mixed. Mixing in the SBR vessel is accomplished by a jet aeration/mixing pump arrangement, or by fixed fine or coarse bubble aeration equipment combined with surface aeration.

**Settle** In this treatment step, the SBR tank has no equipment running and no valves are open. This step permits the solids in the tank to settle before entering decant.

**Decant** During the decant phase, clarified water is removed from the upper levels of the SBR tank for further treatment. To allow sufficient time for the other steps in the process, the decant step is typically rapid. The flow rate from the SBR vessel during decant is typically 5 to 10 times the design average and peak flow into the SBR vessel.

Similar to conventional activated-sludge treatment, sludge wasting and effluent decanting are performed during different phases of treatment. If biological nutrient removal (BNR) is desired, the steps in the individual phases are adjusted to provide anoxic or anaerobic periods within the standard cycle times. Because the SBR process uses the same tank for all phases of operation, there is no need for a return activated-sludge system.

## Recycle—Zero Liquid Discharge

Following the metals removal and SBR processes, the pretreated industrial wastewater is introduced into the recycle/ZLD (zero liquid discharge) treatment system. ZLD comprises a number of additional unit processes that further treat the wastewater through coagulation–clarification, multimedia filtration, and membrane filtration to remove suspended and dissolved contaminants from the wastewater before the water is recycled to the manufacturing facility. Additionally, each step in the ZLD system "rejects" some contaminants in an aqueous stream.

### *Mixed Media Filtration*

Mixed media filtration consists of multiple filter vessels filled with a multimedia material (e.g., anthracite coal, sand, and gravel of varying gradations) that removes suspended solids, algae, and organic matter. The media size varies with depth. This inverse size gradation allows for an increased solids storage capacity of the filter and allows suspended matter to be captured and held throughout the depth of the filter. See Figure 6.9.

Water enters the top of the filter vessel and flows down through the media where the filtration takes place. The filtered water exits through

**Figure 6.9 Typical multimedia filtration arrangement.**

the bottom distribution and outlet piping. Normally, two vessels operate simultaneously until the differential pressure across the filter bed goes above a predetermined set point. At this point, the first unit begins backwash, and the second unit assumes a higher flow output. At the completion of the backwash sequence, the backwashed unit enters service and provides the treatment, while the other unit enters backwash cycle.

The filter vessel's backwash mode dislodges trapped particles and cleans the filter media. The backwash cycle consists of an air scour, clean liquid backwash, and rinse step. Following multimedia filtration, the industrial wastewater stream has been sufficiently pretreated to allow for reverse-osmosis treatment before the water is recycled.

*Reverse Osmosis*

Several parameters are important to determine the effective operation of the reverse-osmosis (RO) system, and they dictate the extent of pretreatment required. These parameters include silt density index (SDI), turbidity, total dissolved solids (TDS), and precipitate forming constituents (such as calcium, silica, carbonate, and sulfate).

**SDI** is strictly a physical analysis completed in accordance with the American Society for Testing and Materials (ASTM) standard designation D4189-95 that correlates the duration to filter a sample using a 0.45-$\mu$m filter with the potential to foul an RO membrane. An SDI value of 3 or less is required in the feed water to the RO. The SDI value can be lowered through standard filtration or microfiltration processes.

**Turbidity** is measured in nephelometric turbidity units (NTU) and indicates the clarity of the feed water. A high turbidity value correlates to a murky water, and a low value correlates to clear water. RO membranes typically require a turbidity value $<1$ NTU. The turbidity can be lowered through filtration processes.

**TDS concentration** governs the quantity and quality of product water that can be generated from the RO system. A high concentration of TDS increases the required quantity of water rejected as concentrate and results in a decreased amount of usable water being generated as permeate. The reverse is true for feed water with a low concentration of TDS.

The concentration of precipitate-forming constituents also contributes to the quantity and quality of product water that can be generated from the RO system. When these constituents are present in significant quantities, the concentration is increased by the membranes that drives the precipitation reaction to occur. This precipitation on the membrane surface can prematurely foul the system, decreasing the efficiency and increasing the required cleaning frequency.

Osmosis itself is the movement of a solvent through a membrane that is impermeable to a solute (dissolved solids). A solution consists of two components:

1. The solvent is the component of a solution that determines whether the solution exists as a solid, liquid, or gas.
2. The solute is the component that is dissolved in solution.

Regarding osmosis, the solvent always flows from the more dilute to the more concentrated solution across the semipermeable membrane. The solvent passes through the membrane in both directions but passes more rapidly in the direction of the concentrated solution, resulting in an increased hydrostatic pressure on the more concentrated side of the membrane.

The excess pressure required to be applied to the more concentrated side of the membrane to produce equilibrium is known as *osmotic pressure*. RO involves the application of hydrostatic pressure on the concentrated solution side of the membrane to overcome the osmotic pressure in addition to overcoming the natural tendency of flow direction of the solvent from more dilute solution to more concentrated solution. Thus, the flow direction is reversed, and the solvent is forced through the semipermeable membrane while the more concentrated solution and the majority of the solute remains on the other side of the membrane.

The RO membrane is the heart of the system. It is responsible for rejecting up to 98% of the total dissolved solids in the water and operates in the following manner. Pressure is applied to the solution that contains the higher concentration of dissolved solids, forcing water through the membrane and leaving behind the dissolved ions and suspended solids. The water that remains behind the membrane along with the dissolved and suspended solids is referred to as *concentrate, brine*, or *reject water*; the water that passes through the membrane is referred to as *permeate* or *product water*. At the Toluca facility, as with most applications of this technology, the permeate is used as the recycle stream; that is, it is returned to the manufacturing facility for reuse. The rate of product water passage (or productivity) through the membrane is referred to as the *flux rate* and is generally expressed in gal/day/ft$^2$ of membrane surface or in gal/day per RO cartridge. The flux rate of a particular membrane is generally limited by several factors including temperature, operating pressure, and the surface flushing action to keep the membrane surface free of suspended solids. The rate of recovery of feed water converted into product water is generally expressed in the form of a percentage; the ratio of product water is expressed as the recovery rate. The percent recovery of RO feed water is

$$\% \text{ recovery} = 100 \times (\text{product flow rate})/(\text{feed flow rate})$$

For example, if the feed flow rate to the RO unit is 200 gal/min and the product rate from the RO unit is 150 gal/min, the rate of recovery would be expressed as 75%. The remaining 50 gal/min did not pass through the membrane and is referred to as the *concentrate water*.

Membrane rejection, an expression of the ability to restrict the passage of dissolved ions through the membrane, is generally expressed as a percentage. Salt rejection is expressed based on the relative total dissolved solids (TDS) concentration, as follows:

$$\% \text{ salt rejection} = 100 \times (\text{feed water TDS} - \text{permeate TDS})/(\text{feed water TDS})$$

For example, if the feed water to the membrane contains 100 ppm of dissolved solids, and the resulting product water contains only 2 ppm dissolved solids after processing, the resultant rejection rate is 98%.

Two major types of membranes are used in industry today: thin-film composite (TFC) and cellulose acetate (CA). Typical salt rejection for a polyamide TFC RO membrane is approximately 98%; a CA membrane rejects approximately 95%. The correct membrane type for a specific application is selected based on the nature of the wastewater, recovery required, and operating conditions. Recovery from the RO system is ultimately controlled by the feed water TDS; however, the concentration of the precipitate-forming constituents (calcium, silica, carbonate, sulfate, etc.) limit the overall recovery. These constituents can either be removed using pretreatment processes or controlled using chemical addition of a sequestering agent. The chemical sequestering agent inhibits precipitate formation of insoluble species allowing the precipitate-forming constituents to become supersaturated, thus improving the overall efficiency and performance of the system.

The quality of the water is typically quantified by measuring its conductivity. Conductivity is measured with units of $\mu$mhos/cm or $\mu$S/cm (microSiemens). The inverse of conductivity is resistivity and is measured with units of megaohm-cm (M$\Omega$-cm). The conductivity of water can often be approximated by 60% of the value of the TDS as measured in mg/L.

The basic equipment necessary for RO treatment is a semipermeable membrane material that is packaged in a suitable membrane element. The membrane material itself is usually a polyamide thin-film composite, a cellulose acetate or triacetate, or some type of plastic-based material. The membranes generally are in the form of one of four element configurations: flat plate, tubular, hollow fiber, or spiral wound. Hollow fiber and spiral wound are the most common in today's technology. The spiral wound membrane configuration is constructed from a flat sheet membrane that is first folded and sealed to form an envelope with one open end. Porous backing material, placed inside the envelope, separates the membrane sheets and forms a flow channel between them. The open end of the envelope is then attached and sealed around a plastic product tube that is perforated to allow the product water, or permeate, to pass into the product tube (Figure 6.10). For compactness, the envelope of membrane material is then wrapped around the product tube in a spiral wound fashion with a coarse plastic screen. This screen is referred to as a *brine channel screen*; it creates a flow channel between the surfaces of the membrane where the feed water enters the element and the concentrate, or brine flow, where the water passes from the element.

Figure 6.10 Typical RO membrane configuration.

The elements are then put into one or more cylindrical pressure vessels that can contain one to six membrane elements, depending on the unit's design.

Pressurized feed water is introduced at one end of the vessel using a high-pressure pump integral to the RO system. Some of the water, driven by the feed pressure through the RO pressure vessel feed port, passes through the membrane and into the product tube. This product water exits the pressure vessel from the pressure vessel product port less most of the dissolved solids and all of the suspended solids. The remainder of the water passes along the surface of the membrane with the concentrated dissolved and suspended solids and passes from the pressure vessel as concentrate, or reject, flow from the pressure vessel concentrate port. A uniform water flow promotes good flushing velocity across the membrane surface and prevents the accumulation of suspended solids on the surface. A buildup of solids fouls the membrane and reduces its productivity. See Figure 6.11.

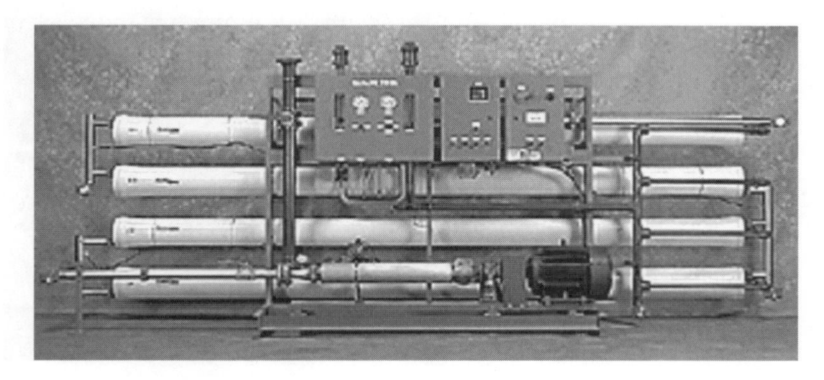

Figure 6.11 Reverse osmosis system.

Toluca has two RO process trains (in series), and the facility is considered state-of-the-art unique because not only is the product water from RO 1 recycled, the reject water from RO 1 undergoes further treatment through a microfiltration unit and then goes to the second RO unit.

The microfiltration unit at the Toluca plant pretreats the RO 1 reject stream before it reaches RO 2. The microfilter softens (removes precipitated calcium and magnesium) and removes suspended silica. Microfiltration is a pressure-driven cross-flow filtration process using a microporous membrane to separate suspended solids from liquids, removing particles >0.1 $\mu$m. A basic filtration unit consists of two reaction tanks, a concentration tank, a single process (recirculation) pump, and a single train of membrane modules. If the solids to be filtered are already in colloidal suspension (as insoluble species), straight filtration is sufficient. If the solids to be removed from the liquid are as the dissolved species, however, the liquid must be treated and precipitated as a metal hydroxide or other insoluble particle to create a colloidal suspension before filtration. Pretreatment includes chemical reactions (reduction, pH adjustment, oxidation, etc.).

After pretreatment in the microfilter, the RO 1 reject stream is treated through RO 2. Permeate from RO 2 is recycled back to the manufacturing facility. The highly saturated reject water from RO 2 then passes to the crystallizer for the final step in the zero liquid discharge process. Product water from the crystallizer is also recycled. (More about the crystallizer below.)

## Crystallizer

A crystallizer reduces highly saturated wastewater to dry solids for disposal. At Toluca, the crystallizer's main purpose is to remove fluoride, metals, and other inorganic material. The crystallizer is a forced-circulation evaporator that uses a mechanical vapor compressor or plant steam as the energy source (Figure 6.12). The details of the crystallizer's

**Figure 6.12 Crystallizer system.**

operation follow. The RO 2 reject is transferred to the degassifier on the front end of the crystallizer. The purpose of the degassifier is to remove carbon dioxide from the solution. Degassification can be accomplished using either a forced-draft decarbonator or a vacuum degassifier. A forced-draft decarbonator consists of a packed tower with counterflow forced air to strip the carbon dioxide from solution. At the Toluca plant, a vacuum degassifier is used. The vacuum degassifier induces a negative pressure on a vessel containing the RO 2 reject. The negative pressure reduces the vapor pressure in the headspace of the vessel and removes the carbon dioxide from the liquid phase. The carbon dioxide must be removed from solution to eliminate the potential for formation of carbonate-based precipitates (most commonly, calcium carbonate) in the tubes of the heat exchanger. Dissolved carbon dioxide exists as a diprotic acid, that, depending on the pH of the solution, can be carbonic acid, bicarbonate, or carbonate. Under certain conditions, carbonate reacts with the cationic species to form a precipitate and substantially reduces the efficiency of the heat exchanger.

The degassified brine is pumped to the crystallizer's vapor body where it is mixed with the circulating fluid already in the vapor body. This fluid is called *liquor*. A recirculation pump constantly circulates the liquor from the vapor body through the crystallizer's heater; this liquor flows on the inside of the heater tubes. Steam is applied to the external surface of the heater tubes and heats the process liquor. The heated liquor returns to the vapor body from the heater and boils. The vapor compressor supplies the energy for evaporation. It converts mechanical energy into heat by increasing the temperature and pressure of the process vapors. The pressurized vapors are then suitable for use as a heat supply to the crystallizer heater.

The vapor released from the vapor body flows to the entrainment separator where water droplets and solid particles are separated from the vapor stream. The vapor then flows to the compressor where it is converted to usable steam. Condensate recovered from the crystallizer heater is pumped through the feed preheater, which heats the feed. The condensate is collected and returned to the manufacturing plant. The liquor in the vapor body is a slurry. This slurry is drawn and sent to a centrifuge where the crystals are separated from the liquid. The solids are discharged to a hopper, and the liquid is drained back to the crystallizer feed tank.

The heat for the crystallizer operation is provided by steam flow to the heater. In normal operation, as evaporation takes place in the crystallizer vapor body, the vapor released flows to the compressor suction. The compressor adds energy to the vapor creating superheated steam by increasing its pressure and temperature. This superheated steam is then at a temperature that is hot enough to be returned to the heater where it is used to drive the evaporation process. Steam control is critical to the evaporation process. If excess steam is not vented, the crystallizer vapor body pressure will rise. If too much steam is vented, the pressure will drop below the set point. The required evaporation rate depends on the flow rate and proportion of solids in the incoming feed to the wastewater evaporator system. As the feed rate increases or the proportion of solids in the feed decreases, the required evaporation rate increases. A change in the proportion of solids of the vapor body can be anticipated if there is a change in the feed rate or the feed composition. As the concentration of dissolved solids increases, this temperature increases, and vice versa. If the liquor is allowed to become too concentrated, the higher liquor temperature causes the compressor to work harder to raise the steam to a higher temperature so that the heat can still be transferred at the same rate. Eventually, the system is limited by the compressor's capability to increase the pressure and temperature; so, capacity would probably have to be reduced to continue operating at the elevated concentration. The relationship controlling the evaporation rate is as follows:

$$\text{BPR} = T + 0.15 \times P^2 - 7.71 \times P - 131.1$$

where BPR = boiling point rise (°F)
$T$ = vapor body temperature (°F)
$P$ = vapor body pressure (psia)

If the suspended solids in the crystallizer are under 25% by weight, the liquid slurry removal flow rate is decreased. If the suspended solids are over 25% by weight, the liquid slurry removal flow rate is increased to reduce the solids. The most soluble species, specifically the nitrates and the chlorides, typically govern the concentration in the vapor body. The removal of soluble species from solution can present operational difficulties and limit the overall capability of the system.

The pH of the crystallizer feed must be maintained between 5.0 and 5.5 to prevent plugging the preheater and degassifier with insoluble solids. The condensate leaving the crystallizer contains a portion of the energy added to the liquid by the steam condensing in the crystallizer heater. The condensate is transferred to the crystallizer feed preheater to recover unused heat.

The heating element in the crystallizer consists of a group of tubes fixed in place at each end by tube sheets. The wastewater solution that is to be concentrated flows inside the tube (tube side). Steam is introduced outside the tube surfaces (shell side) where it condenses and gives up its latent heat value to increase the temperature. As the liquor enters the vapor body, a portion of the water is flashed off, and the solids concentration of the solution increases. The shell-and-tube heat exchanger is modeled using the following equation:

$$Q = U \times A \times \text{LMTD}$$

where LMTD = log mean temperature difference
$Q$ = tubular heat flow (Btu/hr)
$U$ = heater transfer coefficient (Btu/hr-ft$^2$-°F)

$A$ = heat transfer area (tube inside diameter basis or plate surface area; $ft^2$)

The higher the $U$ coefficient, the better is the system's operation. The operator has control over several factors that influence the $U$ coefficient. Note:

$$U = Q/A \times \text{LMTD}$$

where LMTD = log mean temperature difference

$T_S$ = heater shell temperature

$T_1$ = brine in temperature to heater

$T_O$ = brine out temperature from heater

ln = natural log

$$\text{LMTD} = \frac{(T_S - T_1) - (T_S - T_O)}{\ln\frac{(T_S - T_1)}{(T_S - T_O)}}$$

For maximum efficiency and minimum energy use, the system must be maintained. Scale deposits will most likely develop in the crystallizer because of the saturated nature of the solution. Fortunately, the forced circulation design provides high tubular velocities and tube boiling suppression; these factors minimize scale formation. Scale acts as insulation on the tube surface reducing the heat transfer coefficient and increasing the LMTD required for a fixed evaporation rate.

Air and other noncondensable gases cannot be allowed to accumulate in the shell sides of the heaters. These gases, if not removed, form a blanket around the tubes. This prevents steam from reaching the tube wall and decreases the surface area available for condensation. The effective value of the heat transfer coefficient then decreases as the concentration of the noncondensable gases in the shell side increases.

The condensate that forms on the shell side of the heater must be continuously removed to avoid a loss of heat transfer surface area. Higher dissolved concentrations in the liquor cause the boiling point to rise, thus raising the liquor temperature and limiting the efficiency of the entire system. Liquor at a higher temperature liquor requires greater steam pressure to maintain the same temperature across the heater tubes and the same evaporation rate.

The Toluca wastewater treatment plant was designed to recover >95% of the industrial wastewater (approximately 946 $m^3$/day or 250,000 gal/day). Virtually all of the water from the sanitary wastewater treatment system (less that retained with the biological sludge) is reused for irrigation on the facility's grounds. The Toluca facility demonstrates that conventional and advanced wastewater treatment unit operations can be integrated in a synergistic fashion to achieve the goal of zero liquid discharge. Engineers in arid areas have taken the first step forward in implementing this method of green engineering, but as water sources become more costly and discharge regulations become more stringent, recycling provides a cost-effective measure to support manufacturing industry.

The DaimlerChrysler Toluca facility has been a positive addition to the State of Mexico. It employs 7000 workers and includes four separate engine, transmission, stamping, and assembly plants. The sustainability of this manufacturing facility is heavily influenced by environmental conditions in the area it occupies. In addition to helping prevent pollution of the nearby Lerma River system, the most extensive in Mexico, the total reuse of water inside the DaimlerChrysler facility means the manufacturing process draws less water from the Toluca region's dwindling aquifer. Toluca's underground aquifer, which supplies water to residents and industry, is dropping at a rate of more than 6 ft/yr. Total recycling by DaimlerChrysler helps to ensure that production at the plant will not be limited by the lack of water in the future. If other manufacturers in the region adopt a similar pollution prevention and conservation philosophy, the region's aquifer should be able to support manufacturing operations for years to come.

## References

Anastas, P. T. and J. B. Zimmerman. 2003. Design through the 12 principles of green engineering, *Environmental Science & Technology.* **37**(March 1, 2003): 95A–101A.

BMW. 2003a. BMW Spheres of Responsibility, Environment. <http://www.bmwgroup.com/e/nav/index.html?http://www.bmwgroup.com/e/0_0_www_bmwgroup_com/5_verantwortung/5_verantwortung.shtml?5_0>

BMW. 2003b. Environmental Protection. BMW Group Environmental Guidelines. <http://www.bmwgroup.com/e/0_0_www_bmwgroup_com/5_verantwortung/5_2_verantwortung/pdf/BMWGroup_Umweltleitlinien.pdf>

BMW. 2002. Sustainable Value Report 2001/2002. Overview of BMW Green Figures. July 18, 200. <http://www.bmwgroup.com/e/0_0_www_bmwgroup_com/5_verantwortung/5_4_publikationen/5_4_1_umweltbericht/5_4_1_3_downloads/pdf/1_7_clean_produktion.pdf>. Sustained use of water. May 2002. <http://www.bmwgroup.com/e/0_0_www_bmwgroup_com/4_news/4_4_aktuelles_lexikon/pdf/Wasser_17_e.pdf>

DaimlerChrysler. 2003. 360 Degrees: Daimler Chrysler Environmental Report 2003. <http://www.daimlerchrysler.com/index_e.htm?/environ/report2003/2_magazine_15_e.htm>

Endreny, T. A. 2001. Sustainability for New York's drinking water, *Clearwaters.* **31**(Fall 2001):36–40. <http://www.nywea.org/clearwaters/313060.html>

Ford Motor Company. July 14, 2003. Focusing on Water. <http://www.ford.com/en/goodWorks/corporateCitizenship/ourLearningJourney/performance/manufacturingPerformance/focusingOnWater.htm>

Intelimex, Inc. 2002. The Mexican Intelligence Report, Automobile Industry. <http://www.mex-I-co.com/industries/autovrvu.htm>

Ley de Aguas Nacionales. 2003. <http://www.sequia.edu.mx/leyes/aguas-nacionales.html>

Ley General del Equilibrio Ecológico y la Protección al Ambiente. 2003. <http://www.semarnat.gob.mx/marco_juridico/federal/legeepa.pdf>

National Academy of Science, Committee on USGS Water Resources Research, Water Science and Technology Board, Division on Earth and Life Studies, National Research Council. 2002. *Estimating Water Use in the United States.* National Academy Press: Washington, DC. <http://books.nap.edu/catalog/10484.html>

New York City Department of Environmental Protection. 2003. History of Drought and Water Consumption. <http://www.ci.nyc.ny.us/html/dep/html/droughthist.html>

North American Commission for Environmental Cooperation. Publications and Information Resources: 2002. Summary of Environmental Law in Mexico. <http://www.cec.org/pubs_info_resources/law_treat_agree/summary_enviro_law/publication/mx09.cfm?varlan=english>

Paterson, J. W. 1985. *Industrial Wastewater Treatment Technology,* 2nd ed. Butteworth: Boston.

Thompson, S. A. 1999. *Water use: Management and Planning in the United States.* Academic Press: San Diego, Calif.

Toyota North America. 2002. Environmental Report. <http://www.toyota.com/about/environment/news/images/02envrep-all.pdf>

USEPA 2003a (Sector Facility Indexing Project). Sector-Level Data Summary: Automobile Assembly Facilities. <http://www.epa.gov/cgi-bin/sector.cgi> and <http://www.epa.gov/sfipmtn1/access.html> (spreadsheet with TRI reporting, autin.xls).

USEPA. 2003b. TRI Explorer. <http://www.epa.gov/triexplorer/>

USEPA. 2000. Automobile Assembly Facilities—Sector Statistics. <http://www.epa.gov/compliance/resources/publications/assistance/sectors/notebooks/motor.html>

USEPA Office of Compliance Sector Notebook Project. 1995. Profile of the Motor Vehicle Assembly Industry. EPA/310-R-95-009. <http://www.epa.gov/compliance/resources/publications/assistance/sectors/notebooks/motvehsn.pdf>

USEPA. 1992. *Sequencing Batch Reactors for Nitrification and Nutrient Removal,* EPA 832-R-92-002. Washington, D.C.

USEPA. n.d. The Toxics Release Inventory (TRI) and Factors to Consider When Using TRI Data. <http://www.epa.gov/tri/tridata/tri01/press/FactorsToConPDF.pdf>

USFilter. O&M Manuals Specific to Toluca Plant.

Wards Auto.com. 2003. Mexico Vehicle Production Down 13.3 Pct in May [2003]. <http://wardsauto.com/ar/transportation_mexico_vehicle_production_6/>

## CHAPTER 7

# Green Chemistry

MARK R. GREENE

## Introduction

*Green chemistry* is the design of chemical products and processes that reduce or eliminate the use and/or generation of hazardous substances. The science of chemistry itself deals with:

- Composition and properties of substances
- Reactions by which substances are produced from or converted to other substances

Chemicals are the building blocks for products that meet fundamental human needs for food, shelter, clothing, and fuel as well as for products vital to the technology world of computing, telecommunications, and biotechnology. Chemicals are a keystone to manufacturing industries such as pharmaceuticals, automobiles, textiles, furniture, paint, paper, electronics, agriculture, construction, and appliances. In the United States, more than 70,000 different chemical products are registered, and more than 13,000 corporations develop, manufacture, and market chemical products and processes with sales over $415 billion and nearly 17 million workers (U.S. Bureau of Census, 1997). The U.S. government uses 7 codes in the North American Industrial Classification System (NAICS, "nakes") to categorize just the broad areas of chemical manufacturing (see box).

**Major Types of Chemical Manufacturing**

3251 Basic chemical manufacturing

3252 Resin, synthetic rubber, and artificial and synthetic fibers and filaments manufacturing

3253 Pesticide, fertilizer and other agricultural chemical manufacturing

3254 Pharmaceutical and medicine manufacturing

3255 Paint, coating and adhesive manufacturing

3256 Soap, cleaning compound and toilet preparation manufacturing
3259 Other chemical product manufacturing

*Source:* U.S. Bureau of Census, 2003

Products of chemical manufacturing have improved society and the human race. A few of these products and many of the by-products of chemical manufacturing, however, have contaminated the water, air, and land since the beginning of the Industrial Revolution in the 19th century. Green chemistry emphasizes the use of alternative feedstocks; development, selection, and use of less toxic solvents; development of new synthesis pathways; improvements in the sensitivities of reactions; generation of less waste; and avoidance of the use of highly toxic compounds. Green chemistry is often referred to as *environmentally benign chemistry* or *sustainable chemistry*. The focus of green chemistry is to develop products and processes that are both environment friendly and economically viable. Green chemistry is not a discipline in itself but rather a way of applying knowledge in kinetics, catalysis, reaction engineering, materials and interfaces, process design and control, separations, and thermodynamics to lessen the adverse effect that chemical products and processes have on the environment.

This chapter deals with the tools and concepts of green chemistry. Examples of the greening of chemical processes and products are presented. Many green variations have been proposed; however, the current challenge is to make these green options economically viable. When the green methods become less costly to use, green chemistry initiatives will become very popular.

In 1991, the U.S. Environmental Protection Agency (USEPA) established the Green Chemistry Program. It is an initiative under USEPA's Design for the Environment Program (DfE) and responds to the 1990 Pollution Prevention Act (42 U.S.C. 13101 *et seq.*). The Pollution Prevention Act was the first environmental law that established a focus on preventing pollution at the source rather than dealing with remediation, treatment, or capture of pollutants after the chemical process was completed. Green chemistry is a highly effective approach to pollution prevention; it applies innovative scientific solutions to real-world environmental situations. The goal of USEPA's Green Chemistry Program is to promote the research, development, and implementation of innovative chemical technologies that protect human health and the environment in both a scientifically sound and cost-effective manner. The program supports chemical technologies that reduce or eliminate the use or generation of hazardous substances during the design, manufacture, and use of chemical products and processes. The program comprises four major areas:

**Green Chemistry Research** Basic research into the chemical tools and methods necessary to design and develop products and processes that are more environmentally benign than many of those now in use. Research projects under this initiative are jointly funded with the National Science Foundation and typically involve a consortium of industry, university, and government researchers.

**Presidential Green Chemistry Challenge** Annual award that draws national attention to scientifically sound and economically viable technologies that directly reduce risks to human health and the environment. Awards have been made since 1996 to companies, universities, and individuals for their research, development, and implementation of green chemistry and green technologies.

**Green Chemistry Education** Includes the development of materials and courses that incorporate green chemistry principles in the training of professional chemists for industry and the education of students. The objective is to have green chemistry become widely adopted and practiced by integrating the principles in every aspect of chemistry education. USEPA has joined with the American Chemical Society (ACS) in this endeavor. In 1998, ACS and USEPA initiated the Green Chemistry Educational Materials Development Project, which initially generated three tools:

1. Annotated bibliography of green chemistry. See Chemistry.org
2. Green laboratory experiments
3. Real-world cases in green chemistry (Cann and Connelly, 2000).

**Scientific Outreach** Intended to make green chemistry the standard and routine practice in industry, academia, and government. The initiative supports meetings and conferences, such as the National Green Chemistry and Engineering Conference and the Gordon Research Conference on Green Chemistry. Publishing in scientific journals and books as well as developing and disseminating computational tools and databases are supported as part of scientific outreach. An important event in the evolution of green chemistry was the creation of the Green Chemistry Institute in May 1997. This organization was established on the Internet as a virtual not-for-profit entity by representatives from industry, academia, national laboratories, and other organizations. The mission of the institute is to facilitate industry–government partnerships with universities and national laboratories to develop economically sustainable clean-production technologies. In 2001, the Green Chemistry Institute formed an alliance with ACS and subsequently established an office at ACS headquarters in Washington, D.C.

Green chemistry initiatives have not been limited to the United States. Currently, there are 25 green chemistry centers throughout the world. European countries such as Estonia, Finland, Greece, Hungary, Italy, Northern Ireland, Spain, Sweden, and the United Kingdom have green chemistry centers. In South America, Argentina and Brazil have centers, and in the rest of the world, Australia, Canada, India, Japan, Mexico, Nepal, New Zealand, the People's Republic of China, Senegal, South Africa, Taiwan, and Thailand have centers. Green chemistry is an international initiative.

## The 12 Principles of Green Chemistry

**The 12 Principles of Green Chemistry**

Prevention
Atom economy
Less hazardous chemical synthesis
Designing safer chemicals
Safer solvents and auxiliaries
Design for energy efficiency
Use of renewable feedstocks
Reduce derivatives
Catalysis
Design for degradation
Real-time analysis for pollution prevention
Inherently safer chemistry for accident prevention

Paul T. Anastas, an organic chemist with the White House Office of Science & Technology Policy, and chemistry professor John C. Warner of the University of Massachusetts, Boston, developed guidelines to help practicing chemists to evaluate the degree of "greenness" that a synthesis, compound, or technology exhibits (Anastas and Warner, 1998). Anastas' guidelines are called the "12 principles of green chemistry." See box. The philosophical foundation for these principles is the same as that for the 12 principles of green engineering, presented in this book's Introduction.

**Prevention** *It is better to prevent waste than to treat or clean up waste after it has been created.* One objective of green chemistry is to design chemical processes and reactions that consume all the reactants and leave no (or minimal) waste. Waste can be minimized by nongeneration strategies or by reuse strategies. From a prevention standpoint, the nongeneration strategy is preferred. An example of a low waste chemical reaction is the novel Friedel–Crafts acylation based on trifluoroacetic anhydride–phosphoric acid developed by T. Smyth of Ireland and used to produce a pharmaceutical intermediate (Clark and Hodnett, 2001).

**Atom Economy** *Synthetic methods should be designed to maximize the incorporation of all materials used in the process into the final product.* The concept of atom economy is similar to prevention. Waste generation is minimized when all reactants are consumed in the synthesis of new substances; that is, there are no by-products of the chemical reactions. Barry Trost of Stanford University first presented the idea of atom economy (Trost, 1991). An example of this principle is the improvements to the synthesis of the pharmaceutical, Sertraline, which reduced the number of steps and the number of solvents used in the manufacturing process (Clark, 2002).

**Less Hazardous Chemical Synthesis** *Wherever practicable, synthetic methods should be designed to use and generate substances that possess little or no toxicity to human health and the environment.* The objective of this principle is to produce benign by-products when the generation of by-products cannot be avoided and to use solvents that are compatible with the environment. An example of this principle is the production of a pharmaceutical intermediate, NMSM (1-methylamino-1-methylthio-2-nitroethylene), which is used in the manufacture of the drug, ranatidine. This drug is a histamine blocker used to treat ulcers and heartburn. An alternative chemical synthesis process was developed that reduced the raw material consumption and used less hazardous solvents (Krewer et al., 2002).

**Designing Safer Chemicals** *Chemical products should be designed to produce their desired function while minimizing their toxicity.* The objective of this principle is to create substances that accomplish their desired purpose without hazardous side effects or the generation of toxic by-products. An example of this principle is the production of a new wood preservative that is arsenic-free and chromium-free (USEPA, 2002).

**Safer Solvents and Auxiliaries** *The use of auxiliary substances (such as solvents and separation agents) should be made unnecessary wherever possible and innocuous when used.* An example of this principle is the replacement of conventional solvents with inert supercritical fluids. A tool for solvent substitution that was developed at the National Risk Management Research Laboratory of USEPA is PARIS II (Program for Assisting the Replacement of Industrial Solvents) (Technical Database Services, 2003).

**Design for Energy Efficiency** *Energy requirements of chemical processes should be recognized for their environmental and economic impacts and should be minimized; if possible, synthetic methods should be conducted at ambient temperature and pressure.* An example of this principle is an initiative in the electronics industry with respect to the manufacture of microchips where the material and energy inputs are being quantified to identify alternatives for improving energy efficiency in the process (Williams et al., 2002).

**Use of Renewable Feedstocks** *A raw material or feedstock should be renewable rather than depleting, whenever technically and economically* *practicable.* This principle focuses on the substitution of renewable materials for conventional raw materials. Traditionally, many chemical feedstocks have been derived from petroleum-based compounds. An alternative source for some of these substances is agricultural crops—a renewable source. Biodiesel is the name for a clean-burning alternative fuel that is produced from renewable resources. Use of biodiesel in a diesel engine with little or no modification has been shown to provide sufficient power and reduce harmful emissions (Haas et al., 2001). The economic viability of biodiesel has not been proven yet, however. A second example of using renewable feedstocks, which is commercially viable, is the production of polylactic acid (PLA), a polymer used to make fibers, by Cargill Dow LLC (USEPA, 2002).

**Reduce Derivatives** *Unnecessary derivatization (use of blocking groups, protection/deprotection, temporary modification of physical/chemical processes) should be minimized or avoided if possible because such steps require additional reagents and can generate waste.* An example of this principle is the use of noncovalent derivatizations, for example, hydrogen bonds, self-assembly, or molecular recognition instead of covalent techniques to manipulate the physical properties of materials (Warner, 1998).

**Catalysis** *Catalytic reagents (as selective as possible) are superior to stoichiometric reagents.* An example of this principle is a new environment-friendly oxidant activator based on iron, TAML (tetraamidomacrocyclic ligand), which was developed by Terry Collins at Carnegie Mellon University (USEPA, 2002).

**Design for Degradation** *Chemical products should be designed so that at the end of their function they break down into innocuous degradation products and do not persist in the environment.* An example of this principle is the use of thymine as a cross-linking agent in the production of robust polymers. When the time comes, the cross-links can be undone using deoxyribonucleic acid (DNA) photolyase, and the original polymers can be recovered for reuse.

**Real-Time Analysis for Pollution Prevention** *Analytical methodologies need to be further developed to allow real-time in-process monitoring and control before the formation of hazardous substances.* An example of this principle is the use of supercritical fluids where a large degree of control over product selectivity and yield can be achieved by adjusting temperature and pressure of the reactor. The scientific work of Eric Beckman, University of Pittsburgh, has focused on this principle (USEPA, 2002).

**Inherently Safer Chemistry for Accident Prevention** *Substances and the form of a substance used in a chemical process should be chosen to minimize the potential for chemical accidents, including releases, explosions, and fires.* An example of this principle is the substitution of volatile organic compounds with ionic liquids as the solvent for a chemical reaction (Holbrey and Seddon, 1999).

## Expert System

An *expert system* is a computer program that simulates the judgment and performance of a human being or an organization that has expert knowledge and experience in a particular field. The Green Chemistry Expert System (GCES), available free from USEPA as part of the Green Chemistry Program, can be used to select green chemicals and reactions. GCES comes in five modules and assists users in:

- Identifying opportunities to eliminate or reduce the use or generation of hazardous substances during chemical manufacture
- Suggesting molecular modifications to minimize toxicity
- Providing techniques that might reduce hazards to human health and the environment
- Presenting technical information on green synthesis, solvent systems, reaction conditions, and chemical products
- Providing a database of useful green chemistry references

The five GCES modules help users to build a green chemical process, design a green chemical, or survey the field of green chemistry. The system is equally useful for new or existing chemicals and their synthetic processes. The five GCES modules are:

Synthetic Methodology Assessment for Reduction Techniques (SMART)
Green Synthetic Reactions
Designing Safer Chemicals
Green Solvents/Reaction Conditions
Green Chemistry References

SMART incorporates information entered by the engineer to quantify and categorize the hazardous substances used in or generated by a chemical reaction. The module allows the user to modify the chemical reaction and reevaluate the output for improvement or deterioration in the "greenness" of the reaction as a result of the changes. In this manner, the user can minimize the adverse environmental effects of the chemical process. To use the module, three types of information are needed: (1) the quantity of each chemical used and generated in the chemical process, (2) an understanding of the role each chemical plays in the reaction (e.g., product, feedstock, catalyst, solvent), and (3) the basic reaction data, including batches per year, product yield, and annual production volume. SMART calculates the theoretical amount and hazard level of each reaction component on an individual chemical basis and directs the user to the other modules for information on developing a green alternative.

The Green Synthetic Reactions module provides a searchable database of synthetic methods with descriptions, keywords, and references. The module for Designing Safer Chemicals provides guidance on how chemical substances can be modified to make them safer under the categories of chemical class, chemical properties, and chemical use by leading the user through a series of questions and then identifying molecular modifications to minimize hazards. With Green Solvents/Reaction Conditions, engineers can find technical information on alternatives to traditional solvent systems by allowing them to search for substitute solvents based on physiological properties. Several search engines for finding references to additional information are available through the Green Chemistry References module. In practice, these modules may require information beyond that necessary to use the SMART module, including technical information about a chemical product or its manufacturing process.

The five modules of the Green Chemistry Expert System furnish engineers with many tools to assess and identify green alternatives to chemical reactants, chemical products, chemical wastes, solvents, and chemical reaction mechanisms. USEPA is constantly revising and updating this software and the associated databases as new information becomes available. The individual components are described next.

## SMART Module

SMART can be used in parallel with an application for a new chemical under the federal New Chemicals Program (USEPA Office of Pollution Prevention and Toxics), but its usefulness is not limited to new chemicals. The methodology is broadly applicable to any chemical process. The New Chemicals Program is mandated by Section 5 of the federal Toxic Substances Control Act. (TSCA was passed in 1976. It is found at 15 U.S.C. 2601–2671. See Subchapter R.) TSCA gives USEPA broad authority to identify and control newly synthesized substances—therefore, substances newly entering the environment—that may pose a threat to human health or the environment. The New Chemicals Program is tasked to identify conditions, up to and including a ban on production, to be placed on the use of a new chemical before it becomes commercially available. The prospective manufacturer or importer of a new chemical substance for a nonexempt commercial purpose is required by law to notify USEPA of its intent to use the new chemical by filing a premanufacturing notice (PMN) 90 days before beginning any activity. During the review of the PMN, USEPA uses SMART to identify potential green opportunities. The agency then suggests these approaches to the applicant for its voluntary consideration. If the suggestions are feasible and economical, the applicant can incorporate them in its manufacturing process as desired. USEPA suggestions are not binding or mandatory.

SMART's output may include the need to identify benign chemicals as replacements for hazardous starting materials or the goal to change a synthesis to prevent the formation of hazardous or unusable by-products or co-products. The output may also suggest a search for green reaction conditions that eliminate or minimize the use of hazardous solvents or improve the efficiency of the chemical by increasing the conversion of feedstocks and reagents or increasing the product yield or atom economy. A shortcoming of SMART is that it only examines the substances directly involved in the chemical process. The module does not consider other potential environmental concerns, such as the indirect use of an oil burner to heat a chemical reaction with the associated effects on air pollution or as a nonrenewable fuel source.

SMART is a tool. It cannot decide on the relative merits of the changes identified but can only assist in finding potential modifications. The user must interpret SMART's output and decide on the comparative benefits of its suggestions. It is also important to note that alternative substances and processes should be reviewed a second time using SMART for other unforeseen green or nongreen consequences. (A *nongreen* consequence is one more harmful than the alternative.)

SMART can analyze the amount of chemical waste that a manufacturing process produces. The waste can be classified according to its hazards; this feature allows the user to focus on reducing or eliminating specific chemicals from the synthesis. This process is iterative: Changes to the process are entered, and the wastes produced from the modified process are compared to those from an earlier process to decide if the modification is beneficial. Other factors that must be considered are the technical and economic feasibility of the changes. SMART cannot perform technical or economic evaluations; the user must make these assessments. (Total cost assessment, a method is described briefly below, is one tool that can be used to perform the economic evaluations.)

SMART contains mathematical algorithms that perform a series of calculations for a given reaction process. Mass-balance calculations are completed with user-entered data. Classifications of the hazardous nature of each substance are made using established USEPA guidelines. The user enters general information regarding the reaction process, such as the name or identity of the reaction, the typical number of batches per year, and the expected percent yield of the reaction. In addition, specific information for each chemical is entered, including the Chemical Abstracts Service (CAS) Registry Number (if known), the amount per batch charged or formed, the amount per batch recovered, the role of the chemical in the reaction (such as feedstock, product, or solvent), and the number of equivalents required or formed of the reactants and products. If the CAS Registry Number is unknown, SMART uses its lookup table, a reference with information on over 60,000 chemicals listed by the CAS registry. The table can be searched by chemical name, molecular weight, or the chemical structure using Simplified Molecular Input Line Entry System (SMILES) notation. [SMILES uses strings of symbols to represent molecular structures (Weininger, 1988).] If a chemical does not have a CAS Registry Number, the SMART module requires the entry of a unique alphanumeric identifier for the chemical so the system can store information for later use.

Once data have been entered into SMART, the program analyzes all the wastes produced in the reaction and provides the total amount of waste produced, classified into four hazard categories based on the level of concern. The presence of Tier 1 and 2 hazards automatically triggers a connection to another GCES module, such as the Green Synthesis, Green Solvents/Reaction Conditions, or Designing Safer Chemicals modules. (See box.)

> Tier 1 chemicals are few and very hazardous. Their use or generation should be avoided if possible. Examples of Tier 1 chemicals are dioxins and phosgene (carbonic dichloride).
>
> Tier 2 chemicals are USEPA-regulated substances, including those listed in the Emergency Planning and Community Right-to-Know Act [section 302, extremely hazardous substances, and section 313, Toxics Release Inventory; the Clean Air Act section 112(r); and the Hazardous Air Pollutants lists]. This classification includes chemicals with functional groups associated with high toxicity (e.g., acid chlorides, alkoxysilanes, epoxides, or isocyanates).
>
> Tier 3 chemicals are all chemicals of unknown or intermediate toxicity. This classification is broad and includes chemicals not categorized in the other tiers.
>
> Tier 4 chemicals are a few relatively innocuous substances that pose little or no risk of harm under normal conditions, such as water, sodium chloride, and nitrogen.

SMART allows the engineer to list individual chemicals in each tier of the hazard classification levels. SMART itself can provide additional information about each chemical. USEPA-regulated chemicals are identified along with the names of the corresponding rules that govern them.

The algorithms in SMART are designed for single-step reactions that produce a discrete chemical product. It cannot analyze polymers or reactions with multiple products, but individual reactions of a synthetic pathway may be run sequentially by entering the recovered mass of product for each step as the amount of feedstock in the subsequent reaction.

### Example 7.1 Use of SMART

Let us use SMART for the chemical reaction between ethyl amine (CAS RN 75-04-7), phosgene (CAS RN 75-44-5), and bisphenol A (CAS RN 80-05-7). This reaction uses toluene as a solvent and produces a co-product of hydrochloric acid. Two batches per year are going to be produced, and the reaction proceeds with an 87.7% yield. The SMART input data for this reaction are shown in Table 7.1.

The SMART assessment summary for this reaction shows an annual production volume of 13,000 kg and the amount of waste for each hazard classification as a percent of the production volume (Table 7.2).

**Table 7.1** SMART Input Data

| CAS Registry Number | Chemical Name | Role | Equivalents | Amount Charged or Formed per Batch (kg) | Amount Recovered per Batch (kg) |
|---|---|---|---|---|---|
| 75-04-7 | Ethanamine (ethyl amine) | Feedstock | 2 | 3,607 | 0 |
| 75-44-5 | Carbonic dichloride (phosgene) | Feedstock | 2 | 4,946 | 0 |
| 80-05-7 | Phenol, 4,4′-(1-methylethylidine) bis (bisphenol A) | Feedstock | 1 | 4,566 | 0 |
| 108-88-3 | Methylbenzene (toluene) | Solvent | | 10,000 | 8,000 |
| 7647-01-0 | Hydrochloric acid | Co-product | 4 | 2,880 | 0 |
| Product | Product | Product | 1 | 7,409 | 6,500 |

The "green suggestions" for this chemical reaction refer the user to three other GCES modules: (1) the Green Synthesis module to search for an alternative synthesis method that may be applicable to this chemical reaction, (2) the Green Solvents/Reaction Conditions module to search for an alternative solvent to toluene, and (3) the Designing Safer Chemicals module for information on how to reduce the toxicity of the product.

SMART guides the user in estimating the hazard of the chemical process by identifying and ranking the level of concern for the reactants and products. For our example, the highest level of concern is the use of a Tier 1 chemical (phosgene). The next level of concern is the use of USEPA-regulated chemicals (phosgene, ethylamine, bisphenol A, HCl, and toluene) followed by the generation of a relatively large amount of Tier 2 waste. The two lowest levels of concern are the large amount of combined Tier 1 and 2 wastes produced and the large total amount of waste generated by this reaction. The identification and ranking of the sources of concern allow the engineer to develop strategies for eliminating or reducing these concerns. In this example, some possible suggestions are:

- Replace phosgene with dimethyl carbonate, which has been used commercially as a substitute for phosgene. This action removes the

**Table 7.2** SMART Assessment Summary

| Waste Type | Amount of Production Volume | Type |
|---|---|---|
| Tier 1 | 15.22% | Unreacted phosgene |
| Tier 2 | 117.38% | Unreacted ethyl amine, a very small amount of unreacted bisphenol A, unrecovered solvent, and the co-product, hydrochloric acid |
| Tier 3 | None | — |
| Tier 4 | None | — |

Tier 1 chemical and reduces the use of a regulated chemical. (This suggestion is found in the Green Synthesis module.)

- Eliminate the use of toluene as the solvent for this reaction. This action reduces the concern for Tier 2 chemicals and the total amount of chemicals subject to USEPA regulation. The Green Solvents/Reaction Conditions database module could be searched for an alternative solvent by comparing the physiochemical properties of candidates with those of toluene.
- Recover more of the product. This step would reduce the amount of a Tier 2 waste.
- Use less unreacted feedstocks. A reduction in the excess amount of feedstock chemicals—as long as an acceptable yield is maintained—would reduce the amount of Tier 2 waste.
- Recover more of the toluene, if an alternative solvent is impractical. The improved recovery of toluene would also reduce the amount of regulated chemicals wasted by this reaction.

## Green Synthetic Reactions Module

A database—the Green Synthetic Reactions module—contains current references related to selected synthetic processes. The engineer can search the database in the following fields:

- Pollution prevention comments (e.g., safety or less expense)
- Key word
- Status (e.g., bench-scale, pilot plant, or patent)
- Reference citation

The search results are returned with a full reference citation, a description of the process, a list of vendors, the status of the process, and a list of comments about the green advantages of using the process. The user may narrow or expand a search. Word fragments are accepted to allow the user to get information on slight variations (or misspellings) of various items. The search engine does not allow successive searches of intermediate databases. It is a simple once-through tool. The user can perform searches that are more specific by careful selection of search terms, if the output from the initial search is too cumbersome to work with.

**Green Synthetic Reactions Module**

Searchable key words include:

| | | |
|---|---|---|
| acid | catalyst | oxidation |
| alcohol | catalytic | phosgene-free |
| amide | chloride | process |
| amidocarbonylation | chlorine-free | 1,3-propoanediol |
| amine | DMC | rearrangement |
| anhydride | enzymatic | styrene |
| aromatic | hydrolysis | supercritical |
| bromination | isocyanate | THF |
| carbon dioxide | NMP | water |

Pollution prevention comments include:

acetic
acetone
acid
alcohols
aldehyde
alkyl
alternative
amine
ammonium
anhydride
aromatic
arylation
bromination
by-product
carbon
carbonylation
catalysis
catalyst
$CH_3OH$
chlorine
$Cl^+$
concern
conditions
convenient
conventional
co-product
DMC
eliminate
ethylene
feedstock
formation
HCl
HCN
HF
$HNO_3$–$H_2SO_4$
hypochlorite
inexpensive
intermediate
isocyanate
ketones
liquid
nitration
nonacidic
nontoxic
oxidation
phosgene
process
produce
propylene
pure
reaction
reacts
reagents
recycled
removed
route
safety
salt
SC-$CO_2$
secondary
sodium
solvent
source
styrene
sulfate
sulfuric
supercritical
system
traditional
waste
water
yield

Status search terms include:

commercial
commercially
lab
new
patent
pilot
plant
produce
production
R&D
scale
U.S.

Reference citation search terms include:

1992
Encyclopedia
Joseph
market
News
Science
Sheldon
May

In the reference citation search, titles of articles are included, but not all entries actually have titles.

After a search has been performed, the module allows the user to execute another search of the database. This is especially useful if no records are returned or if too many records are returned. The user can adjust the search criteria to obtain a more informative list.

## Designing Safer Chemicals Module

The Designing Safer Chemical module provides the engineer with qualitative information about the toxicities of compounds within certain chemical classes or substances that have specific uses. The classification is based on structure–activity relationships; it is a predictive method to estimate the properties (activity) of a chemical—for example, melting point, vapor pressure, toxicity and ecotoxicity—on the basis of its chemical structure. This module helps to estimate the qualitative toxicity of a particular chemical, to understand the mechanism of toxicity for a particular substance, and to predict which structural changes may reduce the hazard level of the components in the user's chemical reaction. The module is a tool for assessing the relative toxicities between substances; so, the user may make educated decisions on which chemical(s) to include in or which chemical(s) to exclude from a process.

### Example 7.2 Use of Designing Safer Chemicals Module in Choice of Nitriles

The selection of nitriles (organic cyanides containing a CN group) returns three linking options for the user:

- Toxic mechanism and structure–activity relationships of nitrile toxicity, which displays information on toxic responses. The user can then link to additional toxicity information.
- Qualitative toxicity assessment of a planned nitrile, which brings up a series of questions designed to determine the acute lethality or osteolathyrism of a particular nitrile. (*Osteolathyrism* is a form of lathyrism that is characterized by skeletal deformities and the formation of aortic aneurisms. *Lathyrism* is a diseased condition of humans and domestic animals that is characterized by spastic paralysis of the hind or lower limbs.)
- Guidance for designing safer nitriles, which presents known structural modifications that would be expected to reduce nitrile toxicity in the modified substance.

### Green Solvents/Reaction Conditions Module

The Green Solvents/Reaction Conditions module provides databases of examples and physicochemical properties for a large number of common solvents. The information is searchable by physicochemical properties, browsing, and by examples. The module does not provide toxicity information about solvents but does provide regulatory list information, global warming potential, and ozone depletion potential. Global warming potential is a concept that compares the ability of greenhouse gases to trap heat in the atmosphere relative to carbon dioxide. Ozone depletion potential is a comparison between the amount of ozone depletion caused by a substance and that caused by a similar mass of CFC-11 (trichlorofluoromethane). The user can use this information to evaluate qualitatively the health and environmental hazards of various reaction conditions or solvents. The primary physicochemical properties used to search the database are boiling point, melting point, and water solubility.

Information (when available) for solvents in Green Solvents/Reaction Conditions module includes:

molecular weight
specific gravity
log octanol/water partition coefficient
Henry's law constant
ozone depletion potential
water solubility
USEPA regulations
boiling point
vapor pressure
flash point
explosion limits
global warming potential
melting point
dielectric constant

### Green Chemistry References Module

The reference module cites material published in green chemistry and related fields. This list may be browsed by category and subcategory or may be searched by author, title, journal, or key word.

Pertinent key words in the Green Chemistry Reference module:

**Green feedstocks**
biomass
renewable feedstocks

**Green reagents**
photochemical reagents

**Green chemical synthesis/manufacturing**
catalysis
biocatalysis
atom economy

**Green solvents**
aqueous conditions
derivatized/polymeric solvents
ionic liquids
supercritical solvents

**Green reaction conditions**
energy-efficient reactions
solventless reactions

**Green chemical processing**
separations
extractions
purifications
analytes

**Process analytical chemistry**
hazardous chemical reduction/elimination
bioprocessing
reduced solvent
solventless processes
waste reduction/elimination
energy-efficient processes

**Designing safer chemicals**
inherently safer chemicals
less toxic chemicals
biodegradable chemicals
less persistent chemicals

## Assessing Green Approaches to Chemistry

### Life-Cycle Assessment

The effectiveness of green chemistry initiatives can be evaluated using a life-cycle assessment (LCA) technique to quantify the economic and environmental costs of making a product. LCA is a cradle-to-grave quantitative methodology for examining the consequences of making and using products. (Refer to Chapter 2 for a comprehensive presentation on LCA. The brief discussion here just highlights aspects relevant to the chemical industry.) LCA adopts a holistic approach by analyzing the entire life cycle of a product, process, package, material, or activity. In relation to chemical products, five stages are considered in LCA:

- Extraction and processing of raw material
- Manufacturing, transportation, and distribution
- Use, reuse, maintenance
- Recycling and composting
- Final disposition

The analysis proceeds using the four LCA stages elaborated in Chapter 2:

- Goal definition and scoping
- Life-cycle inventory (LCI) analysis
- Impact assessment
- Improvement assessment

The definition of goals and the determination of scope are necessary to provide a boundary for the life-cycle assessment. The results of the life-cycle inventory are needed to perform an impact assessment. The impact assessment has four general categories for classification of effects:

- Environmental or ecosystem quality
- Quality of human life including human health
- Natural resource use
- Social welfare

The *stressor concept* provides a handy way of linking the life-cycle inventory items to subsequent effects. A stressor is defined as any physical, chemical, or biological entity that can affect the environment and may be characterized with the following attributes (AIChE, 1999a):

**Type** physical, chemical, biological

**Intensity** concentration, magnitude, abundance/density

**Duration** acute (short term) versus chronic (long term)

**Frequency** single event versus recurring or multiple exposures

**Timing** time of occurrence relative to environmental and human health parameters

**Scale** spatial extent and heterogeneity in intensity

*Characterization* is determining the magnitude of the potential effects for each category, and *valuation* is assigning a relative value or weight to each effect. The impact assessment provides the basis to perform an improvement assessment. This is a systematic evaluation of the needs and opportunities to reduce the environmental and health burdens associated with the product or process under evaluation and its consumption of raw materials and energy along with its generation of waste emissions for the entire life of the product or process. The results of the improvement analysis may lead to changes in the life-cycle inventory, which could lead to a revision of the impact analysis and a repeat of the improvement assessment. This iterative process continues until a point of diminishing returns is achieved, usually lasting only one or two cycles. See Figure 2.2.

## Total Cost Assessment

Total cost assessment (TCA) has been defined as the identification, compilation, analysis, and use of environmental and environment-related human health (E&H) cost information for internal managerial decision making. In contrast to life-cycle assessment, TCA is an economic method that is used to evaluate the full life-cycle perspective from raw material extraction to the end of the useful life of a product or process. A complete life-cycle inventory is not required for every TCA analysis. The use of "nearest-neighbor" information is sufficient for meaningful TCA results. LCA is more comprehensive, but the information available to the engineer may be insufficient to perform a complete LCA especially with respect to upstream and downstream customers. TCA is a support tool for making informed managerial decisions regarding E&H improvements. Outputs from an LCI can serve as inputs for TCA when they are translated to an economic value, and they can provide focus for cost reduction, risk reduction, or both.

Total cost assessment provides a tool for the quantification of all environmental and health costs, both internal and external, associated with a business decision. This tool enables internal managerial decision making to assess the environmental, health, and safety implications of industrial process and product decisions more accurately, such as:

- Process and product design decisions and performance evaluations
- Comparison of alternative products, processes, and services
- Costing determinations
- Capital budgeting

Total cost assessment is not designed to replace the capital project and product development cost-estimating practices that are already in use. Its role is to enhance these costing exercises further by focusing attention on the potentially hidden E&H costs and effects. The objective of TCA methodology is to provide a disciplined and standardized approach to improve business decisions by incorporating a thorough evaluation of potential E&H costs that may affect the environment and society into the analysis. See box.

**Some Environmental Costs Incurred by Corporations**

*Potentially hidden regulatory costs such as:*

Notification
Reporting
Monitoring and testing
Studies and modeling
Remediation
Recordkeeping

*Up-front potentially hidden costs such as:*

Site studies
Site preparation
Permitting
Research and development
Engineering and procurement
Installation

*Back-end potentially hidden costs such as:*

Closure and decommissioning
Disposal of inventory
Postclosure care
Site survey

*Voluntary (beyond compliance) potentially hidden costs such as:*

Community relations and outreach
Training
Audits
Qualifying suppliers
Recycling
Environmental studies
Habitat and wetland protection
Landscaping
Financial support to environmental groups and/or researchers

The TCA method comprises seven main steps: (1) Define the goal and estimate the scope. (2) Streamline the analysis. (3) Identify potential risks. (4) Conduct financial inventory. (5) Conduct impact assessment. (6) Document results. (7) Send feedback to the company's main decision loop. The first three steps are straightforward and similar to those of an LCA (see Chapter 2). The financial inventory incorporates five different types of costs:

Type I: Direct costs for the manufacturing site

Type II: Potentially hidden corporate and manufacturing site overhead costs

Type III: Future and contingent liability costs

Type IV: Internal intangible costs

Type V: External costs

Type I and II costs represent direct costs attributed to the project or process and allocated corporate overhead. These costs are typically defined by the firm's internal accounting practices or capital project estimation groups. Type III, IV, and V costs are more abstract, and metrics have been prepared from surveys to use as starting points to quantify these costs (AIChE, 1999b). Green chemistry principles have more impact on type III, IV, and V costs than on type I and II costs. The economic viability of a green chemistry alternative, however, is evaluated for its effect on all costs, especially type I and II costs. The proportion of E&H costs associated with a chemical product has ranged around 20% of the total manufacturing cost for the product (AIChE, 1999a).

## Additional Metrics

The Center for Waste Reduction Technologies has made an initial attempt to characterize and quantify some "ecoefficiency" metrics. (See Chapter 1 for further discussion of metrics.) These metrics are intended to cover:

- Material intensity
- Energy intensity
- Toxics dispersion
- Potential for recycling material
- Use of renewable resources
- Product durability
- Service intensity

Material intensity metrics have three ratios: (1) the amount of raw materials plus products plus packaging divided by value added, (2) the amount of raw materials plus products plus packaging divided by the dollar value of the product sold, and (3) the amount of raw materials plus products plus packaging divided by the mass of product sold. Value added is defined as the total value of the product sold minus the purchase cost of the raw materials, packaging, and energy consumed.

Energy intensity metrics use three ratios: (1) the total amount of energy consumed divided by value added, (2) the total amount of energy consumed divided by the dollar value of the product sold, and (3) the total amount of energy consumed divided by the amount of product sold.

Pollutant metrics comprise the following ratios: (1) the *greenhouse gas metric*, which is calculated by dividing the total mass of $CO_2$ equivalents by the dollar value of product sold, (2) the *photochemical ozone creation potential metric*, which is the total mass of $C_2H_4$ equivalent divided by the dollar value of product sold, (3) the *acidification* (air and water separately) *metric*, which is the total mass of $SO_2$ equivalent divided by the dollar value of product sold, and (4) the *eutrophication metric*, which is calculated by dividing the total mass of phosphate equivalent by the dollar value of product sold.

The water usage metric is the sum of the mass of all water used minus the once-through cooling water not from an aquifer plus all rainwater treated in an on-site treatment plant minus any seawater used divided by the dollar value of product sold.

The human health metric is the sum of the products of many emission effects ($Y_i$) times their corresponding mass of emission ($M_i$). The effect of an emission is calculated by:

$$Y_i = (P_i \cdot \mathrm{BCF}_i)/\mathrm{PEL}_i$$

where $Y_i$ = effect of emission $i$
$P_i$ = multimedia weighted half-life ($\mathrm{hr}^{-1}$)
$\mathrm{BCF}_i$ = bioaccumulation factor
$\mathrm{PEL}_i$ = permissible exposure limit

One last metric is the ecotoxicity metric. It is calculated similar to the human health metric, by summing the products of an emission effect times its corresponding emission mass. For this metric, the emission effect is calculated using a toxicity measure such as $LC_{50}$ or $EC_{50}$ using the following formula:

$$Y_i = (P_i \cdot \mathrm{BCF}_i)/\mathrm{LC}_{50j} \text{ or } \mathrm{EC}_{50j}$$

where $Y_j$ = effect of emission $j$
$P_i$ = multimedia weighted half-life ($\mathrm{hr}^{-1}$)
$\mathrm{BCF}_i$ = bioaccumulation factor
$\mathrm{LC}_{50j}$ or $\mathrm{EC}_{50j}$ = lowest value for the species algae (green), daphnid (*Daphnia magna*) and fish (rainbow trout or blue gill)

$LC_{50}$ is the lethal concentration that kills 50% of the test animals in a given time (usually 4 hr). $EC_{50}$ is a statistically derived concentration of a substance in an environmental medium expected to produce a certain effect in 50% of test organisms in a given population under a defined set of conditions.

Additional metrics are under development for land usage, renewable materials, recycled postconsumer materials, extended service life, use and disposal of the product, and social and economic effects.

# Green Chemistry in Practice

## Alternative Feedstocks

Feedstocks are the raw materials used to make chemical products. Historically, they were derived from petroleum products. Alternative feedstocks come from renewable sources, such as agricultural crops. Examples of alternative feedstocks are cellulose, starch, and sugars from plant-based materials. Waste products containing cellulose have been converted to useful chemicals, such as levulinic acid ($C_5H_8O_3$), a building block for making commercial chemicals. Levulinic acid has been used to develop a biodegradable herbicide, as an economical fuel additive to make gasoline burn more efficiently, and to make new biodegradable plastics. The alternative feedstock used to make levulinic acid is municipal solid waste, which contains 60 to 70% cellulose in the form of cardboard, paper, and wood.

An example of using safer chemical feedstocks can be illustrated with the "clock" experiment. These experiments are commonly used to in chemistry demonstrations because the reaction leads to a sudden sharp change in color when it is complete, making it easy to time the rate of reaction (the time from when the reactants are mixed until the chemical reaction is complete). The effect of changing reaction conditions or feedstock concentrations can be readily measured, and optimization strategies to improve reaction times can be easily evaluated. Typically, chemicals such as formaldehyde, mercuric ion, thiosulfates, or bisulfites have been used to demonstrate clock reactions. Alternative, less toxic, reagents for these demonstrations are vitamin C, iodine, hydrogen peroxide, and liquid laundry starch. The demonstration's usefulness and teaching principles are unchanged with the alternative feedstocks.

It is feasible to manufacture plastics using plants instead of fossil fuels, but the amount of energy required to grow the green plant and separate the plastic from the plant far exceeds the amount of energy sacrificed using petroleum-based nonrenewable raw materials (Gerngross and Slater, 2000). For example, the amount of energy required to make one kilogram of polyhydroxyalkanoate (PHA) from genetically modified corn is about 300% greater than the 29 MJ needed to make polyethylene from fossil-fuel-based substances. In other words, there is no benefit from using plants instead of oil for these polymers. Improvements in technology and depletion of oil reserves may narrow this gap. Tools such as total cost accounting and life-cycle analysis will help to focus efforts toward effective improvements in the technology.

One process that has been developed to make PHAs uses food scraps that are decomposed by anaerobic digestion into four types of organic acids: acetic, propionic, butyric, and lactic acids. The organic acids are separated from the suspension by membranes and are used in a fermentation process to produce PHA. The polymer accumulates inside the microorganisms during fermentation, and the polymer content with this process is comparable to that derived from pure glucose fermentation (Du and Yu, 2002).

## Less Harmful Solvents

Cleaner chemical technologies use some of the following reaction media: supercritical carbon dioxide, supercritical water, room temperature ionic liquids, biphasic systems, and solvent-free systems that use surfaces or interiors of clays, zeolites, silica, and alumina.

Supercritical carbon dioxide is probably the most popular supercritical fluid (SCF) that is employed. A supercritical fluid possesses properties that are intermediate between liquids and gases. It is created under the exertion of pressure at high temperatures greater than the *critical point*. The critical point is defined as the physical condition, specified by pressure or density, at which a substance at its critical temperature can exist as a liquid and vapor in equilibrium. Near the critical point, small changes in temperature or pressure can result in significant changes in physicochemical properties—for example, density, diffusivity, or solubility—of the substance. This behavior can be very useful in controlling chemical reactions and is not available using traditional organic solvents.

Other SCFs include methane, ethane, propane, argon, nitrous oxide, and water. These substances have been used for extraction, chromatography, inorganic synthesis, organic synthesis, catalysis, materials processing, and dry cleaning of clothes. The drawback of using SCFs is the high temperature and pressures required for these substances. Technical advances have been made in reactor technology to improve the safety and economy of performing chemical reactions under these extreme conditions. An example is in the manufacture of Teflon by DuPont using supercritical carbon dioxide as a replacement solvent for chlorofluorocarbons. The polymerization of tetrafluoroethylene in carbon dioxide using fluorinated initiators was demonstrated and then incorporated in the processes installed in a new production facility. Another example is the substitution of liquid $CO_2$ for perchloroethylene (PCE) as a solvent in the dry-cleaning process. PCE has been found as a groundwater contaminant and is a possible human carcinogen. The proprietary process with $CO_2$, called Micare, has been incorporated into several dry-cleaning stores using specially designed washing machines to clean clothes with beverage-grade $CO_2$ and specialty surfactants. The 60-lb capacity machines use a conventional rotating drum to agitate the clothes with a 40-minute cleaning cycle. The solvents and about 98% of the $CO_2$ are recovered at the end of the cleaning cycle.

The manufacture of Zoloft (Pfizer), an antidepression drug, was changed to use the solvent, ethanol, instead of toluene, tetrahydrofuran, methylene chloride, and hexane; moreover, three intermediate isolations were combined into one step. These modifications saved hundreds of thousands of dollars and eliminated the use of 440 metric tons of titanium dioxide, 150 metric tons of 35% hydrochloric acid, and 100 metric tons of 50% sodium hydroxide per year.

## New Synthesis Pathways

One example of a new synthesis pathway is the development of paired electrosynthesis where one product is generated at the anode and another is generated at the cathode. This pathway has been used to co-produce aromatic aldehydes (at the anode) and phthalide, a fungicide intermediate (at the cathode).

## Improving Sensitivities in Reactions

Fluorous molecules can improve the yield of an organic chemical reaction and improve the ability to recover high-purity product from the reaction mixture. These molecules are designed to have similar reactivity to other organic molecules and can be separated from standard organic compounds by two- or three-phase liquid extraction or solid–liquid extraction techniques. An example of a fluorous technique is the use of fluorous tin hydrides that have similar reactivity to the classical reagent, tributylin hydride, in the production of a typical Boc-protected amide; that is, an amide attached to *tert*-butoxy carbonyl. The typical Boc-protected amide is difficult to separate from the coupling reagents. The fluorous Boc-protected amide is easy to separate from the coupling reagents by solid–liquid extraction (Curran and Lee, 2001).

## Avoiding Highly Toxic Compounds

In putting out fires, water was historically the only chemical firefighters had at their disposal. In the 1960s, the U.S. Navy developed fire-blanketing chemical foams to suppress fires caused by oil spills. Other foams were developed for combating fires of wood and paper materials. Halon compounds, such as bromotrifluoromethane ($CF_3Br$), have been used to contain fires of electronic equipment. With these substances, the fire-fighting agents have left residual toxic chemicals such as hydrofluoric acid and other fluorine-containing compounds in the environment. These toxic chemicals may contaminate a water supply, cause wastewater treatment systems to fail, and even lead to the depletion of the protective layer of ozone in the atmosphere. Green chemistry techniques have been used to develop alternative fire-fighting agents that are more environment friendly. Pyrocool F.E.F. (fire extinguishing foam) is an example of such a product. The compounds that make up Pyrocool F.E.F. are rapidly biodegradable, are nontoxic, and work at lower concentration doses than earlier foam fire-fighting agents.

The use of harmful chlorofluorocarbons (CFCs) as a blowing agent in the manufacture of polystyrene foam, which is the material used to make such items as coffee cups, meat and poultry trays, and molded packaging, has been connected with the depletion of ozone in the upper atmosphere. Blowing agents are used to transform the hard brittle plastics, such as polystyrene, into lightweight foam materials with suitable strength for use as containers. The blowing agent is forced through the molten plastic, expanding the substance by forming bubbles and gas pockets in the final cooled product resulting in a composition of up to 95% gas and 5% plastic. To avoid the use of CFCs, Dow Chemical Company developed an innovative process that uses 100% carbon dioxide as an environment-friendly blowing agent for polystyrene foam sheets. Carbon dioxide has been implicated in global warming. For this process, however, existing sources either from natural gas or by-products from plants that produce ammonia are used for the supply of carbon dioxide. This means that no new carbon dioxide is generated and concern for global warming is not increased.

Green chemistry initiatives have shown that more environment-friendly chemical processes can be used and less harmful products and by-products can be produced. The challenge is to make these changes economically viable. By taking into account the total cost associated with making or using a chemical substance, the justification for green chemistry methods is likely much easier. A single entity (or business), however, does not bear the total cost of making or using a chemical substance—society bears the total cost. When this perspective is adopted, green chemistry approaches will become very common.

## References

AIChE. 1999a. American Institute of Chemical Engineers, Center for Waste Reduction Technologies. *Total Cost Assessment Methodology*, Appendix 1. AIChE: New York. July.

AIChE. 1999b. American Institute of Chemical Engineers, Center for Waste Reduction Technologies. *Development of Baseline Metrics*. AIChE: New York.

Anastas, P. T. and J. C. Warner. 1998. *Green Chemistry Theory and Practice*. Oxford University Press: New York.

Cann, M. C. and M. E. Connelly. 2000. *Real World Cases in Green Chemistry*. American Chemical Society: Washington, DC.

Chemistry.org. <http://www.chemistry.org/portal/a/c/s/1/acsdisplay.html?DOC=education%5Cgreenchem%5Cindex.html>.

Clark, J. 2002. Impact factors and awards. *Green Chemistry*, August: G38.

Clark, J. and K. Hodnett. 2001. Green chemistry in Ireland. *Green Chemistry*, October: G60.

Curran D. and Z. Lee. 2001. Fluorous techniques for the synthesis and separation of organic molecules. *Green Chemistry*, February: G3–G7.

Du, G. and J. Yu. 2002. Green technology for conversion of food scraps to biodegradable thermoplastic polyhydroxyalkanoates. *Environmental Science & Technology*, **36**(24):5511–5516.

Gerngross, T. U. and S. C. Slater. 2000. How green are green plastics? *Scientific American*, August: 36–41.

Haas, M. J., K. M. Scott, T. L. Alleman, and R. L. McCormick. 2001. Engine performance of biodiesel fuel prepared from soybean soapstock: a high quality renewable fuel produced from a waste feedstock. *Energy Fuels*, **15**(5):1207–1212.

Holbrey, J. D. and K. R. Seddon. 1999. Ionic liquids. *Clean Products and Processes*, **1**:223–236.

Krewer, U., M. A. Liauw, M. Ramakrishna, M. Hari Babu, and K. V. Raghavan. 2002. Pollution prevention through solvent selection and waste minimization. *Industrial and Engineering Chemistry Research*, **41**:4534–4542.

Ryan, M. A. 1999. Benign by design. *ChemMatters*, December:

Technical Database Services, Inc. 2003. <http://www.tds.cc/fsparis2.html>.

Trost, B. M. 1991. The atom efficiency—a search for synthetic efficiency. *Science*, **254**:1471–1477.

U.S. Bureau of Census. 1997. Economic Census. <http://www.census.gov/epcd/ec97/industry/E325.HTM>.

U.S. Bureau of Census. 2003. NAICS codes for chemical manufacturing. <http://www.census.gov/prod/ec97/97numlist/325.pdf>.

USEPA. 2002. Presidential Green Chemistry Challenge, Award Recipients. EPA 744-K-02-002. USEPA: Washington, DC. June.

Warner, J. C. 1998. Pollution prevention via molecular recognition and self assembly: Non-covalent derivatization. In *Green Chemistry: Frontiers in Benign Chemical Syntheses and Process*, Chapter 19. Oxford University Press: Oxford, UK.

Weininger, D. 1988. SMILES, a chemical language and information-system, 1. Introduction to methodology and encoding rules. *Journal of Chemical Information and Computer Sciences*, **28**(1):31.

Williams, E. D., R. U. Ayers, and M Heller. 2002. The 1.7 kilogram microchip: Energy and material use in the production of semiconductor devices. *Environmental Science & Technology*, **36**:5504–5510.

## Resources

### *Introductory Readings in Green Chemistry*

<http://www.chemistry.org/portal/a/c/s/1/acsdisplay.html?id=713ed2eaf02411d6e2f06ed9fe800100>.

Presidential Green Chemistry Challenge: <http://www.epa.gov/greenchemistry/past.html>.

Technical Database Services, Inc. <http://www.tds.cc/fsparis2.html>.

USEPA's Green Chemistry Program: <http://www.epa.gov/greenchemistry>.

### *Reference Information*

Annotated bibliography of green chemistry: <http://www.chemistry.org/portal/a/c/s/1/acsdisplay.html?DOC=education%5Cgreenchem%5Cindex.html>.

Green Chemistry Expert System (download): <http://www.epa.gov/greenchemistry>.

Green chemistry in education from USEPA and American Chemical Society: <http://www.chemistry.org/education/greenchem>.

Green Chemistry Institute: <http://chemistry.org/portal/a/c/s/1/acsdisplay.html?DOC=greenchemistryinstitute/index.html>.

# INDEX